Advances in Human Factors/Ergonomics, 6

Applications of Fuzzy Set Theory in Human Factors

Advances in Human Factors/Ergonomics

Series Editor: Gavriel Salvendy, Purdue University, West Lafayette, IN 47907, U.S.A.

Vol. 1. Human–Computer Interaction (G. Salvendy, Editor)
Vol. 2. Human–Computer Dialogue Design (R.W. Ehrich and R.C. Williges, Editors)
Vol. 3. Expertise Transfer for Expert System Design (J.H. Boose)
Vol. 4. Engineering Physiology (K.H.E. Kroemer, H.J. Kroemer and K.E. Kroemer-Elbert)
Vol. 5. Human Factors Testing and Evaluation (D. Meister)
Vol. 6. Applications of Fuzzy Set Theory in Human Factors (W. Karwowski and A. Mital, Editors)

Advances in Human Factors/Ergonomics, 6

Applications of Fuzzy Set Theory in Human Factors

Edited by

Waldemar Karwowski
Department of Industrial Engineering, Speed Scientific School, University of Louisville, Louisville, KY, U.S.A.

and

Anil Mital
Ergonomics Research Laboratory, University of Cincinnati, Cincinnati, OH, U.S.A.

ELSEVIER
Amsterdam – Oxford – New York – Tokyo 1986

ELSEVIER SCIENCE PUBLISHERS B.V.
Sara Burgerhartstraat 25
P.O. Box 211, 1000 AE Amsterdam, The Netherlands

Distributors for the United States and Canada:

ELSEVIER SCIENCE PUBLISHING COMPANY INC.
52, Vanderbilt Avenue
New York, NY 10017, U.S.A.

ISBN 0-444-42723-6 (Vol. 6)
ISBN 0-444-42396-6 (Series)

Printed in The Netherlands

FOREWORD

The development of the theory of fuzzy sets was motivated in large measure by the need for a computational framework for dealing with humanistic systems, that is, with systems in which human judgement behavior and emotions play a dominant role. Viewed in this perspective, it is hard to understand why so few of the many books on the theory of fuzzy sets deal with its applications to humanistic rather than mechanistic systems. A possible explanation is that the mathematical apparatus of the theory of fuzzy sets is very different in spirit from the traditional mathematics of social and behavioral sciences, which is rooted for the most part in probability, statistics and linear algebra.

As humans, we tend to view with skepticism and suspicion what is strange and unfamiliar. As a case in point, there are very few papers on fuzzy sets in the literature of psychology and cognitive science, even though it is obvious that the theory of fuzzy sets can provide a much better model for human cognition than traditional approaches--approaches in which there is not provision for classes in which the transition from membership to non-membership is gradual rather than abrupt.

By focusing its attention on the application of fuzzy sets in human factors, the present volume serves an important function of presenting to an audience of behavioral and social scientists an authoritative overview which gives a very good idea of what the theory is about and how it can be applied. In this regard, the chapters dealing with the concept of a linguistic variable are of particular relevance. Whereas in dealing with mechanistic systems it is generally reasonable to assume that a variable takes numerical values, this is not the case in humanistic systems, which are usually much more complex and much less well-defined than mechanistic systems. The concept of a linguistic variable, that is, a variable whose values are words or sentences in a synthetic language, provides a natural and effective way of dealing with such systems through the use of linguistic techniques. In this connection, it is of interest to observe that the currently active fields of research in AI centering on qualitative reasoning and qualitative process theory may be interpreted as special cases of the linguistic approach.

What is particularly impressive about the volume is the broad spectrum of applications, ranging from the use of fuzzy methods in the ergonomic diagnostics of industrial production systems to approximate reasoning in risk analysis and the modeling of human-computer interactions in information retrieval tasks. What is equally impressive is that the contributors represent such a wide variety of disciplines and countries of origin.

Clearly, it was not a simple matter to collect and organize the papers which constitute the present volume. Both as editors and as contributors, Professors Karwowski and Mital have done an excellent job. They and their contributors deserve the thanks of all of us for doing so much to advance the theory of fuzzy sets and extend its applications in new and important ways.

Lotfi A. Zadeh
Berkeley, California
July, 1986

PREFACE

The world in which we live and work is not perfect. The problems we attempt to solve are not always clearly defined, simple, or deterministic. Rather, we live in an uncertain and complex world, and the problems we face often have imprecise descriptions or are vague and not easily quantifiable. Indeed, our world is a fuzzy one.

It is living in this technological world that ergonomists attempt to make more comfortable for humans. We are continually challenged by problems of increasing complexity and ambiguity of relationships between people and devices they produce, and the environments in which such devices are used. The ever present uncertainty of our complex world demands appropriate scientific treatment. Recognition of the fact that fuzziness does exist and is critical for human-machine system analyses, although still slow, is getting more and more attention in the scientific community, particularly in the field of human factors/ergonomics. We hope this book will be recognized as a significant step in that direction.

This edited volume contains 25 chapters plus two appendices contributed by authors from Australia (one paper), Canada (one paper), Egypt (one paper), Federal Republic of Germany (one paper), France (two papers), Japan (two papers), People's Republic of China (two papers), Poland (five papers), and the United States (ten papers). The chapters are grouped into five topical parts: The Concept of Fuzziness: An Introduction; Human-Machine Communication; Human Operator: Judgement and Reliability; Design and Analysis of Work Systems; and Mental and Physical Workload and Stress Evaluation.

The theoretical and applied levels of fuzzy models, presented by different contributors, vary considerably from chapter to chapter. Such a mixture, although it may seem to make the material more difficult to comprehend, allows one to develop an appreciation for how much or how little difference scientifically treated fuzziness can make in describing or solving a spectrum of interesting problems in the area of human factors. Even if many of these papers are of exploratory nature and more research may be needed to validate the proposed models, we felt it was important to include them for the benefit of those contemplating application of a fuzzy-approach in their areas of interest.

One of the perceived problems with dissemination of fuzzy theoretical knowledge in the human factors area seems to be not only insufficient mathematical preparation of some of the researchers, but more importantly, the belief that the application of statistical analysis is the only way to analyze experimental data. One must, therefore, learn about fuzzy set theory and accept the fact that fuzziness and randomness, as theory of fuzzy sets and statistics, are complementary, not contradictory tools for theoretical and experimental research. We trust that with this edited book we shall be able to encourage and persuade our colleagues to seek an understanding of fuzzy methodologies and their application in explaining human thought processes and behavior at work, and ultimately in designing work systems to accommodate characteristics expressed in fuzzy terms.

We thank all authors for their efforts to comply with the Publisher's requirements for manuscript preparation. We also wish to acknowledge support of Dr. W. F. Dorrill, Provost of the University of Louisville and Dr. E. R. Gerhard, Dean of the Speed Scientific School, University of Louisville. We owe our sincere appreciation to Dr. K. Quinkert and Mr. C. Davis for their help in overcoming the English language barrier of some of the contributors.

Finally, we must recognize continuous support of our families, especially our wives, Bernardette Karwowski and Chetna Mital, who made our editorial work possible.

Waldemar Karwowski and Anil Mital
Louisville, KY and Cincinnati, OH
July 1986

CONTENTS

Part I.

THE CONCEPT OF FUZZINESS: AN INTRODUCTION

Applications of Fuzzy Set Theory in Human Factors, edited by W. Karwowski and A. Mital
Elsevier Science Publishers B.V., Amsterdam, 1986 — Printed in The Netherlands

A PERSPECTIVE ON MATHEMATICAL MODELING IN HUMAN FACTORS

Gerald W. EVANS and Waldemar KARWOWSKI

Department of Industrial Engineering, Speed Scientific School,
University of Louisville, Louisville, Kentucky 40292, U.S.A.

ABSTRACT

This paper reviews, classifies and discusses mathematical models currently used in human factors research. Emphasis is placed on the applicability of formal modeling techniques in various areas of man/machine/environment systems analysis. The need for adopting theories of fuzzy mathematics in human factors is also discussed.

INTRODUCTION

Human factors (i.e., ergonomics) is concerned with "the consideration of human characteristics, expectations, and behaviors in the **design** of the **things** people use in their work and everyday lives and of the environments in which they work and live" (ref. 1). The broad scope of human factors research is shown in Table 1. "Design" implies decision making -- that is, the choosing of values for the design variables so that the system objectives, in this case fitting the task to a person, are optimized. The "things" that are designed are complex man-machine systems. Hence the use of mathematical models in the design process becomes apparent.

According to Pew and Baron (ref. 2) the ultimate reasons for building models in general, and human performance models in particular, are to provide for:

1) a systematic framework that reduces the memory load of the investigator, and prompts him not to overlook the important features of the problem,
2) a basis for extrapolating from the information given to draw new insights and new testable or observable inferences about system or component behavior,
3) a system design tool that permits the generation of design solutions directly,
4) an embodiment of concepts or derived parameters that are useful as measures of performance in the simulated or real environment,
5) a system component to be used in the operational setting to generate behavior, for comparison with the actual operator behavior to anticipate a display of needed data, to introduce alternative strategies or to monitor operator performance,

TABLE 1

Categories of Ergonomics/Human Factors Studies (based on **Ergonomics Abstracts**)

Psychology
- perceptual input processes
- central processes
- basic motor processes
- perceptual motor performance (PMP)
- factors affecting PMP

Physiology and Anthropometry
- basic sensory physiology
- elementary physiological processes
- basic biomechanical functions
- work capacity
- factors affecting physiological and biomechanical functions
- basic anthropometric data

Technology
- visual displays
- auditory displays
- kinaesthetic and tactile displays
- controls
- special input devices
- display control dynamics
- workspace layout
- equipment design
- physical environmental factors
- protective clothing and equipment

Systems Design and Organization
- assignment of functions
- work design and organization
- training
- selection and classification
- motivation and attitudes

6) consideration of otherwise neglected or obscure aspects of the problem.

The human factors analyst will experiment with a mathematical model of the system, since it may be impractical to experiment with the real system. This impracticality can result from several factors including:

1. The real system is only proposed, and therefore, not in existence.
2. The number and range of feasible values of the design variables are so large that experimenting with the real system would be too costly and too time consuming.
3. Experimentation with the existing system is unsafe, especially when humans are involved.

The use of mathematical models allows the designer a great deal of flexibility in the design process. In this article, we provide a perspective on mathematical modeling in human factors. In particular, we present a classification scheme for mathematical models and a discussion of these various classes. We also give a brief overview of how some specific mathematical models have been employed in human factors, and some suggestions for improving mathematical modeling in this area. Sinclair and Drury (ref. 3) have also given an overview of mathematical modeling in ergonomics.

The next section of the paper gives a definition for a mathematical model and a description of its basic components. The third section of the paper presents a general discussion of the relationship between design theory and models, while the fourth section presents a taxonomy for mathematical models. Section five provides a detailed description of how some specific mathematical models are used in human factors. This section of the paper is divided into five subsections addressing the respective areas of simulation modeling, multiple objectives, experimental design, normative modeling, and fuzzy models. Finally, section six of the paper presents conclusions.

MATHEMATICAL MODELS

A model is just a representation of a "real system." As such, models can be classified as being either iconic models (e.g., pictures or model cars), analog models in which a property of the model is substituted for a property of the real system (e.g., slide rules), or mathematical models. The key feature of a mathematical model is the use of symbols, equations, and other mathematical statements to represent reality. Because of the abstract nature of mathematics, mathematical models can be applied to a much greater variety of situations than either iconic or analog models. This is especially true in decision-making (or design) situations.

There exist many different types of mathematical models. These models can usually be classified as being either normative or descriptive, multiobjective or single objective, dynamic or static, and stochastic or deterministic in nature.

For example, normative models are those types of models which are designed to give "normative solutions." A normative, or prescriptive, solution is one which optimizes the model's objective. One example of such a normative model is given by Bechtold et al. (ref. 4). This model is a mixed integer quadratic program in which the objective is to maximize labor productivity and the decision variables are the number, duration, and placement of rest breaks. These normative models are typically mathematical programs (e.g., linear programs, nonlinear programs, integer programs, goal programs, etc.)

Descriptive models, on the other hand, are designed to give descriptive solutions. A descriptive solution gives the outcome(s) which will occur given particular values for the decision variables. These outcomes can be such quantities as cost, productivity, profit, throughput time, etc.; in other words, they are analogous to objectives in normative models. Examples of descriptive model used in human factors research include simulation models and regression equations. Descriptive models can be used to find a normative solution to a problem by "running" the model several times.

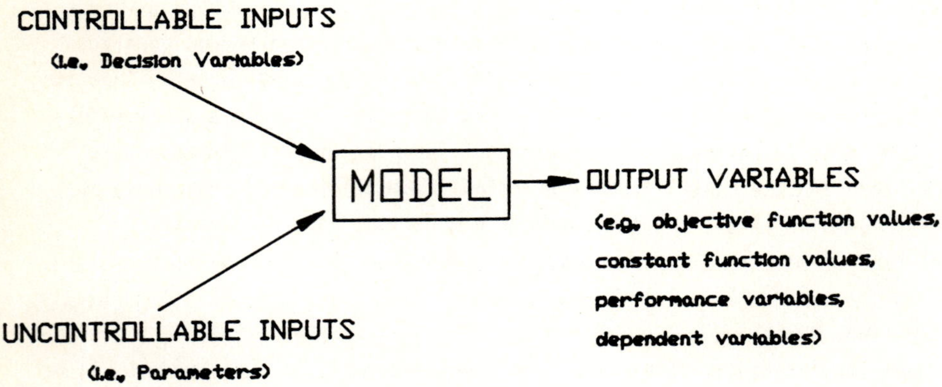

Figure 1. Schematic Representation of a Mathematical Model.

All mathematical models represent a way of converting **inputs** into **outputs** (see Figure 1). The inputs to a model are viewed as being either **controllable** or **uncontrollable**. Controllable inputs are typically called **decision variables** or design variables. A specification of a set of values for the **decision**

variables is termed a **solution**. Uncontrollable variables are often called the model's **parameters**. An example of a model parameter would be the time required for a worker to perform a task.

The "guts" of a mathematical model is a representation of the process (in the real system) which converts the inputs into the outputs. This representation is typically accomplished through the use of functions of the input variables (e.g., objective and constraint functions, or regression equations developed by statistical analysis). The representation can also be accomplished through the use of a computer program (e.g., a simulation model).

According to Alluisi (ref. 21) the selection of a criterion, index, or output to measure will influence not only the conduct of the study, but also the findings and the generalizations that can properly be made on the basis of the results. Likewise, the form of the corresponding mathematical model will depend on such selection of a measurement strategy. In order to optimize this selection, one must consider the following:

1) the purpose of the specific study,
2) the degree of specificity versus generality and
3) the desired area(s) of generalization of the findings of the study.

Pew and Baron (ref. 2) distinguish two basic approaches to human performance modeling, i.e.: 1) psychologically-based models (reliability models, network models, information processing models, and problem solving models), and 2) control-theoretic models for continuous control and control models for signal detection and decision making. Normative models of performance, such as signal detection theory, optimal control, or Bayesian decision making play very important role in workload measurement by specifying precise dimensions of performance and the dimensions along which performance under workload may depart from a model-defined optimal level (ref. 42).

The final report of an experimental psychology group on mental workload (ref. 43) defends the use of formal mathematical models (classical and optimal control theory) in systems where man is an element of a closed loop, and where single channel behavior is known to be probable time line analysis (queuing theory, supervisory theory). As pointed out by Rasmussen (ref. 41), models as well as measures of human workload must account not only for the average performance in frequent tasks but also the operator response in infrequent tasks, which very likely can be related to events implying high risk.

DESIGN THEORY AND MODELS

According to Topmiller (ref. 44), research in human/machine/environment systems poses an important methodological challenge. This is due to the complexity of such systems, and a need for simultaneous consideration of a variety of interacting factors that affect several dimensions of both individual and group performance. The author distingishes three general approaches used to study human/machine/enviroment systems, i.e. 1) direct observations as a means of improving performance through modification, 2) use of various analytical methods of modeling and fast-time computer simulation, especially in the conceptual design stages, and 3) use of real-time simulation and manipulating various system parameters according to experimental design methods. A review of the above approaches (ref. 45) reveal two important methodological issues. First, only a limited number of the existing variety of experimental design and analysis procedures have been used in complex system research involving human performance. Second, there is a lack of methodology development research in human factors which address complex system experimentation problems.

Chapanis argues that (ref. 48) "we do not have adequate methods for finding out all the things we need to know about people. Above all, we need novel and imaginative techniques for the study of man. This is an area in which behavioral scientists can learn much from the engineering and physical sciences." And yet the same author had criticized modeling as applied to human behavior, advocating for theories of human behavior instead (ref. 47).

Research techniques applied in human engineering research about 30 years ago (ref. 48) included the following methods:

1) methods of direct observation (operator opinions, activity sampling techniques, process analysis, etc.),
2) accident study methods (risk analysis, critical-incident technique)
3) statistical methods
4) experimental methods (design of experiments)
5) psychophysical methods (psychophysical scaling and measurement)
6) articulation testing methods.

Comparison of the above with the contemporary methods of human factors engineering (ref. 44) shown in Table 2, indicates that much progress has been made, both in quantity and quality, in the development of new innovative techniques for the study of man. Yet, we are still at the beginning stage of building robust mathematical models for the analysis of complex human-machine systems. This is partially due to lack of appropriate design theory as well as complexity of human behavior at work.

Berntat (ref. 19) points out the nonexistence of a human factors design theory, and gives the main reasons for this situation. These reasons are:

1) the human being is too complex a "system" to be fully understood or describable in all his/her properties, limits, tolerances, and performance capabilities,

2) no comprehensive mathematical tool has been available up to now to describe and integrate all the above mentioned measures and findings about human behavior, and
3) ergonomics is too young a science to have had a real chance to develop the required theory.

Furthermore, mathematical models in human factors suffer from the so called "measuring problem" (ref. 19) which refer to descriptions of the varying task load, the social environment, the state of physical environment, measurement of human workload, design and measurement of the information flow, and status of the machine. At present, only measurements of the state of physical environment and the status of the machine can be relatively easily performed.

According to Sheridan and Ferrell (ref. 49),"Engineering systems can be made compatible with human characteristics and limitations only by means of quantitative analysis and experiment, and only when the behavior of both man and machine can be described in comparable terms." Such comparable terms, applicable to the performance of both machines and human operators, are provided by three classes of models, i.e. information, control and decision models.

Information processing models describe the probabilistic relationships between sets of inputs and outputs, and are based on the Shanon-Winer measure of information. Manual control models, based on control theory, treat the human operator as an integral part of a control loop in order to evaluate total man-machine performance. Such models are either linear or quasi-linear, and depending on the nature of the input to the human operator can be classified as compensatory, pursuit, review or precognitive. Modeling of human decision behavior is needed in order to facilitate the interface between the decision maker and an engineering system and to incorporate subjective knowledge, values and needs into the decision making procedures.

TABLE 2

Classifications of Human Factors Engineering Models

I. EXPERIMENTAL DESIGN METHODS

1. Statistical Procedures

Univariate Methods

Correlations

Simple Regression

Parametric Inferences

Nonparametric Inferences

Multivariate Methods

Multiple Regression
Polynomial Regression

TABLE 2 (continued)

Canonical Analysis
Principal Components
Factor Analysis
MANOVA (Multivariate ANalysis of VAriance)
Discriminate Analysis
Pretesting Methods
ANOVA Designs (ANalysis of VAriance)
DATA Reduction Designs
Blocking Designs
Hierarchical Designs
Fractional-factorial Designs
Central-composite Designs
2. Tailored Methods
Confusion Matrices
Quasi-experimental Designs
Response Surface Methodology
Finite Interaction Test

II. MODELS
1. Biomechanical Models
Architectural Body Models
Dynamic Dan
Combiman
2. Performance Models
Information Theory
Statistical Decision Theory
TSD (Theory of Signal Detectability
Bayesian Decision Making
Estimation theory
Control Theory
Quasilinear Control
FFM (Fixed-Form Models)
OCM (Optimal Control Models)
Queuing Theory
3. Process Models
Short-Term Memory Models

Visual Scanning/Detection Models

GRC (General Research Corporation)

MARSAM II (Multiple Airborne Reconnaissance Sensor

TABLE 2 (continued)

Assessment Model

VISTRAC (VISual Target Recognition and Acquisition)

CRESS/SCREEN (Combined Reconnaissance, Surveillance,

SIGINET/SRI Countersurveillance Reconnaissance Effectiveness

Evaluation)

Autonetics Model

Detect

ASTCAD

REA/BAC

Air Traffic Control Models

Industrial Inspection Models

Attention/Workload Models

Fault-Diagnosis Models

HOS (Human Operator Simulation Model)

III. ENGINEERING DESIGN SIMULATION

1. Human-Machine Integration Engineering Research Simulation

   General Purpose Statis Aircraft Crew Station

   General Purpose Dynamic Aircraft Crew Station

   General Purpose Control Display

   General Purpose Multiperson

2. Human-Machine Integration Engineering Design Simulation

   Static Mockup

   Specific Dynamic Control/Display

   Specific Crew Station

   Outside Dynamic Visual Scene

   Workplace Simulator

   Multiman Workstation

   Command and Control Simulation

   Sensor Simulation

   Computer Simulation

IV. PROCEDURES

1. Systems Engineering Analytic & Management

   1.1. To Aid System Engineering Analysis

      1.1.1. Manual

PERT (Program Evaluation and Review Techniques
TLA-1 (Time Line Analysis-1)
CMP (Critical Path Method)

TABLE 2 (continued)

Expected Value Method
Functional Flow Diagrams
FDI (Functional Description Inventory)
Function Allocation Tradeoffs
Task Analysis
Decision Tree Analysis
Action/Information Requirements
Time Lines
Flow Process
OSD (Operational Sequence Diagrams)
Task Descriptions

1.1.2. Computerized
SW (Siegel-Wolf Model)
SAINT (Systems Analysis of Integrated Networks of Tasks)
PSM (Pilot Simulation Model)
CAPA (Computer Analysis of Personnel Activity)
GERT (Graphical Evaluation and Review Technology)
FOVEA (Field of View Evaluation Apparatus
WSP (Workload Simulation Program)
R&M (Reliability and Maintainability Model)
Sensitivity Analysis

2. Detailed Design Procedures

2.1. Manual
Specification Compliance Summary Sheet
Link Analysis

2.2. Computerized
HECAD (Human Engineering Computer-Aided Design)
CATTS (Continuous Assessment of Task Time Stress)
TBLA (Time-Based Load Analysis
RECEP (RElative Capacity Estimating Process)

CAFES (Computer Aided Function Allocation and Evaluation System)
DMS (Data Management System)
FAM (Function Allocation Model)

WAM (Workload Assessment Model)
CAD (Computer-Aided Crew Station Design Model)
CGS (Crew Station Geometry Evaluation Model)

A TAXONOMY FOR MATHEMATICAL MODELS

As shown in Figure 2, mathematical models in human factors area can be classified along several dimensions, including: normative or descriptive, multiobjective or single objective, dynamic or static, stochastic or deterministic, nonlinear or linear, and fuzzy or nonfuzzy.

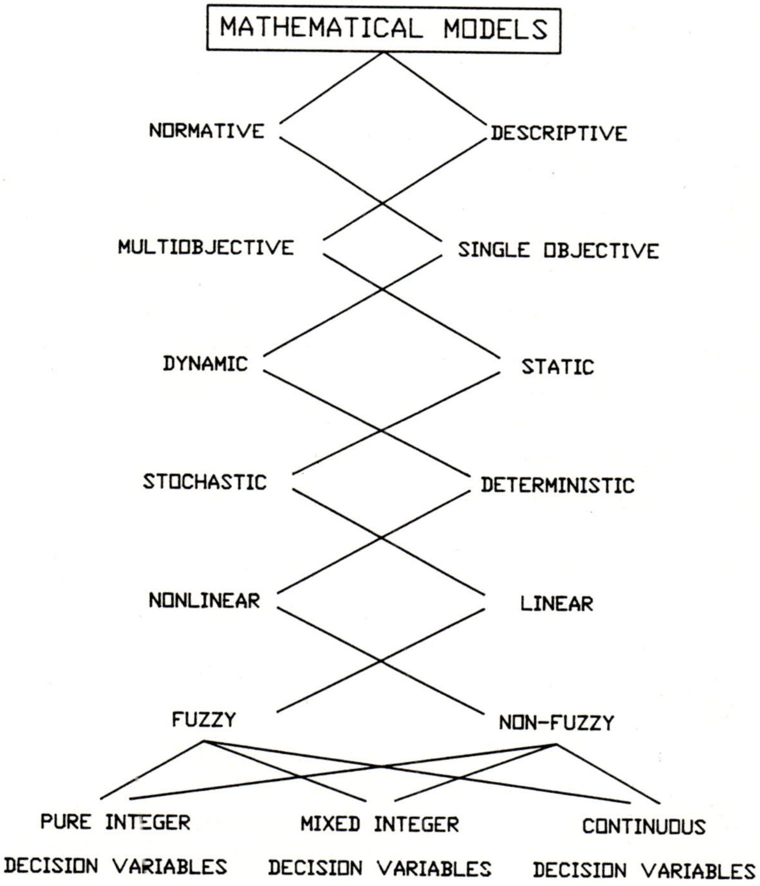

Figure 2. A Taxonomy for Mathematical Models.

In addition, a model can be classified in even greater detail by considering its decision variables. For example, a model can have only continuous decision variables, some continuous and some discrete decision variables, or only discrete decision variables. Typically, a continuous decision variable is one that can take on any value on some interval (or intervals) of the real number line. An example would be the weight of a load to be lifted by a worker. On the other hand, a discrete decision variable is one that is restricted to take on only integral values. An example of a discrete decision variable would be the number of workers to assign to a task.

The method of "solution" for a mathematical model (i.e., the method for determining acceptable or the best values for the decision variables) is dependent upon the form of the model. For example, descriptive models are employed not with the idea of "solving" but with the idea of experimenting with them. That is, several different solutions can be input to the model with the idea of selecting the best one, out of the several tried, in terms of the performance variables of the model.

Prescriptive models typically rely on much more sophisticated approaches for solution. These approaches can either be heuristic or exact in nature. Heuristic techniques will typically give a very good but not necessarily optimal solution to the model. Exact techniques on the other hand guarantee an optimal solution. The drawback of exact techniques has to do with the fact that many models are "too large" (e.g., too many decision variables) to be solved to optimality in a reasonable amount of time even with a computer.

MATHEMATICAL MODELING IN HUMAN FACTORS

Simulation Modeling in Human Factors

Pritsker (ref. 5) notes that due to the nature of the word itself, the term simulation has been "used in different and often conflicting senses." To some, any computer program which represents some system should be thought of as a simulation. When thought of in this way, the concept of simulation modeling covers a very broad spectrum of models indeed. For example, a regression equation implemented on a computer could be viewed as a simulation model.

We have a much more specific concept in mind in this paper when discussing simulation. In particular, we prefer Pritsker's definition since it seems to be relatively easy to understand, yet is much more specific than the basic concept of representation presented above. Pritsker's definition (ref. 6) reads as follows:

> "Simulation is the representation of the dynamic behavior of the system by moving it from state to state in accordance with well-defined operating rules."

A compilation of other definitions of simulation modeling is given in Pritsker (ref. 5). A key concept in the above definition is the idea of a **system state.** The system state is defined in terms of the numeric values assigned to the **attributes** (i.e., descriptors) of the **entities** (i.e., things) in the system. In some cases, these attributes can be viewed as output or performance variables of the system. In other cases the attributes can be viewed as being intermediate variables of the system.

One classification scheme for simulation models/systems concerns the types of attributes in the model/system. For example, the model can be either discrete in nature, continuous in nature, or combined discrete/continuous, depending upon whether the attributes of the model are all discrete variables, all continuous variables, or variables which can change either continuously or discretely, respectively. Most of the simulation models built for manufacturing/production type purposes in general, and ergonomics in particular are discrete in nature.

As an example of an attribute which is a discrete variable, consider the status of a laborer. The laboreer can be either idle, in transit, or working on a particular task. These three worker "states" can be assigned numeric values of 0, 1, or 2 respectively -- hence the discrete nature of the variable.

An **event** in a simulation model is an instant of time at which the system state can change. For example, the arrival of a workpiece to a system is typically modeled as an event, because it is possible for the laborer's status to change from idle to busy at this time.

The simulation process requires that the simulator define a system's entities and the attributes of those entities, as well as the well-defined operating rules under which the entities interact. Following this system definition phase, these operating rules must be programmed so that various policies concerning the system can be experimented with. This programming phase is often aided through the use of special purpose simulation languages, to be discussed later in this section.

Simulation modeling is often viewed as a last resort, i.e., to be employed when no other modeling technique (e.g., analytical models such as those involving queueing theory) is appropriate. This can occur when the relationships between the input variables and the output variables are very complex, such as when these relationships cannot be written down in a functional form, or when there is a great deal of uncertainty in the output variable values for given input variable values.

Another characteristic of simulation models is that they are typically multiobjective in nature; that is, several different performance variable values are output by the model. This is one reason, along with the stochastic nature of most simulation models, why little work has been accomplished in transforming basic simulation models into systematic prescriptive models. A multiobjective prescriptive model requires human (decision maker) interaction in determining te preference structure of the decision maker over the multidimensional outcome space. Instead, simulation models are typically employed in a trial-and-error mode in which various policies are experimented with by "running" them through the simulation model program. The multiple performance measures are therefore, considered in only an implicit, ad-hoc fashion.

Simulation modeling has been applied extensively in ergonomics, especially in the area of manned system design, as noted by Doering and Berheide (ref. 7). They note seven basic design steps that ergonomists must go through in the design of man-machine systems:

1. Allocation of functions to men and machines.
2. Estimation of manning requirements.
3. Design of task structures.
4. Design of workstations.
5. Design and analysis of environment (e.g., lighting, noise, temperature, pollution, etc.).
6. Design of jobs (which consists of the grouping of tasks).
7. Selection and training of personnel.

Simulation modeling can be useful in all of these steps but especially in those numbered two through six. One of the reasons for the usefulness of simulation modeling in these steps is the complexity of the functional relationships (among input, intermediate, and output variables) within the system.

There are many examples of specific applications of simulation modeling in ergonomics, including the works of Wortman et al. (ref. 8) involving the simulation of a remotely piloted vehicle control facility; Wortman et al. (ref. 9) in the simulation of the human interface in a guided missile air defense system; Maltas and Buck (ref. 9) in simulating a hot strip mill; and Kuperman et al. (ref. 10) in developing a simulation model to evaluate the display concepts of a digital avionics information system. See Doering and Berheide (ref. 6) and Pritsker (ref. 5) for some more detailed discussion of these and other applications.

Simulation modeling has become much more popular over the last ten years for complex systems analysis. One of the reasons for this increased popularity is the recent proliferation of flexible software aids (i.e., simulation languages and simulation support software) for the simulation modeler. Another reason is the proliferation of personal computers on which this software can be implemented.

Some examples of these software products include SLAM-II (Simulation Language for Alternative Modeling), TESS (The Extended Simulation System), SAINT (Systems Analysis of Integrated Networks of Tasks), SIMAN (Simulation Analysis Language for Modeling General Systems), BLOCKS, PLAYBACK, CINEMA, and GPSS/PC. SLAM-II and TESS are marketed by Pritsker and Associates, Inc.; SIMAN, BLOCKS, PLAYBACK, and CINEMA are marketed by Systems Modeling Corp.; and GPSS/PC is marketed by Minuteman Software. Much of the recently developed simulation software can be characterized by its ease of use through the application of a process orientation to simulation modeling. The process orientation allows the analyst to build the simulation model in the form of a network, using predefined specifications for the types of nodes and arcs in the network. Hence, no programming is required of the analyst, and computer-implemented models are much easier to build. Pew and Baron (ref. 2) also note the importance of network simulation procedures for modeling human performance.

Along these same lines, the software has even been developed to the point where the analyst can interactively build his network on a computer terminal. The software package will then translate the network into a set of statements which the basic simulation language package can understand. Previously, the analyst had to do this translation himself.

Another characteristic of much of the recently developed software is that it allows the analyst to take an "extended view" of the simulation analysis process. By that, we're referring to the fact that this analysis process is much more than just building a model/program and running it. For example, Pritsker (ref. 5), page 10, defines ten stages of development within the simulation process. Stages seven, eight, and nine, entitled strategic and tactical planning, experimentation, and analysis of results, respectively, require systematic and detailed planning in order to be appropriately accomplished. In particular, use of statistical techniques are appropriate in the analysis of results, since a simulation run, because of the model's stochastic nature, can be viewed as a statistical experiment.

While much has been accomplished recently in extending the utility and flexibility of simulation software, more can still be accomplished in the future. In particular, we foresee the increasing use of software that will be helpful in the prescriptive phase of the decision making process. Since most simulation models/programs are multiobjective and stochastic in nature, this could imply the use of multiattribute utility theory (ref. 12), multiattribute value functions (ref. 13), and multiobjective mathematical programming (ref. 14) in the development of a prescriptive algorithm to be interfaced with the simulation program. This could also imply greater and more systematic involvement on the part of the decision maker in the determination of his preference structure over a multidimensional outcome space. More will be said about this area of multiple objectives in the next section of the paper.

Modeling and Analysis of Multiple Objectives

Multiple objectives arise in almost all decision making situations. One reason for this is simply that most systems have more than one measure of performance. Typically these performance measures are conflicting in nature -- e.g., decreasing a system's cost can be done only with a decrease in reliability, assuming that the system is designed efficiently in the first place.

Sinclair and Drury (ref. 3) allude to the concept of multiple objectives in the modeling of human performance. They note that a recent review of reaction time models stated that there were no models for analyzing the trade-off between speed and accuracy, despite the fact that these two performance measures are known to be related. Sinclair and Drury went on to state that the author should have said that any model not accounting for both of these performance measures is useless.

Other examples of ergonomic studies involving at least the implicit study of multiple performance measures include the works of Williges and Williges (ref. 15), and Wewerinke (ref. 16). Williges and Williges developed statistical models (regression equations) of the human operator in computer-based data entry. Many of the dependent variables could be viewed as performance measures in a multiobjective model. Wewerinke developed a model of a human decision maker observing a dynamic system for predicting the decision speed/accuracy tradeoff.

Even though multiobjective analysis has been acknowledged as an important aspect of system design for a long time, a systematic approach to such analysis has been lacking in ergonomics. In fact, only within the last 15 years has the field of operations research/management science addressed this important aspect of modeling in a systematic fashion.

Typically, multiobjective analysis is addressed in an ad-hoc fashion. For example, most mathematical programming models have a single objective. Other objectives are considered only implicitly through the use of constraints. In particular, consider a system which has associated with it the performance measures of cost and reliability. The objective in a mathematical program to be used in designing the system might be the minimization of cost, while the other performance measure might be only implicitly considered through the use of a constraint restricting system reliability. This constraint will usually be binding, and hence by relaxing the reliability standard, the analyst can achieve a lower minimum cost. By adjusting the standard in a trial and error fashion and running the model through several iterations, the analyst can sometimes achieve satisfying values for both cost and reliability.

More systematic approaches to multiobjective analysis require an explicit representation of the decision maker's preferences over the multiple objectives (i.e., the decision maker's preference structure). One way to categorize the various approaches to multiobjective analysis is to consider the timing of the elicitation of the decision maker's preference structure with respect to the optimization process employed. For example, this elicitation process can take place prior to (prior articulation of preferences), during (progressive articulation of preferences), or after (a posteriori articulation of preferences) the optimization procedure.

Examples of approaches involving a priori articulation of preferences include goal programming (ref. 17) and the use of a multiattribute value (MAV) function (ref. 13). Approaches involving a progressive articulation of preferences usually require some interactive human/computer approach. An a aposteriori articulation of preferences can require that an algorithm be developed to output all of the efficient solutions (ref. 18) to a problem. (A "rational" decision maker will always prefer an efficient solution to a solution that's not efficient.) Following the generation of all (or at least most) of the efficient solutions, the decision maker is expected to choose one of these as his preferred solution. See Evans (ref. 14) for a more detailed discussion of these three general approaches to multiobjective optimization.

Experimental Design Models

Experimental design models, which include statistical procedures and tailored methods (see Table 2) have been and are at present the most frequently used models in human factors research. Examples of various applications of such models are given below.

Williges (ref. 45) reviewed several design alternatives (experimental models) for efficient human performance data collection in man-machine systems simulation. These included data reduction design techniques (single-observation factorials, hierarchical designs, blocking designs, fractional-factorial designs), empirical model building (based on the second-order polynomial approximation), central-composite design, sequential design procedures (response surface methodology), and their multivariate extensions.

Williges and Williges (ref. 15) reviewed modeling needs in the design of human-computer tasks, and concluded that such design requires a quantitative data base that describes operator behavior as a function of various independent variables. They used polynomial regression procedures to generate functional relationships between different classes of metrics and independent variables representing system delay, display rate, keyboard echo rate and buffer length. The use of multivariate response surfaces allowed for a more general interpretation of the human-computer interface.

Williges and Mills (ref. 25) provided support for the contention that the response surface methodology (RSM) central composite design is an efficient way to generate relatively stable and valid prediction of human performance.

Levine and Samet (ref. 28) investigated the effects of information conflict, the frequency with which the information is updated, and the reliability of the information sources upon information seeking and decision behavior in a computer-controlled task. Analysis of variance (ANOVA) based upon a fixed-effects model were carried out on a selected performance measure.

Howard and Ballas (ref. 33) developed a stochastic model to describe the perceptual processes that underlie the classification of complex multidimensional acoustic patterns, such as passive sonar sounds. The model is based on the assumptions that: 1) complex sounds can be represented psychologically as points in a multidimensional perceptual space, 2) different categories are represented as distributions in this space, 3) the category distributions determine the conditional stimulus probabilities and have multivariate Gaussian form, and 4) individual dimensions in the perceptual space may be weighted differentially to reflect a listener's attentional process. A multidimensional scaling analysis was performed in order to identify the potential mapping between the physical characteristics of the stimuli and its psychological dimensions.

According to Carter (ref. 34) most data analyses assume independence of successive observations, and most theories deal with simultaneous values of several variables. When nondynamic modeling techniques such as regression analysis, analysis of variance or factor analysis are applied to human systems, they can be completely misleading. This is due to the fact that most human systems are dynamic in nature where the past is one of the best predictors of the present and future behavior of individuals and organizations. Hess (ref. 22) points out that the dynamic models for human pilots (and vehicle handling qualities) have evolved as linear (or quasi-linear) mathematical structures which tend to mimic the input-output characteristics of actual human pilots when acting as feedback controllers in well-defined tracking tasks.

Descriptive Modeling

Noro (ref. 29) has proposed a descriptive model of visual search on inspection task based on corneal reflection technique. Morawski et al. (ref. 30) have proposed two different models of predicting visual search performance for multiple targets, i.e.: 1) random search, and 2) systematic search. Such procedures were based on the use of the cumulative distribution functions.

Bechtold et al. (ref. 31) developed analytical solution procedures for mixed-integer quadratic programming models for determining the optimal number, duration and placement of rest breaks. In the proposed models the decay of work output and recovery of work rate potential during rest breaks are modeled as a linear functions of time, allowing one to optimize labor productivity over a simple, finite time horizon.

Ayoub et al. (ref. 23) have developed and tested a biomechanical model for the upper extremity using optimization techniques. They concluded that while linear and geometric programing techniques were not suitable for providing a solution for the model, the dynamic programming was the most favorable algorithm to be used for that purpose. The authors also indicated that stochastic approaches to model building in connection with human motion should be considered.

Herrin and Newhardt (ref. 34) derived an empirical, non-linear model for automobile driver horizontal curve negotiation based on a driver aspiration velocity, maximum lateral acceleration tolerance, and an expedience parameter related to the driver's willingness to trade velocity for lateral acceleration.

According to Drury and Addision (ref. 27) the theory of signal detection (TSD) provides a useful description of the performance of groups of industrial on-line inspectors. The authors recommended not to derive individual values of TSD parameters but to use the theory on a macro scale as a model for the inspection system and then to determine its parameters.

In many situations, in order to account for and identify the dynamic relationships among system components multivariate time series models (ref. 35) of human factors data must be used. Carter (ref. 34) also points out that dynamic and nondynamic models are complementary. Nondynamic models assume that successive data are independent, while the dynamic models explain the dependency among successive observations. The analyst must be familiar with both types of research techniques. The potential areas of application of dynamic models in human factors include: analysis of data pathology, intervention analysis and control tasks (forecasting the future states of humanistic systems).

Fuzzy Models

Referring to the field of engineering psychology, Van Cott (ref. 20) argues that there exist reasonably good methods for the analysis of physical tasks, but not for the description or analysis of cognitive tasks. In addition, there is a need to understand how people aggregate the small chunks of information from different sources into larger concepts. The theory of fuzzy sets can contribute to addressing the above needs.

W. Rohmert (ref. 60) proposed a human work taxonomy which includes five different levels ranging from primarily physical tasks to primarily information processing tasks. These include:

1) producing force (primarily muscular work),
2) continuously coordinating sensory-motor functions (like assembling or tracking tasks),
3) converting information into motor actions (e.g. inspection tasks),
4) converting information into output information (e.g. required control tasks), and
5) producing information (primarily creative work).

It is our view that regardless of the level of human work, three types of fuzziness are present and should be accounted for in man/machine/environment systems, i.e.: 1) fuzziness stemming from our inability to acquire and process adequate amounts of information about the behavior of a particular subsystem (or the whole system), 2) fuzziness due to vagueness of the relationships between people and their working environments, and complexity of the rules and underlying principles related to such systems, and finally, 3) fuzziness inherent in human thought processes and subjective perceptions of the outside world. Figure 3 illustrates the above thesis. Traditional man-machine interfaces, which include: 1) information sensing and receiving, 2) information processing, 3) decision making, 4) control actions, and 5) environmental and situational variables, are represented in the two blocks, i.e., human interpretation block and a complex work system block.

Even though fuzzy methodologies have only recently been developed, there are a few examples of applications in human factors. For example, Terano et al. (ref. 36) introduced a fuzzy set approach into fault-tree analysis, and studied the fuzziness of a human-reliability concept from the man-machine systems safety point of view. Kramer and Rohr (ref. 37) developed a fuzzy model of driver-behavior based on simulated visual pattern processing in lane control. Saaty (ref. 38) distinguishes two types of fuzziness in layman perception (for example, perception of illumination intensity) and fuzziness in meaning, advocating that fuzziness is a basic quality of understanding. Hirsh et al. (ref. 39) used a fuzzy dissimilitude relation to describe human vocal patterns.

Simcox (ref. 32) used the linguistic approach to devise a method for pragmatic communication in graphic displays. The proposed model involves determining a compatibility function that describes the degree of correspondence between an implied attribute of the display and the linguistic category that summarizes values of this attribute. He concluded that compatibility functions can be used to guide the graph-construction process.

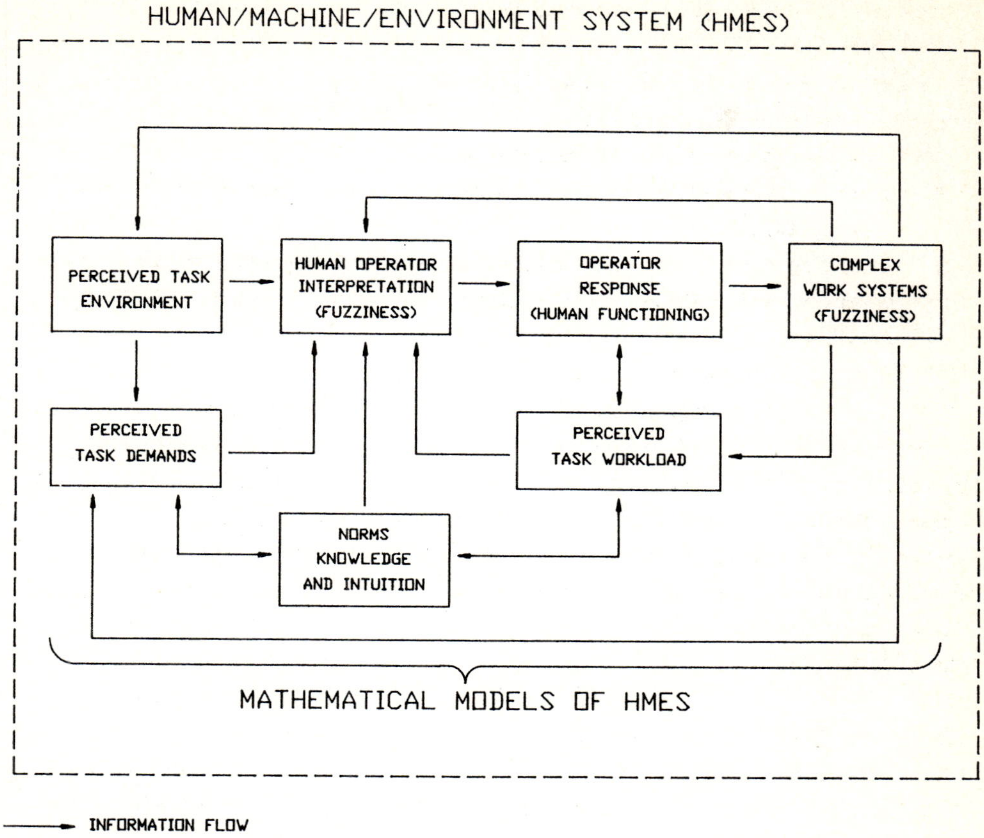

Figure 3. Fuzziness in the complex human/machine/environment systems.

Uncertainty, (looked upon in the context of mental workload) which causes unpredictability in one's stimulus and/or response, enters a work situation from several sources (ref. 51). These are:

1) external disturbance model,
2) varying parameters of the system structure external to the human operator,
3) human produced noise in observing the task stimuli,
4) lack of good internal model of the external system,
5) human-produced distortions in interpreting the externally stipulated criterion of performance, and
6) human produced motor noise.

In view of the above, the theory of fuzzy sets offers a useful approach when the task demands are vague, with the main advantage being its ability to model imprecise task situations and, therefore, a potential to develop a framework for implementation of workload measures.

Rouse (ref. 52) defines a model of mental workload as a procedure for predicting workload based on prior conditions, and evaluates three categories of models used in this area of research, i.e., 1) information theory models, 2) control theory models, and 3) queuing theory models. The author argues that information models are too structured to apply to many realistic tasks. The reason being that information theory measures rely on conditional probabilities that are highly context and individual dependent, and which are difficult to assess in all but simple tasks. The control theory models seem adequate only for a fairly small set of tasks where tracking of outputs of linear dynamic systems are of interest. On the other hand, the queuing theory models are applicable to more robust tasks than information theory or control theory models. However, queuing models disregard task performance other than the task completion time. The above difficulties can be partially reduced by combining different classes of models into one.

CONCLUSIONS

According to Harre (ref. 55) there are two major purposes of models in science: 1) logical: to enable certain inferences which would not otherwise be possible to be make; and 2) epistemiological: that is to express and enable us to extend our knowledge of the world. Thus, models are helpful for explanation and theory formation, as well as simplification and concretization.

Zimmermann (ref. 56) classifies models into three groups: 1) formal models (purely axiomatic systems with purely fictitious hypotheses, 2) factual models (conclusions from the models have a bearing on reality and they have to be verified by empirical evidence), and 3) prescriptive models (which postulate rules according to which people should behave.

The quality of a model depends on the properties of the model and the functions for which the model is designed (ref. 56). In general, good models must have three major properties: 1) formal consistency (all conclusions follow from the hypothesis), 2) usefulness, and 3) efficiency (the model should fulfill the desired function at a minimum effort, time and cost).

Zimmerman (ref. 56) indicates that although the usefulness of the mathematical language for modeling purposes is undisputed, there are limits of the possibility of using the classical mathematical language which is based on the dichotomous character of set theory to models in particular systems and phenomena. Such restriction applies especially to the human factors area. This is due to vagueness of the natural language and the fact that in empirical research natural language cannot be substituted by formal languages. Furthermore, formal languages are rather simple and poor, and are useful only for specific purposes. Mathematics and logic as research languages widely applied in natural sciences and engineering are not very useful for modeling purposes in behavioral sciences and especially in human factors studies. Rather, a new methodology, based on the theory of fuzzy sets and systems is needed to account for the ever present fuzziness of man/machine/environment systems

As suggested by Smithson (ref. 57), the potential advantages for applications of a fuzzy approach in human sciences, and therefore, human factors are: 1) fuzziness,itself, may be a useful metaphor or model for human language and categorizing processes, and 2) fuzzy mathematics may be able to augment conventional statistical techniques in the analysis of fuzzy data. The author reviewed several alternative methods for the analysis of fuzzy data (like measure of intercategory overlap, equivalence, etc.) and concluded that fuzzy methods are useful supplements for statistical techniques such as reliability analysis and regression, and structurally oriented methods such as hierarchical clustering and multidimensional scaling. More theoretical and experimental research should be carried out to explore potential applications of fuzzy set methods, especially in combination with other analytical techniques.

As early as 1959, six years before Zadeh published his first paper on fuzzy sets (ref. 61), Helmer and Rascher (ref. 58) had indicated a need for the new reasoning procedures that would be tailored to the domain of the inexact sciences, i.e., sciences where reasoning is informal, terminology at times exhibits inherent vagueness, and reasoning may rely on reference to intuitively perceived facts. Clearly, inexact sciences, which do not make predictions with great precision include human factors studies. Fuzzy set theory, concerned with mathematical representation and manipulation of degree vagueness, is a powerful tool for the analysis of human work systems. Such systems are complex, their underlying structure and governing relations are not precisely known, its descriptions are generally linguistic in nature, and definitions of many variables and several concepts are vague. Human factors researchers must look into the enormous potential offered by fuzzy methods and fully explore their applications in the analysis of ergonomics systems.

REFERENCES

1 E. J. McCormick, Human Factors Engineering, McGraw Hill, New York, 1970.
2 R. W. Pew and S. Baron, Automatica, 19 (1983) 663-676.
3 M. A. Sinclair and C. G. Drury, Applied Ergonomics, 10.4 (1979) 225-234.
4 S. E. Bechtold, R. E. Janaro, and D. L. Summers, Management Science, 30 (1984) 1442-1458.

5 A.A.B. Pritsker, Simulation, 33 (1979) 61-63.
6 A.A.B. Pritsker, Introduction to Simulation and SLAM II, Systems Publishing Corporation, West Lafayette, Indiana, 1984.
7 B. Doering and W. Berheide, in Manned Systems Design: Methods, Equipment, and Applications, Plenum Press, New York, 1981, p. 91.
8 D. B. Wortman, S. D. Duket, and D. J. Seifert, in Proceedings of the Human Factors Society 19th Annual Meeting, Dallas, 1975, p. 342.
9 D. B. Wortman, A. F. Hixson, and C. C. Jorgensen, in Proceedings of the 1978 Winter Simulation Conference, Miami Beach, Florida, 1978, p. 879.
10 K. L. Maltas and J. R. Buck, in Proceedings of the Human Factors Society 19th Annual Meeting, Dallas, 1975, p. 193.
11 G. G. Kuperman, R. L. Hann, and K. M. Berisford, in Proceedings of the Human Factors Society 21st Annual Meeting, San Francisco, 1977, p. 305.
12 P. H. Farquhar, in M. K. Starr and M. Zeleny (Eds.), Multiple Criteria Decision Making, TIMS Studies in the Management Sciences, 6 (1977).
13 R. L. Keeney and H. Raiffa, Decisions with Multiple Objectives: Preferences and Value Tradeoffs, John Wiley and Sons, Inc., New York, 1976.
14 G. W. Evans, Management Science, 30 (1984) 1268-1282.
15 R. C. Williges and B. H. Williges, Human Factors, 24 (1982) 285-299.
16 P. H. Wewerinke, Automatica, 19 (1983) 693-696.
17 S. M. Lee, Sloan Management Review, 14 (1973) 11-24.
18 T. Gal, in Proceedings of the 5th International Conference on Multiple Criteria Decision Making, Springer Verlag, Berlin, 1982.
19 R. Berntat, in H. Schmidtke (Ed.) Ergonomic Data Equipment Design, Plenum Press, New York, 1984.
20 H. P. Van Cott, Human Factors, 1984 (26) 115-122.
21 E. A. Alluisi, Human Factors, 1975 (17) 309-320.
22 R. A. Hess, Human Factors, 1977 (19) 459-475.
23 M. A. Ayoub, M. M. Ayoub and A. G. Walvekar, Human Factors, 1974 (16) 585-594.
24 G. D. Herrin and J. B. Newhardt, Human Factors, 1974 (16) 129-133.
25 R. C. Williges and R. Mills, Human Factors, 1973 (15) 349-354.
26 M. W. Bartlett and L. A. Smith, Human Factors, 1973 (15) 1-7.
27 C. G. Drury and J. L. Addison, Human Factors, 1973 (2) 159-169.
28 J. M. Levine and M. G. Samet, Human Factors, 1973 (15) 407-419.
29 K. Noro, A descriptive model of visual search, Human Factors, 1983 (25) 93-101.
30 T. Morawski, C. G. Drury and M. H. Karwan, Human Factors, 1980 (22) 707-718.
31 S. E. Bechtold, R. E. Janaro and D. L. Sumners, Management Science, 1984 (30).
32 W. A. Simcox, Human Factors, 1984 (26) 483-487.
33 J. M. Howard and J. Ballas, Perception of Simulated Propeller Cavitation, Human Factors, 1983 (25) 643-655.
34 R. C. Carter, Human Factors, 1984 (26) 83-95.
35 G. E. P. Box and G. M. Jenkins, Time Series Analysis, Forecasting and Control, Holden-Day, San Francisco, 1970.
36 T. Terano, Y. Murayama and N. Akijama in G. Johannson and J. E. Rijnsdorp (Eds.) Analysis, Design and Evaluation of Man-Machine Systems, Pergamon Press, Oxford, pp. 347-354.
37 U. Kramer and G. Rohr, 1982, in G. Johannsen and J. E. Rijnsdorp (Eds.) Analysis, Design and Evaluation of Man-Machine Systems, Pergamon Press, Oxford, pp. 31-35.
38 Saaty, Fuzzy Sets and Systems, 1978 (1) 57-68.
39 G. Hirsh, M. Lamotte, M. T. Mass and M. T. Vigneron, Fuzzy Sets and Systems, 1981 (5) 267 - 275.
40 N. Moray, in Mental Workload. Its Theory and Measurement, N. Moray (Ed.), Plenum Press, New York, 1979, pp. 13-21.
41 J. Rasmussen, in Mental Workload. Its Theory and Measurement, N. Moray (Ed.), Plenum Press, New York, 1979, pp. 29-40.
42 C. D. Wickens, in Mental Workload. Its Theory and Measurement, N. Moray (Ed.), Plenum Press, New York, 1979, pp. 79-99.
43 Johanssen et al., in Mental Workload. Its Theory and Measurement, N. Moray

(Ed.), Plenum Press, New York, 1979, pp. 101-114.
44 D. A. Topmiller, in Manned Systems Design: Methods, Equipment and Applications, J. Moraal and K. F. Kraiss (Eds.) Plenum Press, New York, 1981, pp. 3-31.
45 R. C. Williges, in Manned Systems Design: Methods, Equipment and Applications, J. Moraal and K. F. Kraiss (Eds.) Plenum Press, New York, 1981, pp. 59-87.
46 W. W. Wierville, in Manned Systems Design: Methods, Equipment and Applications, J. Moraal and K. F. Kraiss (Eds.) Plenum Press, New York, 1981, pp. 201-220.
47 A. Chapanis, Men, Machines and Models. American Psychologist, 1961 (6) 113-131.
48 A. Chapanis, Research Techniques in Human Engineering, The John Hopkins Press, Baltimore, 1959.
49 T. B. Sheridan and W. R. Ferrell, Man-Machine Systems: Information, Control, and Decision Models of Human Performance, The MIT Press, Cambridge, 1981.
50 V. D. Hopkin, D. L. Parks, W. Rohmert, A Rault, T. Soede, and H. Schmidtke in N. Moray (Ed.) Mental Workload, Its Theory and Measurement, Plenum Press, New York, 1979.
51 R. J. Audley, W. Rouse, T. Senders, and T. Sheridan in N. Moray (Ed.), Mental Workload, Its Theory and Measurement, Plenum Press, New York, 1979, pp. 269-285.
52 W. B. Rouse in N. Moray (Ed.), Mental Workload. Its Theory and Measurement, Plenum Press, New York, 1979, pp. 255-262.
53 R. Curry et al., 1979, in N. Moray (Ed.) Mental Workload, Its Theory and Measurement, Plenum Press, New York, 1979.
54 W. B. Rouse in N. Moray (ED.) Mental Workload, Its Theory and Measurement, Plenum Press, New York,1979, pp. 255-267.
55 R. Harre, The Philosophies of Science. Oxford University Press, London, 1972.
56 H. J. Zimmermann, Mathematical Modeling, 1 (1980) pp. 123-139.
57 M. Smithson, Mathematical Social Sciences, 2(1982) pp. 257-274.
58 O. Helmer and N. Rascher, Management Science, 6(1959) 25-52.
59 L. L. Pipino and J. P. Van Gigch, Cybernetics and Systems, 12 (1981) pp. 21-35.
60 W. Rohmert, in N. Moray (Ed.) Mental Workload, Its Theory and Measurement, Plenum Press, New York, 1979, p. 481.
61 L. A. Zadeh, Information and Control, 8 (1965), 338-353.

Applications of Fuzzy Set Theory in Human Factors, edited by W. Karwowski and A. Mital
Elsevier Science Publishers B.V., Amsterdam, 1986 — Printed in The Netherlands

AN INTRODUCTION TO FUZZY SET THEORY

RONALD R. YAGER

Machine Intelligence Institute, Iona College, New Rochelle, NY 10801 (U.S.A.)

ABSTRACT

In this paper we present a short introduction to the basic elements of fuzzy set theory with the purpose of providing an understanding useful for the remaining articles.

INTRODUCTION

The concept of a fuzzy subset was first introduced by L. A. Zadeh in 1965 (ref.6). It is especially useful for the representation of imprecise knowledge of the type which is prevalent in human concept formulation and reasoning.

A fuzzy set is a generalization of the ideas of an ordinary or crisp set. A fuzzy subset can be seen as a predicate whose truth values are drawn from the unit interval, I = [0,1] rather than the set {0,1} as in the case of an ordinary set. Thus the fuzzy subset has as its underlying logic a multi-valued logic.

The fuzzy subset allows for the description of concepts in which the boundary between having a property and not having a property is not sharp.

FUNDAMENTAL CONCEPTS OF FUZZY SUBSETS

In general a fuzzy subset is always defined in the framework of some crisp set, called the universe of discourse.

Def: Assume X is a set corresponding to the universe. A fuzzy subset A of X is a subset in which the membership grade of any element $x \in X$ is drawn from the unit interval, I.

Example: Assume:

$$X = \{x_1, x_2, x_3, x_4\} \tag{1}$$

$A = \{.7/x_1, .3/x_2, 1/x_3, 0/x_4\}$ is an example of a fuzzy subset of X.

In the framework of the theory of fuzzy subsets terms of the form a_1/x_1 are understood to indicate that the element x_1 has membership grade a_1 in the fuzzy subset A. Using this interpretation we see that in the above example that x_2 has membership grade .3 in the fuzzy subset A. We note the larger the membership grade of an element the more strongly it is a member of the fuzzy subset. If A is a fuzzy subset of X we shall generally use A(x) to indicate the membership function of A, with $A(x_1)$ indicating the membership grade of x_1 in the set A. Using this notation we shall see terms like $A(x_1)/x_1$. The membership function of a fuzzy subset can be seen to be a predicate in this sense $A(x_1)$ indicates the degree to which x_1 has the property represented by a fuzzy subset A.

Any ordinary set can be represented as a fuzzy subset.

Example: Assume $X = \{x_1, x_2, x_3, x_4\}$. The crisp set $B = \{x_1, x_2\}$ can be expressed in the previous notation as:

$$B = \{1/x_1, 0/x_2, 1/x_3, 0/x_4\}. \tag{2}$$

Thus crisp sets are special cases of fuzzy sets in which the membership grades are just either zero or one.

Fuzzy subsets are particularly useful for representing concepts corresponding to imprecise predicates. Assume our universe of discourse is a set of people X. If we are interested in representing the subset of "tall" people the use of a fuzzy subset in this situation to represented "tall people" frees us from the restriction of having to categorize each person as being a member or not of this set but allows us to make more subtle distinctions by using the grades of membership. This ability to more naturally represent this imprecise concept is particularly important in designing and studying human concept driven systems in that it allows us to represent more closely the types of concepts which humans use, such as old, young, short, fat, etc. If our universe of discourse is the set of real numbers concepts such as small, large, "close to 5" etc. are more naturally represented as a fuzzy subset rather than as a crisp subset with its abrupt transition from membership to non-membership.

We should note that the value of the membership grade for a fuzzy subset is in many cases a subjective context dependent value as is the idea of belonging and not belonging to a crisp subset. Furthermore, in many instances it is the shape of the membership function that is of significance rather than the actual values.

Most of the definitions and operations associated with fuzzy sets are straight forward extensions of corresponding definitions from ordinary set theory. However in some cases, because of the fact that membership is drawn from the unit interval rather than simply {0,1} we shall make definitions and define operations which are nonexistent in ordinary set theory. These new definitions generally become vacuous when applied to crisp sets. Furthermore, those definitions and operations which are extensions of ordinary set definitions usually collapse to their ordinary counterpart when restricted to fuzzy subsets with membership grades of zero or one (crisp sets).

Def: A fuzzy subset A of X is called normal if there exists at least one element $x \in X$ such that $A(x) = 1$. A fuzzy subset that is not normal is called subnormal.

We note all crisp subsets except for the null set are normal.

Def. The height of a fuzzy subset A is the largest membership grade of any element in element in A. It is denoted height(A). Hence:

$$\text{height}(A) = \max_{x \in X} A(x) \tag{3}$$

Thus a normal fuzzy subset has height equal to one.

Def: Assume A is a fuzzy subset of X the support of A, denoted Supp(A), is the crisp subset of X whose elements all have nonzero membership grades in A. Hence:

$$\text{Supp}(A) = \{x \mid A(x) > 0 \text{ and } x \in X\}. \tag{4}$$

Example: Assume $X = \{a, b, c, d, e\}$ where $A = \{1/a, .3/b, .2/c, .8/d, 0/e\}$ and $B = \{.6/a, .9/b, .1/c, .3/d, .2/e\}$. A is a normal fuzzy subset and B is a subnormal fuzzy subset of X.

height(A) = 1

height(B) = .9

$$\text{Supp}(A) = \{a, b, c, d\} \tag{5}$$

$$\text{Supp}(B) = \{a, b, c, d, e\} \tag{6}$$

We now extend the definitions of containment and equality among sets to fuzzy sets.

Def: Assume A and B are two fuzzy subsets of X. A is said to be a subset of (or contained in) the fuzzy subset B, denoted $A \subset B$ if $B(x) \geq A(x)$ for each $x \in X$.

Example: Assume $A = \{.1/a, .6/b, 1/c\}$ and $B = \{.3/a, .9/b, 1/c\}$ then $A \subset B$. To see that this definition extends the ordinary set idea we look at the following example:

Example: Let $X = \{a, b, c, d, e\}$. Assume $A = \{a,b\}$ and $B = \{a, b, c\}$. It is obvious that $A \subset B$. In fuzzy set notation:

$$A = \{1/a, 1/b, 0/c, 0/d, 0/e\} \tag{7}$$

$$B = \{1/a, 1/b, 1/c, 0/d, 0/e\} \tag{8}$$

Since $B(x) \geq A(x)$ for each $x \in X$ it follows that $A \subset B$.

Def: Assume A and B are two fuzzy subsets of X. A and B are said to be equal, denoted A = B, if $A \subset B$ and $B \subset A$. We note that A = B if A(x) = B(x) for $x \in X$.

Def: The null fuzzy subset of X is defined as the fuzzy subset ϕ of X such that $\phi(x) = 0$ for each $x \in X$. We note that for any fuzzy subset A of X, $\phi \subset A$. Furthermore, since the universe X can be seen to be the fuzzy subset for which X(x) = 1 for all $x \in X$ then for any A, $A \subset X$.

Def: Assume A is a fuzzy subset of the set X, where $X = \{x_1, \ldots x_n\}$. The power of the fuzzy set A, denoted Power[A], is defined as:

$$\text{Power}[A] = \sum_{i=1}^{n} A(x_1). \tag{9}$$

We note that for crisp sets the power becomes equivalent to the cardinality of the set. While the idea of the cardinality of a fuzzy set is a complex concept in some cases the power is used as a measure of the cardinality. In these cases it is called the sigma-count of the fuzzy subset and is denoted Σ-Count(A).

OPERATIONS ON FUZZY SETS

In this section we extend the set theoretic notions of ordinary set theory to fuzzy sets. In most cases there exist various possible ways for extending these operations. A very lively literature in the field is concerned with this issue. The definitions we usually provide are the so-called default definitions which are most often used. As we noted earlier all those operations which are extensions of crisp concepts default to their usual meaning when the fuzzy subsets have membership grades that are drawn from {0,1}. For this reason when extending operations to fuzzy sets we use the same symbol as in set theory. We also introduce some operations which are not usually defined for crisp sets.

Def: Assume A and B are two fuzzy subsets of X. Their union is a fuzzy subset C of X, denoted $C = A \cup B$, such that for each $x \in X$

$C(x) = \text{Max}\,[A(x), B(x)] = A(x) \vee B(x)$.

Example: Let X = {a, b, c, d, e}. Assume A = {1/a, .7/b, .3/c, 0/d, .9/e} and B = {.2/a, .9/b, .4/c, 1/d, .4/e}, then C = {1/a, .9/b, .4/c, 1/d, .9/e}.

Def: Assume A and B are two fuzzy subsets of X. Their intersection is a fuzzy subset D of X, denoted $D = A \cap B$, such that for each $x \in X$

$$D(x) = \text{Min}\,[A(x), B(x)] = A(x) \wedge B(x) \tag{10}$$

Example: Using same values for X, A and B as in the previous example we note that D = {.2/a, .7/b, .3/c, 0/d, .4/e}.

Theorem: Assume A and B are two fuzzy subsets of X. If $C = A \cup B$ and $D = A \cap B$

then:

(1) $D \subset C$ (11)

(2) $A \subset C$ and $B \subset C$ (12)

(3) $D \subset A$ and $D \subset B$ (13)

Proof: (1) For each $x \in X$, $D(x = \text{Min}\ [A(x), B(x)]$ and $C(x) = \text{Max}\ [A(x), B(x)]$ hence $C(x) \geq D(x)$ and thus $D \subset C$

(2) Since $C(x) = \text{Max}\ [A(x), B(x)]$, then $C(x) \geq A(x)$ and $C(x) \geq B(x)$ for all $x \in X$ hence $A \subset C$ and $B \subset C$

(3) Since $D(x) = \text{Min}\ [A(x), B(x)]$ then $D(x) \leq A(x)$ and $D(x) \leq B(x)$ thus $D \subset A$ and $D \subset B$.

The following is a list of properties relating the intersection and union of fuzzy subsets. They are derived by arguments similar to those used in the above theorem. These properties completely replicate those existing in ordinary set theory.

Assume A, B, and C are fuzzy subsets of X. The following properties hold regarding union and intersection.

Commutativity $A \cup B = B \cup A$; $A \cap B = B \cap A$ (14)

Idempotency $A \cup A = A$; $A \cap A = A$ (15)

Associativity $A \cup (B \cup C) = (A \cup B) \cup C = A \cup B \cup C$ (16)

$A \cap (B \cap C) = (A \cap B) \cap C = A \cap B \cap C$ (17)

Distributivity $A \cap (B \cup C) = (A \cap B) \cup (A \cap C)$ (18)

$A \cup (B \cap C) = (A \cup B) \cap (A \cup C)$ (19)

$A \cup \phi = A$; $A \cup X = X$; $A \cap \phi = \phi$; $A \cap X = A$ (20)

$A \subset A \cup B$; $A \supset A \cap B$; $A \cap B \subset A \cup B$ (21)

If $A \subset B$ then $B = A \cup B$; $A = A \cap B$ (22)

$A \subset B$ and $B \subset C$ then $A \subset C$ (23)

Note The operations of Max and Min which play such a fundamental role in fuzzy set theory can be written in algebraic terms:

$$\text{Max}\ (a,b) = (a + b + |a - b|)/2 \tag{24}$$

$$\text{Min}\ (a,b) = (a + b - |a - b|)/2 \tag{25}$$

We now extend the operations of relative complement and complement (or negation) to fuzzy subsets.

Def: Assume A and B are two fuzzy subsets of X. The relative complement of A with respect to B, denoted $E = A - B$, is defined as the fuzzy subset E where for each $x \varepsilon X$, $E(x) = \text{Max}\,[0, B(x) - A(x)]$.

Example: Assume A = {1/a, .6/b, .3/c, 0/d, .5/f} and
B = {1/a, .3/b, .7/c, .4/d, 1/f} then E = B - A = {0/a, 0/b, .4/c, .4/d, .5/f}.

Def: Assume A is a fuzzy subset of X. The complement or negation of A, denoted $\bar{A}$ is defined as the fuzzy subset $\bar{A} = X - A$, hence for each $x \varepsilon A$
$\bar{A}(x) = 1 - A(x)$.

Example: A = {1/a, .6/b, .3/c, 0/d, .5/f}, $\bar{A}$ = {0/a, .4/b, .7/c, 1/d, .5/f}.
We note that in general $\bar{A} \cap A \neq \phi$, and $\bar{A} \cup A \neq X$.

The following is a list of properties involving complements and relative complements of fuzzy subsets A and B of X.

Double Negation $(\bar{\bar{A}}) = A$ (26)

De Morgans Law $(\overline{A \cup B}) = \bar{A} \cap \bar{B}$; $(\overline{A \cap B}) = \bar{A} \cup \bar{B}$ (27)

$\bar{\phi} = X$; $\bar{X} = \phi$ (28)

If $A \subset B$ then $\bar{A} \supset \bar{B}$ and $A - B = \phi$ (29)

$A - A = \phi$ (30)

$A - \phi = A$; $\phi - A = \phi$ (31)

We should note that the definitions for relative complement and complement provided here are equivalent to the classic definitions for crisp subsets.

Def: Assume A is a fuzzy subset of X and α is any non-negative number then by A^α we mean the fuzzy subset of X such that $A^\alpha(x) = (A(x))^\alpha$

Example: If A = {1/a, .6/b, .3/c, 0/d, .5/f}
A^2 = {1/a, .36/b, .09/c, 0/d, .25/f}

We note that if $\alpha > 1$, then $A^\alpha \subset A$ and if $\alpha < 1$, $A^\alpha \supset A$. Furthermore if A is a crisp subset and $\alpha > 0$ then $A^\alpha = A$.

Def: Assume A and B are two fuzzy subsets of X, the bounded sum of A and B, denoted $D = A \oplus B$, is a fuzzy subset of X such that
$D(x) = \text{Min}\,[1 \wedge (A(x) + B(x))]$.

Def: Assume A is any fuzzy subset of X and α is any number in the unit interval. We define the fuzzy subset $F = \alpha A$ as $F(x) = \alpha A(x)$.

Def: Assume A is a fuzzy subset of X, the α-level set of A, denoted A_α, is the crisp subset of A consisting of all the elements in X for which $A(x) \geq \alpha$, hence $A_\alpha = \{x \mid A(x) \geq \alpha, x \varepsilon X\}$.

The α -level sets play a role in the decomposition of any fuzzy subset. In particular any fuzzy subset A of X can be written as:

$$A = \bigcup_{\alpha \in I} \alpha A_{\alpha} . \tag{32}$$

More particularly if X is finite and K is the finite subset of K consisting of the set of all membership grades of elements in A then:

$$A = \bigcup_{\alpha \in K} \alpha A_{\alpha} . \tag{33}$$

FUZZY RELATIONSHIPS

In this section we extend the concept of relationship to fuzzy sets. We recall that if X and Y are two crisp sets their cartesian product is the set X x Y consisting of all pairs (x,y) where $x \in X$ and $y \in Y$.

Def: A fuzzy relationship over the pair X, Y is defined as a fuzzy subset of the cartesian product X x Y.

Example: Let X = {a, b, c} and Y = {1, 2} then:
A = {.1/(a,1), .6/(a,2), .9/(b,1), 1/(b,2), 0/(c,1), .2/(c,2)} is a fuzzy relationship. More generally if $X_1, X_2, \ldots X_n$ are a collection of sets n-nary fuzzy relationship over $X_1 \times X_2 \ldots \times X_n$ is a fuzzy subset over the cartesian product of $X_1, X_2, \ldots X_n$.

Many relationships are usually expressed as fuzzy relationships which allow for the capturing of the imprecision inherent in these relationships. For example if X is a set of people and Y is a set of cities the relationship of a person being close to a city is capturable by a fuzzy relationship R over X x Y, where R(x,y) indicates the degree to which x lives near city y.

Since a fuzzy relationship is simply a fuzzy subset over the base set defined by the cartesian product then all the mechanism we have developed for handling fuzzy subsets can be used to manipulate fuzzy relationships. We now provide a mechanism for obtaining fuzzy relationships from fuzzy sets.

Def: Assume A and B are two fuzzy subsets of X and Y respectively their cross product A x B is a fuzzy relationship T on the set X x Y in which, T = A x B where T(x,y) = Min [A(x), B(y)].

Example: Assume A = {1/a, .6/b, .3/c} and B = {1/1, .5/2, 0/3} then:
A x B = {1/(a,1), .5/(a,2), 0/(a,3), .6/(b,1), .5/(b,2), 0/(b,3), .3/(c,1), .3/(c,2), 0/(c,3)} (34)

More generally if $A_1, A_2, \ldots A_n$ are fuzzy subsets of $X_1, X_2, \ldots X_n$ respectively then the cross product $A_1 \times A_2 \ldots \times A_n$ is a fuzzy subset of $X_1 \times X_2 \ldots X_n$, denoted

$T = A_1 \times A_2 \ldots A_n$ where:

$$T(x_1, x_2, \ldots x_n) = \underset{i=1,n}{\text{Min}}\ A_1(x_1), \text{ and } x_1 \in X_1. \tag{35}$$

Assume A is a fuzzy subset of X. Let Y be another set. The <u>cylindrical extension</u> of A to X x Y is the special cross product denoted $A^\wedge$. In particular $A^\wedge = A \times Y$ hence $A^\wedge(x,y) = A(x) \wedge Y(y) = A(x) \wedge 1 = A(x)$ for each $(x,y) \in X \times Y$.

<u>Def</u>: Assume A is a fuzzy relationship on X x Y. The projection of A onto X is a fuzzy subset of X which we denote $\text{Proj}_x A$ such that:

$$\text{Proj}_x A(x) = \underset{y \in Y}{\text{Max}}\ A(x,y). \tag{36}$$

FUZZY NUMBERS AND THE EXTENSION PRINCIPLE

In many situations people are only able to characterize numeric information imprecisely. For example people use terms like "about 5," "near zero," "more or less than 10." These are examples of what are called fuzzy numbers. Using the theory of fuzzy subsets we can represent these fuzzy numbers as fuzzy subsets of the set of real numbers. However, in order to use these fuzzy numbers in any intelligent system we must be able to perform arithmetic operations on these numbers. In particular we must be able to add, subtract, multiply, etc. with fuzzy numbers.

We shall first introduce the extension principle and then use it to provide for these arithmetic operations on fuzzy numbers.

The extension principle plays a fundamental role in enabling us to extend many point operations to operations involving fuzzy sets.

<u>Extension Principle</u>: Assume X and Y are two sets. Let f be a mapping from X into Y, i.e. $f: X \rightarrow Y$ such that for each $x \in X$, $f(x) = y \in Y$.

Assume A is a fuzzy subset of X, we can define f(A) as a fuzzy subset of Y such that $f(A) = \underset{x \in X}{\cup} \{A(x)/f(x)\}$.

<u>Example</u>: Assume $X = \{1, 2, 3\}$ and $Y = \{a,b\}$ and f is defined by $f(1) = a$, $f(z) = a$, $f(3) = b$. Let $A = \{1/1, .3/2, .7/3\}$ then: $f(A) = \{1/a\} \cup \{.3/a\} \cup \{7/b\} = \{1/a, .7/b\}$.

It can be easily seen that if we denote $B = f(A)$ then B is a fuzzy subset of Y such that for each $y \in Y$,

$$B(y) = \underset{\substack{\text{all } x \\ \text{such that} \\ f(x) = y}}{\text{Max}}\ A(x). \tag{37}$$

We note that the extension principle extends the mapping f(A). Consider the point $x \in X$, this can be represented as the fuzzy subset A; where $A(x^*) = 1$, $A(x) = 0$ for all $x \neq x^*$. Then $f(A) = B$ where $B = \{1/f(x^*)\} = f(x^*)$.

We can use the extension principle to define the degree of membership of one fuzzy subset in another.

Assume A is a fuzzy subset of X. The membership function of A can be considered a mapping from X into the unit interval $A: X \rightarrow [0, 1] = I$ such $A(x) \in I$ equals the membership grade of x in A.

Let B be another fuzzy set of X, the membership grade of B in A, denoted A(B) can be obtained via the extension principle as the fuzzy subset over I where

$$A(B) = \bigcup_{x \in X} \{B(x)/A(x)\}. \tag{38}$$

The extension principle can be generalized. Let $X_1, X_2, \ldots X_n$ and Y be a family of sets. Assume f is a mapping from the Cartesian product $X_1 \times X_2 \ldots x X_n$ into Y that is for each n tuple $(x_1, x_2, \ldots x_n)$ such that $x_1 \in X_1$ we have $f(x_1, \ldots x_n) = y \in Y$. Let $A_1, \ldots A_n$ be fuzzy subsets of $X_1, \ldots X_n$ respectively then the extension principle allows for the evaluation of $f(A_1, A_2, \ldots A_n)$. In particular $f(A_1, \ldots A_n) = B$, where B is a fuzzy subset of y such that

$$B(y) = \underset{\substack{\text{over all} \\ (x_1 \ldots x_n) \in X_1 x \ldots X_n \\ \text{such that} \\ f(x_1, \ldots x_n) = y.}}{\text{Max}} [A_1(x_1 \wedge A_2(x_2 \wedge \ldots A_n(x_n)] \tag{39}$$

We note that if there exists no tuple $(x_1 \ldots x_n)$ such that $f(x_1 \ldots x_n) = y$ then $B(y) = 0$. That is:

$$B = \bigcup_{\substack{x_1 \ldots x_n \\ \in \\ X_1 \ldots X_n}} \{\text{Min } [A_1(x_1)]/f(x_1, \ldots x_n)\} \tag{40}$$

A special application of this form of the extension principle allows us to define arithmetic operations on fuzzy numbers.

Assume R is the set of real numbers. The operation of addition can be seen as a mapping, P from R x R into R. That is for each pair of $x, y \in R$, $P(x,y) = z$, whenever $x + y = z$. Thus addition is a mapping and hence we can use the extension principle to allow for fuzzy numbers.

Thus if A and B are two fuzzy numbers, their sum $A + B = C$, where C is a fuzzy number (fuzzy subset of the real number), where for each $z \in R$,

$$C(z) = \underset{\substack{\text{all } x, y \\ \text{such} \\ x+y=z}}{\text{Max}} [A(x) \wedge B(y)] \tag{41}$$

More generally assume $\perp$ is any binary arithmetic operation, addition (+), subtraction (-), multiplication (*), division (/) or exponentiation (^) and A and B are two fuzzy numbers. Then AxB=C, where C is a fuzzy number such that

$$C = \bigcup_{\substack{\text{all} \\ x,y \in R}} \{A(x) \wedge B(y))/(x \perp y)\} \qquad (42)$$

or equivalently for all z ε R

$$C(z) = \underset{\substack{\text{all } x,y \\ \text{such that} \\ x \perp y = z}}{\text{Max}} [A(x) \wedge B(y)] \qquad (43)$$

In some cases the membership grade of a fuzzy subset itself may be imprecise. For example, we may have membership grades like "high," "close to 1," "about .5," that is the membership grade is a fuzzy subset of the unit interval. Fuzzy subsets having these types of membership functions are called type II fuzzy subsets.

Assume A and B are two type II fuzzy subsets of A. Their intersection $C = A \cap B$ and union $D = A \cup B$ are also fuzzy subsets of type II such that

$$C(x) = A(x) \wedge B(x) = \text{Min } [A(x), B(x)] \qquad (44)$$

$$D(x) = A(x) \cup B(x) = \text{Max } [A(x), B(x)] \qquad (45)$$

However since A(x) and B(x) are fuzzy subsets we must be able to obtain the min and max of fuzzy numbers. The extension principle provides a mechanism for accomplishing this task.

Assume R and S are two fuzzy subsets of the unit interval, I, then Min (R, S) = T where T is a fuzzy subset of I such that for each y ε I

$$T(y) = \underset{\substack{\text{all } u,v \\ \text{such that} \\ y=\text{Min}(u,v)}}{\text{Max}} [R(u) \wedge S(v)]. \qquad (46)$$

If T = Max [R, S] then T again is a fuzzy subset of I such that for each y ε I

$$T(y) = \underset{\substack{\text{all } u,v \\ \text{such that} \\ y=\text{Max}(u,v)}}{\text{Max}} [R(u) \wedge S(u)]. \qquad (47)$$

We shall provide another form of extension associated with fuzzy sets. Let X and Y be two crisp sets. A set mapping G from X into Y associates with each subset of X a point in Y. That is, if ρ (X) is the power set of X, the set of all subsets of X, then $G: \rho(x) \to Y$ where for each $A \subset X$ $(A \varepsilon \rho(X))$ $G(A) = y \varepsilon Y$.

Examples of this type of function are probability measures and belief and plausibility measures used in the Shafer theory of evidence.

In order to extend G to act on fuzzy subsets of X, we proceed as follows. Let B be any fuzzy subset of X. B can be expressed as $B = \bigcup_{\alpha \in I} \alpha B_\alpha$. where B_α is the α-level set of B.

Then $$G(B) = \bigcup_{\alpha \in I} \{ \alpha / G(B_\alpha) \}. \qquad (48)$$

We note that since B_α is a crisp subset of X the operation $G(B_\alpha)$ is well defined.

REFERENCES

1 D. Dubois and H. Prade, Fuzzy Sets and Systems Theory and Applications, Academic Press, New York, 1980.
2 J. Kacprzyk, Multistage Decision Making Under Fuzziness, Verlag TUV Rheinland, Cologne, 1983.
3 A. Kandel, Fuzzy Mathematics, Addison-Wesley, Reading, MA, 1985.
4 A. Kaufmann, Theory of Fuzzy Subsets, Academic Press, New York, 1975.
5 C.V. Negoita and D. Ralescu, Applications of Fuzzy Sets to Systems Analysis, John Wiley & Sons, New York, 1975.
6 L.A. Zadeh, Fuzzy Sets, Information and Control 8, 1965, pp. 338-353.
7 H.J. Zimmermann, Fuzzy Set Theory - and Its Applications, Kluwer-Nijhoff Publishing, Hingham, MA, 1985.

Applications of Fuzzy Set Theory in Human Factors, edited by W. Karwowski and A. Mital
Elsevier Science Publishers B.V., Amsterdam, 1986 — Printed in The Netherlands

FUZZY CONCEPTS IN HUMAN FACTORS/ERGONOMICS RESEARCH

WALDEMAR KARWOWSKI[1] and ANIL MITAL[2]

[1]Department of Industrial Engineering, University of Louisville, Louisville, Kentucky 40292 (U.S.A.)

[2]Ergonomics Research Laboratory, University of Cincinnati, Cincinnati, Ohio 45221-0072 (U.S.A.)

(...."One should never try to be more precise
than the problem situation demands."
K. Popper)

ABSTRACT

The perspectives and potential impact of the concept of fuzziness in human factors studies are explored. It is advocated that fuzzy methodologies should be applied in this field of research to a much greater extent than has been done in the past.

INTRODUCTION

Human factors/ergonomics, defined operationally as an interdisciplinary study aimed at optimization of work systems with respect to physical and psychological characteristics of a person, investigates complex and usually ill-defined (imprecise) relationships between people, machines, and physical environments. The main goal of such investigation is to identify and remove (or reduce) the existing incompatibilities between human capacities and task requirements, and to make the workplace safe, healthy, and productive, as well as a comfortable and satisfying one.

According to Oborne (ref. 1), the human factors/ergonomics discipline arose as a response to the need to consider how the human operator manages to cope with his environment. From the very beginning, however, this objective has not been easy to fulfill for at least two main reasons, i.e.: (1) natural imprecision and uncertainty inherent to complex human-centered systems, and (2) lack of research methodology which would allow one to account for, rather than disregard, the human- and system-based uncertainties in the analysis process.

Human-centered systems, which are the objects of ergonomics research, are very complex and therefore difficult to analyze. Furthermore, an ergonomist must deal effectively with at least three different types of uncertainty inherent to such systems; i.e., inaccuracy, randomness, and vagueness (ref. 2). Uncertainties due to inaccuracy are related to observations and measurements (representations), while those due to randomness (of events) are independent from observations and constitute an objective property of some real process (ref. 2). Uncertainty due to vagueness (or fuzziness) has to do with the complexity of the system under investigation and the human thought and perception processes (ref. 3). The last category of uncertainty is of utmost importance to human factors/ergonomics studies, and has to be taken into account more carefully.

Contemporary research techniques in the human factors/ergonomics area are based on the premise that if the uncertainty exists, one must restrict the model to eliminate, rather than incorporate, as much of it as possible, then precisely explain whatever uncertainty is left. Such a premise is a consequence of adapting quantitative methods of analysis directly from physical sciences (ref. 4). As pointed out by Lord Kelvin in the XIX century, "...a first essential step in the direction of learning any subject is to find principles of numerical reckoning and practicable methods for measuring some quality connected with it."

Traditional scientific thinking, based primarily on the Aristotelian logic is oriented towards exact quantitative methods of analysis. Such methods (and corresponding models) equate uncertainty with randomness only and fail to recognize the human- and system-based uncertainties due to vagueness. Also, according to Zadeh´s principle of incompatibility (ref. 3), at a high level of complexity, precision and significance (of the statements about the system´s behavior) become almost mutually exclusive characteristics. Therefore, an attempt to make precise and yet significant statements about the complex relationships between people, machines, and environments may be an illusive task, and the traditional modeling methods may not have much relevance here.

A new methodology in the area of human factors/ergonomics is needed to account for imprecision and vagueness of such relationships. Zadeh (ref. 5) points out that "Although the conventional mathematical techniques have been and will continue to be applied to the analysis of humanistic systems, it is clear that the great complexity of such systems call for approaches that are significantly different in spirit as well as in substance from the traditional methods -- methods which are highly effective when applied to mechanistic systems, but are far too precise in relation to systems in which human behavior plays an important role." Furthermore, as the author points out, "in order to be able to make significant assertions about the behavior of humanistic systems, it may be necessary to abandon the high standards of rigor and precision that we have become conditioned to expect of our mathematical analyses of well-structured mechanistic systems, and become more tolerant of approaches which are approximate in nature"

The above points of view can be defended by early school of thought shared by such philosophers of science as J. Schwartz and K. Popper. According to

Schwartz (ref. 6), "an argument, which is only convincing if it is precise, loses all its force if the assumptions on which it is based are slightly changed, while an argument, which is convincing but imprecise, may well be stable under small perturbations of its underlying axioms." Popper (ref. 7) had reinforced such position by advocating that both precision and certainty are false ideals which are impossible to attain, and therefore often misleading if accepted as guides. According to the philosopher, "The quest for precision is analogous to the quest for certainty, and both should be abandoned." Zimmermann (ref. 8) indicates that real situations are very often not crisp and deterministic and they cannot be described precisely.

Most of the traditional human factors methodologies show an intended disregard for system complexities, and assume that the formal properties of mathematics (usually statistics) correspond to some existing relationships characteristic to the system under investigation (ref. 5). For example, an uncertainty due to vagueness is often modeled (if not disregarded) as being of stochastic nature. One should notice that such treatment appears to defeat the purpose of any formal system's analysis and modeling efforts.

THE CONCEPT OF FUZZINESS

Fuzziness relates to the specific kind of vagueness having to do with gradations in categories, i.e. degree vagueness (ref. 9). Uncertainty measured by fuzziness refers to the gradation of membership of an element in some class (category). Although such uncertainty arises at all levels of cognitive processes (ref. 10 and 11), people have the abilities to understand and utilize vague and imprecise concepts which are difficult to analyze within the framework of traditional scientific thinking. Therefore, awareness of vagueness and inexactness, implicit in human behavior, should be the basis of any human factors/ergonomics studies. Furthermore, ergonomists should learn and apply the mathematical tools for dealing with vague and imprecise concepts.

According to Zadeh (ref. 12), the theory of fuzzy sets represents an attempt for constructing a conceptual framework for a systematic treatment of vagueness and uncertainty due to fuzziness in both quantitative and qualitative ways. Such framework is much needed in the human factors/ergonomics area. As pointed out by Singleton (ref. 13) "most human characteristics have very complex contextual dependencies which are not readily expressible in tabulations of numbers even in multivariate equations." Yet there is growing evidence that people comprehend vague concepts, such as concepts of a natural language, as if those concepts were represented by fuzzy sets, and manipulate them according to the rules of fuzzy logic (ref. 14 and 15). Moreover, recent research in semantic memory and concept formation (ref. 16) indicates that natural categories are fuzzy sets with no clear boundaries separating category members from nonmembers.

One can certainly understand the meaning of such concepts as "excessive workload," "low illumination," "heavy weight," "high level of stress," and "tall man," to name a few commonly used descriptors of the human-environment relationship.

As noted by Singleton (ref. 13), "no one has yet developed a comprehensive set of crude and approximate but simple and inexpensive techniques finding solutions to ergonomics problems." Fuzzy set theory, which allows interpretation and manipulation of imprecise (vague) information and recognition and evaluation of uncertainty due to fuzziness (in addition to randomness), may be the closest solution, to the above stated need available to ergonomists today. For the above reasons, a formal treatment of vagueness is an important and necessary step toward more realistic handling of imprecision and uncertainty due to human and behavior thought process at work. It is our view that the theory of fuzzy sets will prove successful in narrowing the gap between the world of the precise or "hard" sciences and the world of the cognitive or "soft" sciences. This can be achieved by providing a mathematical framework in which vague conceptual phenomena where fuzzy descriptors, relations, and criteria are dominant (ref. 8) can be adequately studied and modeled.

CONVENTIONAL VERSUS FUZZY SET THEORY AND LOGIC

In a conventional (classical) set theory, an element x either belongs or does not belong to a set X, and the characteristic (membership) function f_x can be represented as follows:

$$\begin{cases} 1 & \text{if } x \in X \quad \text{(truth value = 1: true)} \\ 0 & \text{if } x \notin X \quad \text{(truth value = 0: false)} \end{cases}$$

The concept of fuzzy set extends the range of membership values for f_x, and allows graded membership, usually defined on an interval [0,1]. Consequently, an element may belong to a set with a certain degree of membership, not necessarily 0 or 1. The "excluded middle" concept is then abandoned, and more flexibility is given in specifying the characteristic function.

In view of the above, the mathematical logic can also be modified. Interestingly, the classical logic was actually extended as early as 1930 by Lukasiewicz, who proposed the infinite-valued logic. As stated by Giles (ref. 17), "Lukasiewicz logic is exactly appropriate for the formulation of the "fuzzy set theory" first described by Zadeh; indeed, it is not too much to claim that is related to fuzzy set theory exactly as classical logic is related to ordinary set theory."

Fuzzy logic and fuzzy reasoning are described by Baldwin (ref. 18).

FUZZY METHODOLOGIES - BASIC DEFINITIONS

Although the formal introduction to fuzzy set theory was given in the previous chapter, for the convenience of the reader selected definitions of fuzzy set theory are recalled here to facilitate further discussion.

According to Zadeh (ref. 3), a fuzzy subset A of a universe of discourse U is defined by a membership function f_A: $U \rightarrow [0,1]$ which associates with each element u of U a number $f_A(u)$ in the interval [0,1], where $f_A(u)$ represents the grade of membership of u in A. Formally, A can be written as:

$$A = \{(u, f_A(u)), u \in U\} \qquad (1)$$

To simplify the notation, a fuzzy subset A with discrete membership function can be expressed as follows:

$$A = f_1/u_1 + f_2/u_2 + f_3/u_3 + \ldots + f_n/u_n \qquad (2)$$

where f_i, $i = 1,2,3,\ldots n$, is the grade of membership of u in A, and

$U = u_1 + u_2 + u_3 \ldots + u_n$

U is the universe of discourse, where + denotes "union" rather than the arithmetic sum.

For the given fuzzy subsets A and B one can perform basic operations as follows:

Complement: $\bar{A} \leftrightarrow \bar{f}_A = 1 - f_A$ (3)

Union of two fuzzy subsets A and B, denoted

$A \cup B$: $A \cup B \leftrightarrow f_{A\,B} = f_A \vee f_B$ (4)

Intersection of two fuzzy subsets, denoted

$A \cap B$: $A \cap B \leftrightarrow f_{A\,B} = f_A \wedge f_B$ (5)

where V and $\wedge$ denote MAX and MIN operators, respectively.

Product of A and B: $A \times B \leftrightarrow f_{AxB} = f_A f_B$ (6)

Concentration of A:

$$CON(A) \rightarrow A^2 \leftrightarrow CON(f_A) \rightarrow f^2_A \qquad (7)$$

Dilation of A:

$$DIL(A) \rightarrow A^{0.5} \leftrightarrow DIL(f_A) \rightarrow f^{0.5}_A \qquad (8)$$

Contrast intensification of A, denoted by INT(A):

$$INT(f_A) \leftrightarrow \begin{cases} 2f^2_A & \text{for } 0 \leq f_A \leq 0.5 \\ 1-2(1-f_A)^2 & \text{for } 0.5 \leq f_A \leq 1.0 \end{cases} \qquad (9)$$

is the operation that increases these values of f_A which are above 0.5 and decreases those which are below that point.

A graphical illustration of the fuzzy operations defined by formulas 7 through 9 are given in figure 1.

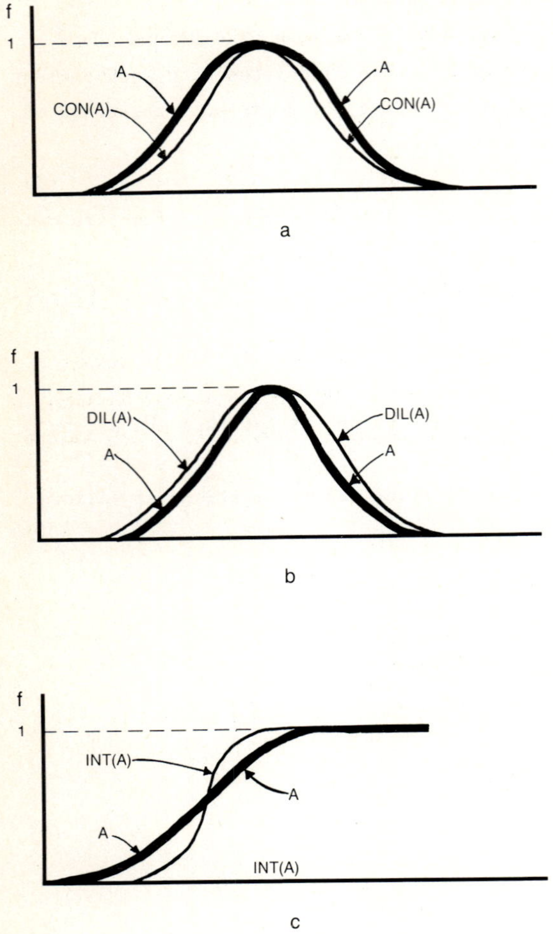

Figure 1. Graphical representation of the fuzy operators of: a) concentration, b) dilation, and c) contrast intensification.

THE LINGUISTIC APPROACH

The linguistic characterization (ref. 4) uses a concept of a linguistic variable with values which are not numbers but words (or sentences) of a natural (artificial) language. A linguistic value (e.g. ´high´) is interpreted as a label for a fuzzy restriction on the values of the base variable. The fuzzy restrictions are characterized by the compatibility functions. Each such function associates with each value of the base variable a number in the interval

[0,1] representing the compatibility with the fuzzy restriction.

Typical values of linguistic variables contain not only the primary terms such as 'low' or 'average,' but also hedges such as 'very' or 'more or less,' fuzzy connectives such as 'and' and 'or,' and the negation 'not.' The hedges, connectives, and negation are used as modifiers of the operands (primary terms) in a context-dependent situation. The modification of the meaning of primary terms (e.g. for f(u)) can be done as follows:

very low $\Leftrightarrow$ $low^2 \rightarrow f^2(u)$
not low $\Leftrightarrow$ $1-low \rightarrow 1-f(u)$
more or less high $\Leftrightarrow$ $high^{0.5} \rightarrow f^{0.5}(u)$
extremely high $\Leftrightarrow$ very(very high) = $high^4 \rightarrow f^4(u)$

Figure 2 depicts the result of an experimental work by Hersh and Caramazza (ref. 10) on experimental verification of the predictions of fuzzy set theory, and in particular on the perception of natural categories as being fuzzy, rather than crisp or well defined sets.

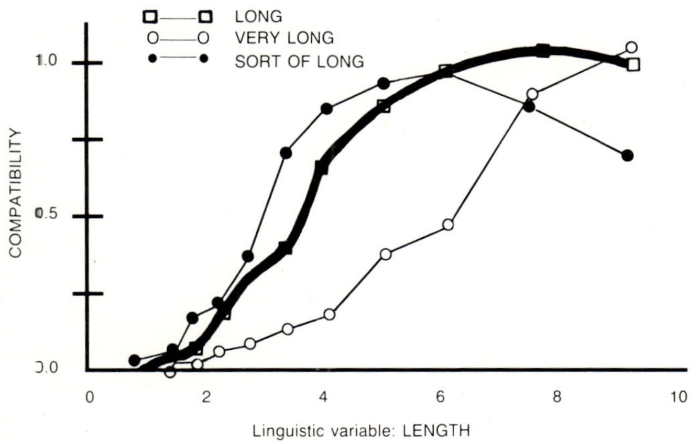

Figure 2a. Experimentally derived linguistic representations of the linear length (after ref. 10).

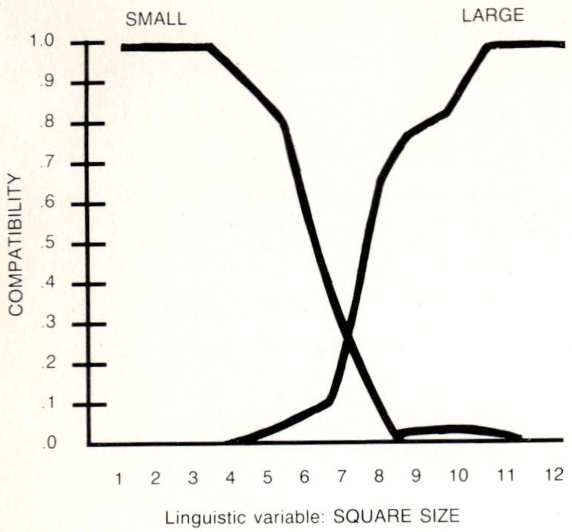

Figure 2b. Experimentally derived linguistic representations of the square size (after ref. 10).

APPROXIMATE REASONING

Approximate reasoning (ref. 3) utilizes fuzzy conditional statements and compositional rules of inference. In a fuzzy conditional statement, "If A then B," or "A→B," antecedent (A) and consequent (B) are fuzzy sets rather than propositional variables.

If A is a fuzzy subset in a universe of discourse U, and B is a fuzzy subset in a universe of discourse V, then the cartesian product of A and B, A x B, is defined as a fuzzy relation R from U to V:

$$A \times B = \sum_{R=U\times V} (f_A(u) \wedge f_B(v))/(u,v) \qquad (10)$$

where R is usually given in the form of a matrix.

A fuzzy conditional statement can be represented as follows:

$$A \rightarrow B = \text{IF A THEN B} \leftrightarrow (A \times B + \bar{A} \times V) \qquad (11)$$

where + denotes a union of fuzzy relations.

In case of a conditional statement of "IF A THEN B ELSE C" we will have the following:

$$\text{IF A THEN B ELSE C} \leftrightarrow (A \times B + \bar{A} \times C) \qquad (12)$$

If the fuzzy relation R from U to V is known, and A is a fuzzy subset of U, then the fuzzy subset B of V is induced by A, and is given by the composition of R and A:

$$B = A \circ R \tag{13}$$

where B is given by the max-min product of A and R:

$$B = A \cdot R \leftrightarrow f_B(v) = \sup_v (f_A(u) \wedge f_R(u,v))/(v) \tag{14}$$

The conceptual framework for approximate reasoning is based on fuzzy logic and the notion of linguistic variables proposed by Zadeh (ref.3-5) who advocates that "approximate reasoning underlines the remarkable human ability to understand natural language, decipher sloppy handwriting..., and make rational decisions in complex and/or uncertain environments."

FUZZY NUMBERS

A fuzzy number (ref. 19) is a fuzzy subset of the real line defined by its membership function $f: R \rightarrow [0,1]$. Any normal fuzzy number can be represented as a 4-tuple (a,b,α,β) where α- is called the 'left bandwidth,' and β the 'right bandwidth,' and $[a,b]$ is the closed interval on which the membership function is equal to 1.0. The shape of the left and right slopes of the membership function can be limited to an even function S, such that $S(-x) = S(x)$, and $S(0) = 1$. Several operations such as addition, subtraction, multiplication, division and exponentiation, allow one to perform calculations on fuzzy numbers similar to those performed on non-fuzzy numbers in crisp algebra. A graphical illustration of an addition of two fuzzy numbers (ref. 20) is given in figure 3.

THE CONCEPT OF FUZZY EVENTS

Let X be a set of events $(x_1, x_2, \ldots, x_n)$ with probabilities $p(x_i)$. A fuzzy event A (ref. 21) is a fuzzy set on X which membership function f_A is measurable and in a discrete case:

$$P(A) = \sum_{i=1}^{n} f_A(x_i)\ p(x_i) \tag{15}$$

where P(A) is called the probability of fuzzy event A. Such defined P(A) can be interpreted as the expectation of the membership function of a fuzzy event.

The amount of uncertainty associated with fuzzy event can be represented as the entropy measure H(A) of a fuzzy subset A of the finite set $(x_1, x_2, \ldots, x_n)$ with respect to a probability distribution $P = (p_1, p_2, \ldots, p_n)$ as follows:

$$H^F(A) = -\sum_{i=1}^{n} f_A(x_i)\ p_i \log p_i \tag{16}$$

This entropy measure could be used in system analysis as an assessment of the value of information describing given condition(s).

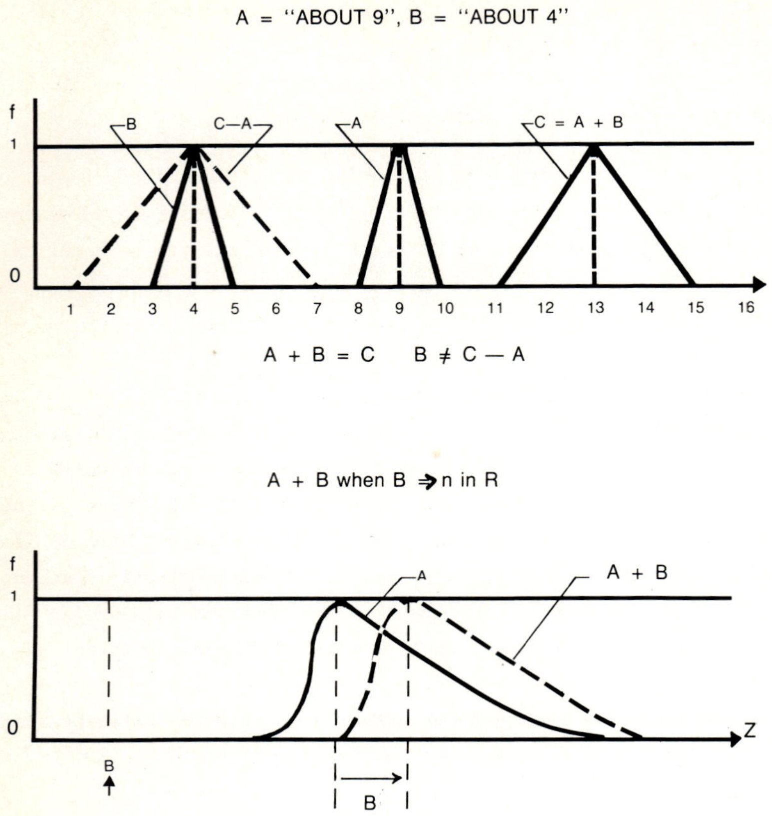

Figure 3. Graphical representation of addition of the two fuzzy numbers, and of sum of the real and fuzzy numbers (after ref. 20).

THEORY OF POSSIBILITY

According to Zadeh (ref. 23) a variable may be associated with both a possibility distribution and a probability distribution. The possibility/probability consistency principle states that what is probable must also be possible but not vice versa.

If B is a fuzzy subset of a universe of discourse $U = \{u\}$, and B is characterized by the membership function f_B, then the proposition of the form ´X is B,´ where X is a variable taking values in U, induces a possibility distribution Π_x which equates the possibility of X taking the value u to $f_B(u)$, i.e. the compatibility of u with B, which is denoted by

$$\Pi(X = u) = \Pi_x(u) \Leftrightarrow f_B(u) \tag{17}$$

Set B can be treated as a fuzzy restriction on X since it acts as a constraint on the values that may be assigned to X.

If R(X) denotes a fuzzy restriction associated with X, then one can write the above in the form of a relational assignment equation as follows:

$$R(X) = B \tag{18}$$

The possibility distribution Π_x satisfies the equation $\Pi_x = R(X)$ and the possibility that X = u, $\Pi_x(u)$, is equal to $f_B(u)$.

Consider the following example (ref. 23). Let X be a variable in the universe of discourse U = [1, 2, 3, 4,...] of natural numbers. A statement, "Jim eats X eggs for breakfast," can be interpreted in view of the possibility/probability consistency principle as shown in table 1. X can be associated here with both $\Pi_x(u)$- possibility distribution of X specifying the degree of ease with which Jim can eat U eggs for breakfast," and $P_x(u)$- probability distribution of X, indicating the probability of Jim eating U eggs for breakfast." Notice that $\Pi_x(u)$ has non-distributional property ($\Pi_x(u) > 1.0$), and $\Pi_x(u) \geq P_x(u)$.

TABLE 1. Possibility versus probability distributions of X.

U	1	2	3	4	5	6	7	8	9
$\Pi_X(U)$	1	1	0.9	0.7	0.5	0.3	0.1	0.0	0.0
$P_X(U)$	0.1	0.7	0.2	0	0	0	0	0	0

Example: $\Pi_X(U = 3) = 1$, $P_X(U = 3) = 0.2$

EXAMPLES OF EARLY APPLICATIONS

Since its birth in 1965 (ref. 24), approximately 10,000 papers and more than 20 books have been published concerning fuzzy set theory. Currently, two journals: "Fuzzy Mathematics" (published in China), and "Fuzzy Sets and Systems: An International Journal" (North-Holland), and a quarterly: "Bulletin on Fuzzy Sets and their Applications" (France), are being read all over the world. In addition, the International Journal of Man-Machine Studies and other journals (Information Sciences, Journal of Cybernetics, and IEEE Transactions on Systems, Man and Cybernetics) regularly publish papers dealing with applications of fuzzy sets theory in the specific areas of interests. At least three major "fuzzy" conferences are being held every year in the U.S. and Europe. The First World Congress on Fuzzy Sets and Systems met in Europe in 1985.

The theory of fuzzy sets has been successfully applied in the modeling of ill-defined systems in a variety of disciplines (cognitive psychology, information processing and control, decision-making sciences, biological and medical sciences, sociology and linguistics, image processing and pattern recognition, and artificial intelligence). Yet it has so far attracted only a limited number of researchers in the area of human factors/ergonomics.

Willaeys and Malvache (ref. 25) investigated the perception of visual and vestibular information in a "watch-and decision" or industrial inspection (control) tasks. The imprecise nature of the human problem solving procedures was related to the "shaded" strategy of the operator´s perception and to the "hard--to-predict" environment of the man-machine environment. The labels of fuzzy sets used by the operator to describe different physical variables of the task were identified, and the fuzzy model of the process-control task was formulated. It was concluded that the fuzzy treatment of the operator´s subjective information allowed the modeling of this complex system.

Benson (ref. 26) developed an interactive computer graphics program for analytical tasks which are not well defined or utilize imprecise data. Color scales were used to model subjectively defined categories under investigation. Such fuzzy categories were then presented to the analyst. The use of a linguistic approach allowed the identification of membership for different categories of description by visual inspection. The perceptual properties of color proved to be useful in selective focus attention and in distinguishing or disregarding variations between imprecisely defined categories.

Karwowski (ref. 27) proposed a fuzzy approach in modeling the relationship between the biomechanical, physiological and the psychophysical stress responses of a human operator performing manual materials lifting tasks.

Mital and Ulgen (ref. 28) used fuzzy sets theory to model information regarding the decision task and the operator´s mental and clinical status. A model was developed to aid the decision-making process in a situation where the operator responsible for performing critical tasks is under the influence of mental stress and his ability to continue on the job is temporarily compromised. Three different decision alternatives were analyzed and the unique data base for individual operators were considered.

Karwowski and others (ref. 29-31) developed a fuzzy set based model to assess the acceptability of stresses in manual lifting tasks. Measures of acceptability were expressed in terms of membership functions which described the degrees to which the combined effect of biomechanical and physiological stresses were acceptable to the human operator. The combined acceptabilities of a lifting task were similar to the subjective estimations of the overall task acceptability established by the subjects in a psychophysical experiment.

Hunt and Rouse (ref. 32) proposed a fuzzy rule-based model of human problem solving in the fault diagnosis tasks. Two types of procedural rules (in the form of: IF (situation) THEN (action)): the context-specific rules, and the typographic or context-free rules were used to derive the model. The model was then validated using a simulated fault diagnosis task (FAULT) for trouble-shooting in the functional network diagrams of six different automotive and aircraft systems. The fuzzy model matched 50% of human subjects' actions exactly while using the same rules approximately 75% of the time.

CONCLUSIONS

Since fuzziness plays an essential role in human cognition and performance, more research is needed to fully explore the potential of this concept in the area of human factors. It is believed that the theory of fuzzy sets and systems will allow one to account for natural vagueness, nondistributional subjectivity, and imprecision of human-centered systems which are too complex or too ill-defined to admit the use of conventional methods of analysis.

REFERENCES

1 D. T. Oborne, Ergonomics at Work, John Wiley and Sons, New York, 1982.
2 J. Bezdek, Pattern Recognition with Fuzzy Objective Function Algorithms, Plenum Press, New York, 1981.
3 L. A. Zadeh, IEEE Transactions on Systems, Man and Cybernetics, SMC-3 (1973 28-44
4 L. A. Zadeh, Information Sciences, 8 (1975) 199-249.
5 L. A. Zadeh, ERL Memo M-474, University of California Berkelay, 1974.
6 J. Schwartz, in E. Nagel, P. Suppes and A. Tarski (Eds.), Logic, Methodology and Philosophy of Science, Stanford, 1962.
7 K. Popper, in P. A. Schilpp (Ed.), The Philosophy of Karl Popper, La Salle, Illinois, 1974.
8 H. J. Zimmermann, Fuzzy Set Theory and Its Applications, Kluwer-Nijhoff Publishing,Boston, 1985.
9 M. Smithson, Mathematical Social Sciences, 2(1982) 257-274.
10 M. M. Hersh and A. Caramazza, Journal of Experimental Psychology: General, 105 (1976)254-276.
11 U. Kramer, in Proceedings of the Third European Annual Conference on Human Decision Making and Manual Control, Roskilde, Denmark, 1983, p. 313.
12 L. A. Zadeh, Information and Control, 8 (1965) 338-353.
13 W. T. Singleton (Ed.), The Body at Work. Biological Ergonomics, University Press, Cambridge, 1982.
14 G. C. Oden, Journal of Experimental Psychology: Human Perception and Performance, 3(1977) 565-575
15 H. H. Brownell and A. Caramazza, Memory and Cognition 6(1978) 481-490.
16 M. E. McCloskey and S. Glucksberg, Memory and Cognition, 6(1978) 462-472.
17 R. Giles, in E. H Mamdani and B. R. Gaines (Eds.), Fuzzy Reasoning and Its Applications, Academic Press, London, 1981.
18 T. F. Baldwin, in E. H. Mamdani and B. R. Gaines (Eds.), Fuzzy Reasoning and Its Applications, Academic Press, London, 1981.
19 D. Dubois and H. Prade, Fuzzy Sets and Systems: Theory and Applications, Academic Press, London, 1981.
20 Sanchez, E., Fuzzy Sets and Systems, 12 (1984) 237-248.
21 L. A. Zadeh, Journal of Mathematical Analysis and Applications, 23 (1968) 421.

22 T. Okuda, H. Tanaka and K. Asai, Information and Control, 38 (1978) 135.

23 L. A. Zadeh, Fuzzy Sets and Systems, 1 (1978) 3.

24 L. A. Zadeh, Information and Control, 8 (1965) 338-353

25 D. Willaeys and N. Malvache, in M. M. Gupta, R. K. Ragade and R. R. Yager (Eds.), Advances on Fuzzy Set Theory and Applications, North Holland, Amsterdam, 1979.

26 W. H. Benson, in R. R. Yager (Ed.), Fuzzy Sets and Possibility Theory, Pergamon Press, New York, 1982.

27 W. Karwowski, Unpublished Doctoral Dissertation, Texas Tech University, Lubboch, TX, 1982.

28 A. Mital and O. M. Ulgen, in Proceedings of the Human Factors Society Annual Meeting, Santa Monica, California, 1982, pp.474.

29 W. Karwowski, in K. Coombes (Ed.), Proceedings of the Ergonomics Society Conference, Taylor and Francis, London, 1983, pp. 95-100.

30 W. Karwowski and M. M. Ayoub, Ergonomics, 27 (1984) 641-649.

31 W. Karwowski, M. M. Ayoub, L. R. Alley and T. L. Smith, Fuzzy Sets and Systems,14 (1984) 65-76.

32 R. M. Hunt and W. B. Rose, IEEE Transactions on Systems, Man and Cybernetics, SMC-14 (1984) 112-120.

Applications of Fuzzy Set Theory in Human Factors, edited by W. Karwowski and A. Mital
Elsevier Science Publishers B.V., Amsterdam, 1986 — Printed in The Netherlands

MEASUREMENT OF MEMBERSHIP FUNCTIONS

I.B. TURKSEN

Department of Industrial Engineering, University of Toronto, Toronto, Ontario
M5S 1A4, Canada.

ABSTRACT

Empirical measurement of membership functions of fuzzy sets are considered with the fundamental axioms of measurement theory. An experimental construction of fuzzy set membership leads to a realization of stochastic fuzziness and a type II fuzzy set representation. Axioms of measurement can be validated with a probabilistic interpretation.

INTRODUCTION

In fuzzy set theories, the set calculus is carried out with membership functions and/or values that characterize imprecision, i.e., vagueness, ambiguity and uncertainty, conveyed by linguistic variables, A, B, etc., that describe attributes of elements $\theta \in \Theta$. A membership function is a rule that assigns to each element in the universe of discourse a number in the closed interval [0, 1] indicating the degree to which that element is a member of a particular set defined by linguistic variables. If the set is well defined, i.e., no imprecision is associated with the set as in, for example, "men of at least 1.80 meters in height," then all membership values are either 0 or 1. On the other hand, if the set is not well defined, i.e., there is some imprecision associated with the set due to linguistic variables, as in, for example, "tall men", then the membership values can take on any value in the interval [0, 1].

Shortly after an introduction to the theory, most students of fuzzy set theory ask: "How does one determine membership functions and/or values?" Inquiries of this sort have at times questioned the scientific merit of the fuzzy set theories. This has lead to a controversy which is much subdued at present after some twenty years of research and development. In the early days of the controversy some researchers even went so far as to state, for example, that "...There is no way of determining the values of the memberhsip function, either rationally or empirically... (ref. 1). At present, it is a common knowledge, at least, among the fuzzy set researchers, that membership functions can be determined either normatively or empirically.

There are, however, two distinct schools of thought amongst the fuzzy set researchers themselves: normative versus empirical. The normative researchers essentially follow L.A. Zadeh's original approach (ref. 2). They point out that the imprecision conveyed by linguistic variables is subjective and thus need to be defined directly by their users as a function, for example, an S function (ref. 3).

The empiricists generally follow some aspect of measurement theory (ref. 4) resting their argument on the objectivity of the experimental procedure of the scientific method. Although much has been written about the measurement of imprecision, empirical work has been relatively sparse. There are essentially two methods that have been reported so far.

(i) Binary Direct Rating:

In this method, a subject answers "yes or no" according whether the linguistic term describes the element (stimulus). The fraction of "yes" responses over all responses are taken as the degree of membership for that element (stimulus) in the imprecise concept represented by the linguistic variable (refs. 5, 6).

(ii) Continuous Direct Rating:

In this method, the subjects rate elements (stimuli) on a continuous scale from "definitely in the concept" to "definitely not in the concept". There are a number of variants of this method:

(a) Oden (ref. 7) had subjects rate propositions on a continuous scale form "absolutely true" to "absolutely false", (b) Turksen and Norwich (ref. 8) had subjects respond to the following question: "how A is θ?" where A is a linguistic variable, θ is an element (stimulus). Norwich and Turksen (refs. 9, 10, 11, 12) have experimented with other equivalent questions as well; such as "how much membership does θ possess in the fuzzy set A?" or "what degree of belonging θ has in the fuzzy set A?" etc.

With experimental results at hand from both of these methods, Norwich and Turksen (ref. 6) among others (ref. 13) criticized the binary direct rating method for several reasons.

It can be argued that the binary direct rating method:

(a) confounds measures of fuzziness with response variability due to experimental procedure;

(b) is subsumed by continuous direct rating; and

(c) misrepresents the subject's perception of a linguistic variable by making it appear as a crisp attribute; i.e., by decreasing the amount of imprecision associated with the linguistic variable.

An extensive discussion of these criticisms may be found in the references indicated above. However, let us briefly point out, for the criticism (a) that the two components of measurement, i.e., membership and response error, can be separated in stochastic measurement experiments as will be shown in the sequel. Now for the other two criticisms, (b) and (c), consider two measurement experiments, one for each method, to determine, let us say, the membership function of the set of "tall men". The fuzzy concept "tall men" partitions the set Θ = {different men}, and its associated numerical domain $X(\Theta)$ = {numerical heights of men in Θ} = $[0,\infty]$ meters, for each subject into three subsets. It is observed and experimentally found that the fuzzy subset F^B due to the binary direct rating is contained in the fuzzy subset F due to the continuous direct rating, i.e., $F^B \subseteq F$ or $|F^B| \leq |F|$ where f^B and f are the average membership responses in both of the experiments, respectively (Figure 1). Thus it is observed that the fuzzy subset F gives a better representation of the imprecision due to a linguistic variable. As a result, the continuous direct rating methods are generally prefered over the binary direct rating methods. In addition, in the continous direct rating methods, the responses of subjects may be embedded within a theory of measurement and some of their scale properties may be analysed. Various checks could be executed on data to determine if certain axioms of measurement theory are satisfied with the implied result on the scale strength of data; i.e., ordinal, interval, ratio, or absolute scale.

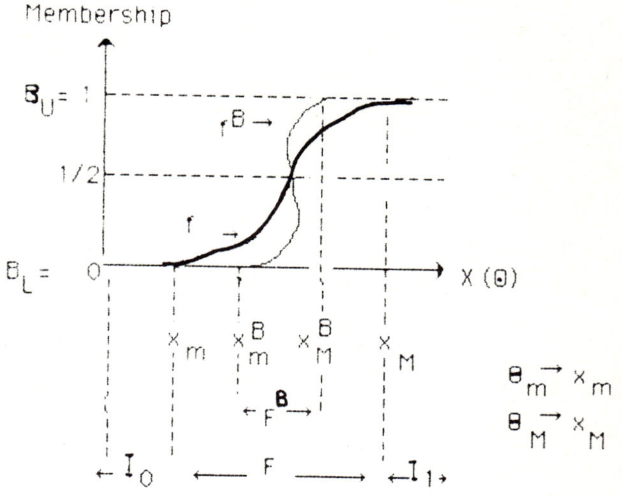

Figure 1 A comparison of the binary and continuous direct rating methods.

Validation studies are rather important because there exists yet another controversy among the empirical researchers of fuzzy set theories. For example, Thole, Zimmermann and Zysno (ref. 14) have asserted that membership is on an absolute scale, while Saaty (ref. 15) has exposed a ratio scale. Sticha, Weiss and Donnell (ref. 16) claim interval scale strength but Goguen (ref. 17) has stated that ordinal scale may be all that can be obtained for membership. It appears that different membership scale strengths may be obtained in different domains of application. Hence every new domain of application requires a validation study of its own.

But such validation studies require a probabilistic and/or, at times, possiblistic interpretation of the axioms of measurement theory in order to account for variabilities due to experimental procedure and inconsistencies due to subjects responses. This in turn requires extensive data analysis with implied costs.

AXIOMATIC FOUNDATIONS

The fundamental measurement of membership may be regarded as the construction of homomorphisms from an empirical relational structure to a numerical relational structure. In particular, if $\langle \Theta, \succsim \rangle_A$ is an empirical relational structure of a fuzzy concept (attribute) A over a set of objects Θ and $\langle Re, \geqslant \rangle$ is a numerical relational structure, then a real valued function f, called 'membership function', on Θ is a homomorphism if it takes the empirical relation $\succsim$ into the numerical relation $\geqslant$ (refs. 8, 10, 11, 12, 18, 19).

Almost all of measurement theoretic studies include a set of axioms that attempt to clarify: (i) behavioural aspects of subject's reponses and (ii) procedures for constructing numerical assignments. The behavioural axioms are generally stated as weak order and weak monotonicity, whereas the structure axioms are known as solvability and Archimedian conditions. For example, P. Suppes (ref. 20) presents axioms for "rational subjective choice" and separates them into two categories: (i) rationality axioms, and (ii) structure axioms. It is assumed that the "pure" rationality axioms, deterministic as they are, should be satisfied by any rational, reflective person in a decision making situation. On the other hand, the "pure" structure axioms impose limitations on the kind of situations to which an analysis may be applied.

Since there is no really "satisfactory" definition of rationlity, these theories have a somewhat tenious quality for empirisists. The experimental literature concerned with the validity of these theories are indeed very few and only test certain limited aspects of these theories. In measurement experiments of membership based on an axiomatic formalism, subjects almost always exhibit inconsistencies among and uncertainties within their responses to given stimuli (refs. 8, 9, 10, 11, 12, 18). These inconsistent and ambiguous response behaviour of subjects display a certain sense of "rationality" that needs to be qualified. Thus historical debate on "rationality" and scarce analysis of experimental data strongly suggest that axiomatic theory of measurement needs to be interpreted in a new way. Such interpretations have to account for "inexactness and randomness" on the one hand and the insufficiency of data, on the other, that are inescapable components of every empirical study of human subjects. In the rest of this paper it will be argued that "inexactness and randomness" can be separated into fuzziness and stochasticity, respectively, and the insufficiency of data suggest a type II fuzziness.

Interval Scale

Our research findings point out that membership functions could at best be on an interval scale when they are obtained from the responses of human subjects. To discuss every aspect of our research findings would require a rather lengthy paper. Instead, certain essential issues are chosen for discussion here.

First of all, it is required that a subject be able to compair any pair of elements $\theta_i, \theta_j \in \Theta$ with regard to a linguistic variable A in order to establish a 'weak-order of elements' $\succsim$ on Θ. That is to say, if a subject judges that "θ_i is at least as A as θ_j", then we have $\theta_i \succsim \theta_j$, $\forall\ \theta_i, \theta_j, \in \Theta$.

Definition

Let Θ be a finite non-empty set of elements and $\succsim$ be a binary relation on Θ with regard to a linguistic variable A, i.e., $\succsim$ is a subset of $\Theta' = \Theta \times \Theta$. The relational structure $\langle \Theta, \succsim \rangle_A$ is a 'weak-order' and will be called a 'multi-valued membership structure for a linguistic variable A', if and only if, for all $\theta_i, \theta_j, \theta_k \in \Theta$, the following two axioms are satisfied:

1. Connectedness: Either $\theta_i \succsim \theta_j$, or $\theta_j \succsim \theta_i$,
2. Transitivity: if $\theta_i \succsim \theta_j$, and $\theta_j \succsim \theta_k$, then $\theta_i \succsim \theta_k$,

It should be clear that this weak-order of elements establishes a relexive, antisymmetric and topological ordering of elements. But it does not say anything about the distance between these elements.

Definition

A membership structure $\langle \Theta, \succsim \rangle_A$ is called 'bounded' if there exist elements θ_M and θ_m, such that $\theta_M \succsim \theta$ and $\theta \succsim \theta_m$, $\forall\, \theta \in \Theta$. θ_M is an element which is judged to be 'definitely A' by the subject and θ_m is an element which is judged to be 'definitely not A'. That is, a subject assigns maximum membership to θ_M and to its equivalence class $[\theta_M]$ and minimum membership to θ_m and to its equivalence class $[\theta_m]$ in the fuzzy set representation of A.

It can be shown that if a membership structure $\langle \Theta, \succsim \rangle_A$ is a bounded weak-order, then there exists a bounded real-valued membership function f on Θ such that, $\forall\, \theta_i, \theta_j \in \Theta$ $\theta_i \succsim \theta_j$ iff $f(\theta_i) \geq f(\theta_j)$

Moreover, f' is unique to a strictly increasing (monoton) function ϕ, i.e., if f' has the same properties as f, then $f'(\theta) = \phi[f(\theta)]$, $\forall\, \theta \in \Theta$.

i.e. the membership function f has an ordinal scale property.

Secondly, it is required that a subject be able to compare any pair of intervals specified by any four elements $\theta_i, \theta_j, \theta_k, \theta_l \in \Theta$ in order to establish a 'weak-order of intervals' $\succsim'$ on Θ'. That is to say, if a subject judges that " θ_j is A-er than θ_i by at least as much as θ_l is A-er than θ_k ", then we have

$$\theta_j\, \theta_i \succsim' \theta_l\, \theta_k, \forall \theta_i, \theta_j, \theta_k, \theta_l \in \Theta.$$

Unlike the weak-order of elements $\succsim$ the "weak-order of intervals" $\succsim'$ is algebraic. Such responses of a subject are needed for the axiom of 'weak-monotonicity'.

Definition

Let Θ be a finite non-empty set of elements and $\succsim'$ be a quaternary relation on Θ' with regard to a linguistic variable A. The relational structure $\langle \Theta', \succsim' \rangle_A$ is said to have 'weak-monotonicity' and will be called a 'bounded, multi-valued, difference-comparable membership structure for a linguistic variable A', if and only if, for all $\theta_i, \theta_j, \theta_k, \theta_i', \theta_j', \theta_k', \in \Theta$, the following is satisfied:

If $\theta_j\, \theta_i \succsim' \theta_j'\, \theta_i'$ and $\theta_k\, \theta_j \succsim' \theta_k'\, \theta_j'$ then $\theta_k\, \theta_i \succsim' \theta_k'\, \theta_i'$

Let us next suppose that a bounded, multi-valued, difference-comparable membership structure satisfies the remaining three axioms of 'algebraic-difference structure' stated by Krantz et. al. (ref. 4), i.e., sign reversal, solvibility and the Archimedian conditions. A detailed explanation of these axioms may be found in Turksen (ref. 21) in the context of the measurement of membership. All of these axioms require that Θ be order-dense with respect to a linguistic variable A in a relation $\succsim$. This property will now be assumed.

Consider now, any empirical domain Θ that can be represented by an associated numberical domain $X(\Theta)$. Whenever Θ is order-dense, $X(\Theta)$ is a continuum. The introduction of $X(\Theta)$ serves the dual purpose of rendering a graph of membership over X continuous (note that the graph is discrete over Θ) and in accordance with the standard practice of fuzzy set literature, making membership a function of a numerical base variable rather than of empirical elements.

When a subject's responses with regard to a linguistic variable A over an order-dense domain Θ

satisfy the axioms of an algebraic difference structure, then the following numerical representation of membership is implied.

It can be shown that if a bounded, multi-valued, difference-comparable membership structure $\langle \Theta', \succsim' \rangle_A$ is an algebraic difference structure, then there exists a bounded real-valued function f on Θ such that, for all $\theta_i, \theta_j, \theta_k, \theta_l \in \Theta$,

$\theta_j \; \theta_i \succsim' \theta_l \; \theta_k$ iff $f(\theta_j) - f(\theta_i) \geq f(\theta_l) - f(\theta_k)$

Moreover, f is unique up to a positive linear transformation, i.e., if f' has the same properties as f, then there are real constants α, β with $\alpha \geq 0$, such that

$f' = \alpha f + \beta$,

i.e., the membership function f has an interval scale property.

This means that a membership function incorporates two parameters: an arbitrary origin and an arbitrary unit. All θ- in the equivalence class of θ_m, $[\theta_m]$, define an interval I_0 on $X(\Theta)$, i.e. $\theta- \in [\theta_m] \triangleq I_0$, and take on an arbitray value B_L, the lower bound of f and all θ+ in the equivalence class of θ_M, $[\theta_M]$, define an interval I_1 on $X(\Theta)$, i.e., $\theta+ \in [\theta_M] \triangleq I_1$, and take on an arbitrary value B_U, the upper bound of f. B_L and B_U may be chosen as 0 and 1, repectively. Once specific values are chosen for B_L and B_U, then the membership values of all other elements in Θ are totally relative to these bounds. The intervals I_0, I_1, B_L, B_U and other symbols are displayed in Figure 1. In the following, membership values are denoted with $y \in [B_L, B_U]$ in order not to confuse them with a membership function f.

EXPERIMENTAL METHOD

Both the continuous direct rating and the continuous reverse rating procedures were used in our experiments. A brief explanation is now given for these procedures.

Direct Rating

In our continuous direct rating procedure randomly selected elements (or objects) $\theta \in \Theta$ with values $x(\theta) \in X(\Theta)$ are presented to a subject in a random manner. The subject is asked to respond to the following question: "how A is θ $(x(\theta))$". $\theta \in \Theta$ is displayed to the subject but $x(\theta) \in X(\Theta)$ is known to the analyst. The subject's response is a value $y \in [B_L, B_U]$, where B_L and B_U are two marks on a simple piece of wood stick and y is determined by a movement of an indicator that slides on the wood stick. The subject is also asked randomly to identify θ_m $(x_m(\theta))$ corresponding to B_L and θ_M $(x_M(\theta))$ corresponding to B_U in repeated experiments in order to identify the range $[x_m, x_M]$ of the base variable. A subject is presented with the same θ $(x(\theta))$ a number of times, say n=10, in between other random presentations of $\theta \in [\theta_m, \theta_M]$ $(x \in [x_m, x_M])$ in order to avoid memorization.

The subject's responses are recorded as observed values y for given $x(\theta)$, specified as x^*, i.e., $y|x^*$. Thus, in every direct rating experiment, the observed values $y|x^*$ generate a conditional distribution (Figure 2). Let $\phi_{Y|x^*}(y|x^*)$ be the pdf, of the conditional cummulative distribution function, cdf, $\Phi_{Y|x^*}(y|x^*)$, of the random variable Y generated in a continuous direct rating experiment at a given value x^* of X. The properties of this distribution turns out to be either a Beta distribution or a Gaussian error distribution, $N(\mu, \sigma^2)$, with the mean $\mu_{Y|x^*} = E(Y|x^*)$ and the variance

$\sigma^2_{Y|x^*} \equiv Var(Y|x^*)$ which represents the inherent inconsistencies in human perception in identifying the grades of membership $y \in [B_L, B_U]$. The mean $\mu_{Y|x^*}$ and the variance $\sigma^2_{Y|x^*}$ are unknown parameters.

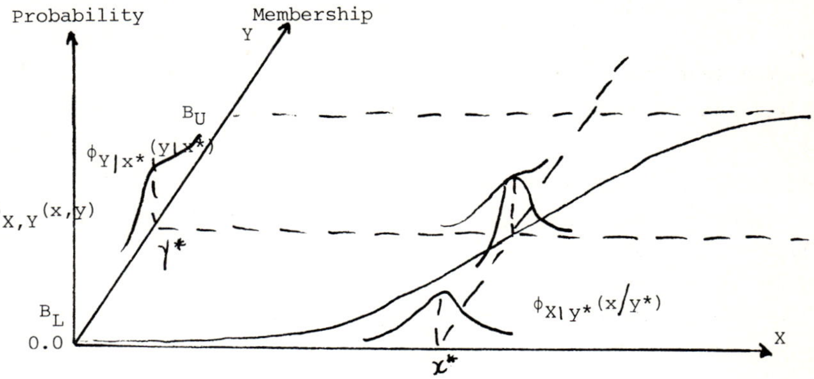

Figure 2. Experimental distributions over the base variable X and the membership variable Y

However, unbiased estimates of these parameters can be obtained in the usual manner:

$\bar{y}|x^* = (1/n)\Sigma(y_i|x^*)$, $s^2_{Y|x^*} = (n-1)^{-1}\Sigma(y_i|x^* - \bar{y}|x^*)^2$

These computations are meaningful for the class of membership functions that have the interval scale property which was established in the previous section.

It was noted in the introduction and later in the previous section that inexactness and randomness are two different components of uncertainty. With the stochastic fuzziness described here the components of membership and errors due to experimental procedures are thus separated in this experimental setting. The mean value $\mu_{Y|x^*}$ is assumed to be equivalent to the deterministic membership value if $f(x^*)$ of a fuzzy set $[x^*, f(x^*)]$ in current literature. Thus in this model, if sufficiently large number of responses are obtained in an experiment, then a probabilistic interpretation would be appropriate with the second order variation $\sigma^2_{Y|x^*}$, due to inconsistencies in human perception, around the first order effect $\mu_{Y|x^*} \equiv f(x^*)$. On the other hand; if sufficiently large amount of data could not be obtained in an experiment, then the spread should be intepreted as to the type II fuzziness. This suggests a consideration of, at least, interval valued membership functions (ref. 22). This would give rise to a possibilistic analysis.

EXAMPLE 1:

An example of direct rating for "Tall man" is shown graphically in Figure 3 and the values of the replicated experiment are listed in Table 1 for subject #3 (ref. 9).

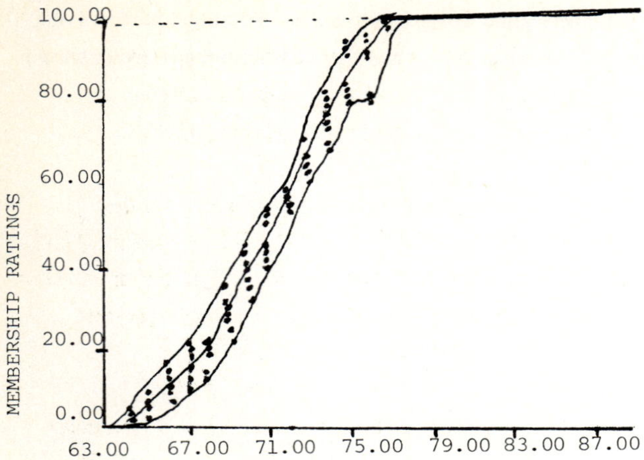

Fig. 3. Direct ratings for TALL man (Subject #3)

Table 1. Data for Figures 3 (Subject #3) (TALL man)

Stimulus Inches	Replicated direct ratings*									Mean of Ratings	Std. Dev. of Ratings
63	0	0	0	0	0	0	0	0	0	0.00	0.00
64	0	3	0	4	0	0	2	0	0	1.00	1.58
65	2	4	9	7	8	4	8	12	6	6.67	3.04
66	6	10	15	8	11	10	13	8	9	10.00	2.74
67	11	17	20	12	18	12	19	15	15	15.44	3.28
68	20	14	29	22	23	15	22	18	23	20.67	4.58
69	30	37	29	32	20	30	33	26	31	29.78	4.74
70	35	37	42	45	43	41	32	42	37	39.33	4.27
71	39	42	52	46	44	50	51	43	45	45.78	4.41
72	48	54	59	58	57	58	55	56	59	56.00	3.46
73	77	66	67	63	67	70	74	71	68	69.22	4.29
74	81	82	78	77	67	70	79	73	81	76.44	5.29
75	92	94	88	82	85	85	84	88	85	87.00	3.91
76	97	92	96	96	79	94	82	92	95	91.44	6.48
77	100	100	100	100	96	100	100	98	100	99.33	1.42
78	100	100	100	100	100	100	100	100	100	100.00	0.00

We have arbitrarily set $B_L = 0$, $B_U = 100$. Each rating was actually produced by the pointer being moved to one of the 101 gradations from the left end (i.e. B_L) to the right end (i.e. B_U) of memembership dimension.

Reverse Rating

There is also a corresponding reverse rating procedure where randomly selected membership values y are presented to a subject in a random manner. The subject is asked to respond to the following question: "identify θ or $(x(\theta))$ that possesses yth grade of membership in the fuzzy set A." The subject's response is a $\theta \in \Theta\ (x(\theta) \in X(\Theta))$.

The subject is randomly presented with the same degree of membership y a number of times, say m = 10 times in between the other random presentations of $y \in [B_L, B_U]$ in order to avoid memorization. The subject's responses are recorded as the observed values of x for given y, specified as y^*, i.e., $x|y^*$. Thus in every reverse rating experiment, the observed values $x|y^*$ generate a conditional distribution (Figure 2). Let $\phi_{X|y^*}(x|y^*)$ be the pdf of the conditional cdf $\Phi_{X|y^*}(x|y^*)$ of the random variable X generated by a reverse rating experiment at point y^* of Y.

This conditional distribution turns out to be the Gaussian error distribution $N(\mu, \sigma^2)$ with mean $\mu_{X|y^*} = E(X|y^*)$ and a variance $\sigma^2_{X|y^*} = Var(X|y^*)$ which again represents the inconsistencies in human perception in identifying $\theta \in \Theta\ (x(\theta) \in X(\Theta))$. The mean $\mu_{X|y^*}$ and the variance $\sigma^2_{X|y^*}$ are unknown parameters. Their unbiased estimates can be obtained again in the usual manner:

$$\overline{x}|y^* = (1/m) \Sigma (x_i|y^*), \quad s^2_{X|y^*} = (m-1)^{-1} \Sigma (x_i|y^* - \overline{x}_i|y^*)^2$$

These computations are, however, meaningful so long as the values of the base variable have the properties of "interval" scale as a minimum..

EXAMPLE 2

An example of reverse rating for "Tall man" is shown in Table 2 (ref. 9) to determine the region $[x_m, x_M]$ (all values are in inces). It should be noted that the fuzzy interval regions indicated in Figure 3 are induced by the indirect rating procedure, which does not explicitly measure x_m, x_M.

TABLE 2.
Data to determine the fuzzy region $[x_m, x_M]$.

Subject Number	Region $[x_m, x_M]$	$x_M - x_m$
1	[70.73, 81.62]	10.89
2	[60.03, 77.78]	17.75
3	[62.90, 76.88]	13.98
4	[50.29, 78.68]	28.39
5	[64.51, 80.82]	16.31
6	[51.40, 84.72]	33.32
7	[50.21, 74.38]	24.17

Correspondence Principle

In a long run experiment it is natural to expect that the direct and reverse ratings be consistent. That is, the central tendencies of the two conditional distributions correspond to each other.

Postulate:

$E(Y|x^*) = y^* = \mu_{Y|x^*}$ if and only if $E(X|y^*) = x^*$ of X and y^* of Y.

This postulate can be validated experimentally.

MEASUREMENT AND VALIDATIONS

Our initial mesurement experiments include the following concepts: "middle aged men", "middle aged women", "warm weather", "ordinary size stamp", "medium size stamp", "high note (with reference to one's musical experience)", "closeness of people (to each other in elevators)", "redness (of colours)", "young person", "tall men (undergraduate students)", etc.

Within the last few years, we conducted measurement experiments for the mental workload analyses. (ref. 22). For example, membership functions were determined for such linguistic terms as 'difficulty' and 'effort' within the context of rule-based mental workload studies of the following type:

"If the 'bandwith' is high and the operator uses moderate 'effort', then the task will seem moderately 'difficult and 'performance' will be poor".

In Figure 4, scatter diagrams of a subject's responses are displayed for 'difficulty' (Figure 4(a)) and for 'effort' (Figure 4(b)) for four particular experimental settings of a task as 1, 2, 3, and 4. As it can be seen, a type II fuzzy set is realized in these diagrams.

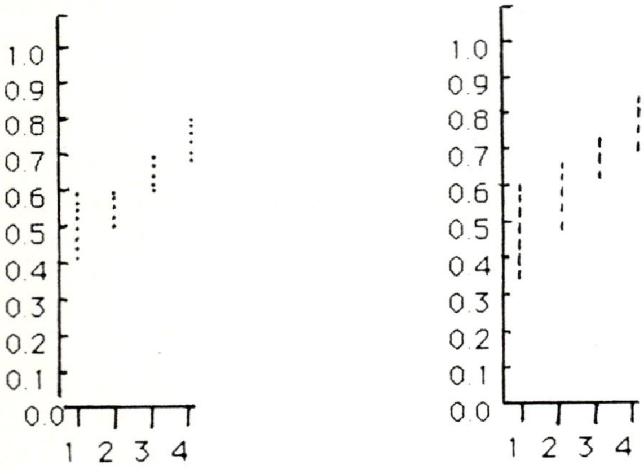

Figure 4 (a) Membership values of 'difficulty' (b) Membership values of 'effort'

Validation of Axioms

In the introduction, it was stated that the axioms of measurement could be validated with an appropriate interpretation of responses from a subject. This will be explained now for the weak monotonicity axiom discussed earlier. Similar arguments may be put forth for the other axioms.

It should be recalled that a bounded, multi-valued, difference-comparable membership structure for a linguistic variable A, $\langle \Theta', \succsim' \rangle_A$ has the property of weak-monotonicity if and only if, for all θ_i, θ_j,

θ_k, θ_i', θ_j', $\theta_k' \in \Theta$, the following is satisfied:

If $\theta_j \theta_i \succsim \theta_j' \theta_i'$ and $\theta_k \theta_j \succsim \theta_k' \theta_j'$, then $\theta_k \theta_i \succsim \theta_k' \theta_i'$

In order to check the validity of this property, the argument of 'weak stochastic monotonicity' is invoked within the experimental setting of stochastic measurement discussed in the previous section. Therefore, it must be experimentally determined that if a subject judges $\theta_j \theta_i \succsim \theta_j' \theta_i'$ with at least probability p and also judges $\theta_k \theta_j \succsim \theta_k' \theta_j'$ with at least the same probability p, then the same subject must judge $\theta_k \theta_i \succsim \theta_k' \theta_i'$ again with at least the same probability p. That is if $P(\theta_j \theta_i \succsim \theta_j' \theta_i') \geq p$ and $P(\theta_k \theta_j \succsim \theta_k' \theta_j') \geq p$ then $P(\theta_k \theta_i \succsim \theta_k' \theta_i') \geq p$.

In other words, whenever a subject is presented with two intervals:

(1) $\theta_j \theta_i$ that corresponds to a membership interval (difference) d_a, and

(2) $\theta_j' \theta_i'$ that corresponds to a membership interval (difference) d'_a, such that

$d_a \geq d'_a$, in a replicated experiment, then if the subject's correct responses form a frequency of response, at least, equivalent to p, i.e., (see Figure 5)

$P(d_a \geq d'_a) \geq p$.

And similarly a frequency of, at least, p is obtained for the other pair $(\theta_k \theta_j \succsim \theta_k' \theta_j')$, i.e., $P(d_b \geq d'_b) \geq p$, (see Figure 5)

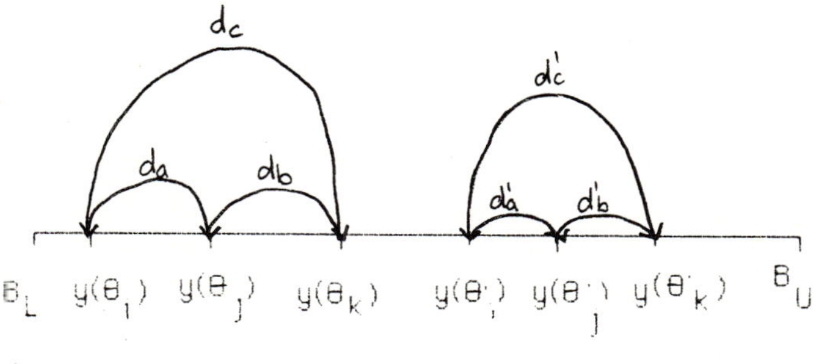

Figure 5 Weak Stochastic Monotinicity

Then, for the axiom to be satisfied, the same subjects correct responses for the pair $(\theta_k \theta_i \succsim \theta_k' \theta_i')$ must, at least, produce a fequency p. i.e., $P(d_c \geq d'_c) \geq p$, (see Figure 5)

In such a replicated experiment for all θ_i, θ_j, θ_k, θ'_i, θ'_j, $\theta'_k \in \Theta$, if a subjects responses provide all these frequencies, then it can be said that the weak stochastic monotonicity is satisfied for this subject in a measurement of membership experiment for a given linguistic variable A.

Furthermore, such tests of validity has to be done for the other axioms in order to determine a stochastic scale strength of data collected in such measurement experiments. As can be appreciated such tests of validity are rather involved and are very costly. It is for these reasons, it was stated at the beginning that membership functions might at best be on an interval scale more specifically a "stochastic interval scale". It may be that a membership function with ordinal scale property may be sufficient for a study that may than be able to use only the "max-min" operators of the fuzzy set calculus.

Most students of fuzzy set research initially are introduced to max-min operators based on their reading of L.A. Zadeh's paper. And a large number of novices in the field are not aware that there are many fuzzy set operators. Notably, there are a large class of operators known as t-norms and t-conorms (refs. 24, 25, 26). Existence of many operators gives us the opportunity to apply an operators to fit a particular phenomona. However, the users of these new operators have to be extremely careful that the experimental data has the properties required by the arithmetic operations that are part of the definitions of these new operators. For instance, for those operators that require an <u>addition</u>, the analyst must justify the existence of a ratio scale, and for those operators that require <u>multiplication</u> the analyst must justify the existence of an absolute scale. For example, the Bold intersection, i.e.,
$a \underset{\cdot}{\wedge} b = \max(0, a + b - 1)$, and the Bold union, $a \dot{\vee} b = \min(1, a + b)$ require addition and hence a ratio scale justification; whereas the probabilistic sum, i.e., $a \hat{+} b = a + b - ab$ and product $a \cdot b = ab$ require addition and multiplication and hence an absolute scale justification.

Norwich and Turksen (ref. 9) have discussed the question of 'extensive measurement' rather thoroughly. The notion of a "standard function of the membership of θ in A", $\sigma_A(\theta)$, was introduced there in order to get around some of these difficulties. Due to space limitations, the readers are referred to 'the meaningfulness of operations on membership' in Norwich and Turksen (1984).

ACKNOWLEDGEMENTS

The author is grateful for the partial support provided by NASA-AMES NAGR-429 and NSERC-Operating Grant, A-7698.

REFERENCES

1 S. Watanabe, "A Generalized Fuzzy Set Theory", IEEE Trnasactins on Systems, Man, Cybernetics, 8, 1978, 756-760.
2 L.A. Zadeh, Fuzzy sets, Information and Control, 8, 1965, 338-353.
3 L.A. Zadeh, The concept of a linguistic variable and its application to approximate reasoning - I, II, III, Information Sciences, 8, 1975, 199-249, 301-357; 9, 43-80.
4 D.H. Krantz, R.D. Luce, P. Suppes and A. Tversky, Foundations of Measurement, Vol. 1 (Academic Press, New York, 1971).
5 H.M. Hersh and A. Caramazza, A fuzzy set approach to modifiers and valueness in natural language. J. of Experimental Psychology: General, 1976, 105, 254-276.
6 A.M. Norwich and I.B. Turksen, Stochastic fuzziness, in: M.M. Gupta and E.E. Sanchez, eds., Fuzzy Information and Decisions Processes (North-Holland, Amsterdam, 1982).
7 G.C. Oden, Integration of fuzzy logical information, J. of Experimental Psychology: Human Perception and Performance, 1977, 3, 565-575.
8 I.B. Turksen and A.M. Norwich, Measurement of fuzziness, Proceedings of the International Conference on Policy Analysis and Information Systems, August 19-22, 1981, Taipei, Taiwan (Meadea Enterprises Co., Ltd., Taipei Taiwan, 1981) 745-754.
9 A.M. Norwich, and I.B. Turksen, A model for the measurement of membership and the consequences of its empirical implementation, Fuzzy Sets and Systems, 12, 1984, 1-25.
10 A.M. Norwich and I.B. Turksen, The fundamental measurement of fuzziness, in: R.R. Yager, Ed., Fuzzy Set and Possibility Theory (Pergamon Press, New York, 1982).
11 A.M. Norwich and I.B. Turksen, The construction of membership functions, in: R.R. Yager, Ed., Fuzzy Set and Possibility Theory (Pergamon Press, New York, 1982).
12 A.M. Norwich and I.B. Turksen, Meaningfulness in fuzzy set theory, in: R.R. Yager, Ed., Fuzzy Set and Possiblility Theory (Pergamon Press, New York, 1982).
13 D.C. Rubin, On Measuring fuzziness: A comment on 'A fuzzy set approach to modifiers and vagueness in natural language.' Journal of Experimental Psychology: General 1979, 108, 486-489.
14 U. Thole, H.-J. Zimmermann and P. Zysno, On the suitability of minimum and product operators for the intersection of fuzzy sets, Fuzzy Sets and Systems 2 (1979) 167-180.
15 T.L. Saaty, Measuring the fuzziness of sets, Journal of Cybernetics 4 (1974) 43-61.
16 P.J. Sticha, J.J. Weiss and M.L. Donnell, Evaluation and integration of imprecise information, Final

Technical Report PR 79-21-90, Decisions and Designs, In., Suite 600, 8400 Westpark Drive, P.O. Box 907, mcLean, VA 22101 (1979) (also available from JSAS through Order Department, American Pscyhological Association, 1200 17th Street N.W., Washinton, DC 20036).
17 J.A. Goguen, The logic of inexact concepts, Synthese 19 (1969) 325-373.
18 I.B. Turksen, Measurement of linguistic variables in medical diagnoisis, proc. of the 23rd Annual North American Meeting, Society for General Systems Research, January 3-6, Houston, Texas, (1979) 278-284.
19 I.B. Turksen, Stochastic and fuzzy sets. Proc. 2nd World Conference on Mathematics at the Service of Man, June 28-July 3, (1982), Las Palmas, Canary Islands, Spain, 649-654.
20 P. Suppes, The Role of subjective probability and utility in decision making, Proceedings of the Third Berkeley Symposium on Mathematical Statistics and Probability, 195, 61-73.
21 I.B. Turksen, Measurement of fuzziness: an interpretation of the axioms of measurement. Proc. of the Conference on Fuzzy Information Knowledge Representation and Decision Analysis, July 19-22, (1983), Marseille, France, 97-102.
22. I.B. Turksen, Interval Valued Fuzzy Sets Based on Normal Forms, Fuzzy Sets and Systems, (to appear).
23 I.B. Turksen, N. Moray, K. Fuller, A linguistic rule-based expert system for mental workload, H.J. Bullinger, H.J. Warnecke eds., Toward the Factory of the Future, 1985, 865-875.
24 B. Schweizer and A. Sklar, Associative functions and abstract semi-groups, Publicationes Mathematicae Debrecen, 1963, 10, 69-81.
25. B. Schweizer and A. Sklar, Associative functions and statistical traingle inequalities, Publicationes Mathematicae Debrecen, 1961, 8, 169-186.
26 B. Schweizer and A. Sklar, Statistical metric spaces, Pacific J. of Math., 1960, 10, 313-334.

Part II.

HUMAN-MACHINE COMMUNICATION

Applications of Fuzzy Set Theory in Human Factors, edited by W. Karwowski and A. Mital
Elsevier Science Publishers B.V., Amsterdam, 1986 — Printed in The Netherlands

DEALING WITH THE VAGUENESS OF NATURAL LANGUAGES IN MAN-MACHINE COMMUNICATION

Henri FARRENY, Henri PRADE

ENSEEIHT & Université Paul Sabatier

Laboratoire "Langages et Systèmes Informatiques"

118, route de Narbonne, 31062 TOULOUSE CEDEX - FRANCE

ABSTRACT

This paper presents a fuzzy set-based approach which takes into account the vagueness that pervades natural languages in task-oriented dialogues. The communication process involves a matching step where representations of the meaning of the words or expressions under consideration are compared with representations of other words or with pieces of information about the environment stored in a knowledge base. Words which refer to "measurable" attribute values (e.g. 'large', 'old', ...) can be conveniently represented in terms of possibility distributions, which enables us to model the possible fuzziness of their meaning. The context dependency of the meaning of words such as 'large' can also be easily handled with this approach. The use of possibility distributions for representing the meaning attached to the label of more complex attribute values (description of shapes or of abstract notions for instance) is also discussed. The approximate synonymy between labels of attribute values or between names of object categories (possibly characterized by means of several possibility distributions), is estimated in terms of a possibility and a necessity measure, which are valued on the real interval [0,1]. Examples of the treatment of vagueness in man-machine communication, using this approach, are briefly discussed.

KEY WORDS : natural language understanding ; task-oriented dialogue ; vagueness ; synonymy ; fuzzy attribute value ; fuzzy category ; pattern matching ; possibility ; necessity ; fuzzy set.

I - INTRODUCTION

In many respects natural languages are flexible. First different surface structures can be given to a sentence without considerably modifying its intended meaning. Moreover, people (especially when they speak) do not adhere to rigid grammatical behavior and some violations of the grammar can be tolerated without creating much ambiguity in general. Additionally, objects or entities can be referred to in various ways by means of descriptive properties in natural language sentences. What makes natural languages still more flexible is the existence of many words whose meanings are more or less vague and context-dependent.

The idea of a fuzzy set has been from the beginning [29], [30] very much related to the problem of representing a collection of objects whose characteristic property is stated in vague linguistic terms. Then the extent to which an object satisfies a vague predicate is a matter of degree. Even if degrees are difficult to assign, generally there are objects which certainly belong (or certainly do not belong) to the set associated with the vague predicate, while others can only be regarded as possible elements of this set.

In the following we present a fuzzy set-based approach which takes into account the vagueness of words in man-machine systems for natural language communication. The related question of approximate synonymy between words is also discussed. Although the grammatical correctness might be a matter of degree also (see [1], [13]), this problem is not considered in the paper.

Within this paper several ideas are addressed; first, the treatment of vagueness is examined from a man-machine communication point of view. Second the representation of linguistically stated attribute values and of vague categories is discussed within the framework of the possibility theory [32] and finally, the similarity between the representations of two approximately synonymous expressions is estimated by means of a fuzzy pattern matching method. In the last section a discussion of how the treatment of vagueness can be handled in natural language interfaces with knowledge or data bases or in task-oriented communication systems for dialogues with robots, such as the system SYROCO [7], [8], [11] is presented.

II - VAGUENESS AND MAN-MACHINE COMMUNICATION

It is desirable to enable users of robots, data bases or expert systems to communicate in natural languages with these computer systems. Since dialogues with such systems are task-oriented, discourse domains are limited, which makes the problem of understanding natural languages computationally tractable.

The treatment of vagueness may be slightly different according to the intended use of the natural language communication system. In the following, two kinds of systems will be considered. The first type are interfaces with mechanical systems acting in a physical environment and the second are those which enable the users to query knowledge bases.

The communication systems commonly utilized with robots generally have two modules, one for language understanding and another for language generation. The interpretation of natural language sentences is guided by the recognition of semantical entities and the checking of their compatibilities with the frame associated to the verb and the confrontation with the robot's world (see [8]). This confrontation is necessary in order to determine the objects to which the discourse refers and also to solve some possible ambiguities. For instance, the command "put the box on the table against the wall" could be interpreted as either "put the box which is on the table against the wall" or "put the box on the table which is against the wall" depending on the context. Thus the language understanding module has to identify the objects which are designated in the sentence under consideration by comparing their descriptions with the known properties of the objects present in the environment. Either this identification must be certain (i.e. lead to unique results) or the sentence must be recognized as ambiguous. Conversely, the language generation module must be able to build discriminating natural language designations of objects in a given context from the available knowledge concerning the properties and the relationships of the objects present in the environment [9], [11]. Obviously, the meaning of the words which appear in the imperatives, questions or statements addressed to the robot, may be pervaded with vagueness ; moreover the available knowledge concerning the objects in the environment may be incomplete or fuzzy [22]. When this is the case, a fuzzy designation will be solved, using the "maximum meaningfulness" principle [14]. This involves finding the object which fits the description in the best way ; the designation will be more or less ambiguous as soon as another object will also fit it to a possibly different but still similar extent.

The specificity of the natural language communication in robotics is that the discourse refers to objects existing in the environment. The situation is slightly different when we consider interfaces with knowledge bases. Then the problem is to retrieve items which satisfy some condition, or pieces of information which contain parts which are similar, analogous or related in some sense to what is expressed in the request. An item which corresponds to a request is not necessarily unique, unlike the object fitting a non-ambiguous designation which must be unique. Query languages for data bases can be designed in order to allow restrictions on attribute values stated by means of linguistic expressions and fuzzy comparators between attribute values (e.g. "approximately equal", "much greater than") [23]. Query languages are thus brought closer to natural languages. More generally the understanding of a request expressed in natural languages may pose problems of approximate synonymy in order to be able to transform the request (or its representation) into a form which can be answered more easily [15], [24].

In the next section a possibility theory-based approach to the representation of vagueness in natural languages is presented. The estimation of the similarity between the representations of two natural language expressions, or between the representation of a natural language expression and the representation of a piece of information relative to an object is also discussed.

III - MEANING REPRESENTATION ISSUES

In the following we consider three main kinds of words or expressions, i) those which pertains to features of objects and which enable us to specify values of attributes (e.g. the size, the age, the color, the shape, ...) or to compare them ; generally these correspond to adjectives ;

ii) those which are names of categories of objects ; these are nouns ; and

iii) the verbal expressions.

1) Linguistic value of an attribute

A word or an expression which conveys some information about the value of an attribute, when applied to an object, can be represented by means of a possibility distribution [32], [33], [34]. This possibility distribution is a function from the attribute domain to the real interval [0,1], which gives an estimate of the compatibility between the word and each particular value of the attribute domain (also called universe of discourse).

When the attribute underlying the word is measurable, the attribute domain is some interval of the real line (or the set of integers or the set of decimal numbers, included in such an interval). In this case the possibility distribution attached to the word is very often context dependent.

More precisely this possibility distribution depends on i) the attribute domain and ii) the other words used in the taxonomy of the values of this domain. For instance, in order to define the possibility distribution attached to "large", we need to know the universe of discourse (i.e. the minimal possible size a and the maximal possible size b of the objects under consideration) ; then the interval [a,b] will be the attribute domain. Moreover, the meaning of 'large' will be slightly different if the values of the attribute domain are allowed to be classified by means of the three fuzzy categories 'large', 'medium' and 'small' rather than by means of two fuzzy categories 'large' and 'small' only, since the considered categories should provide at least an approximate covering of the attribute domain, as pictured in Figures 1.a and 1.b. However in both

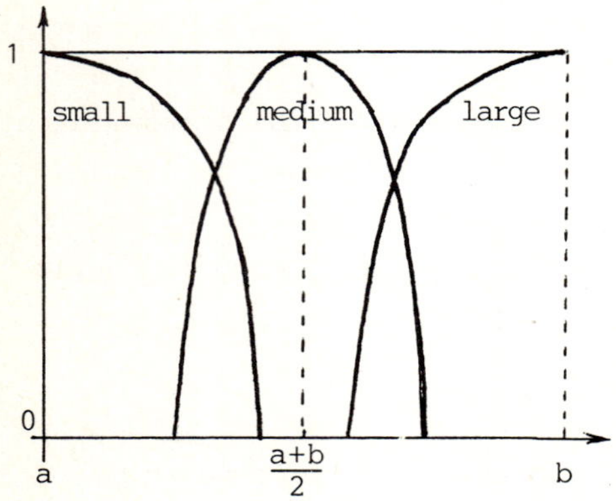

Figure 1.a

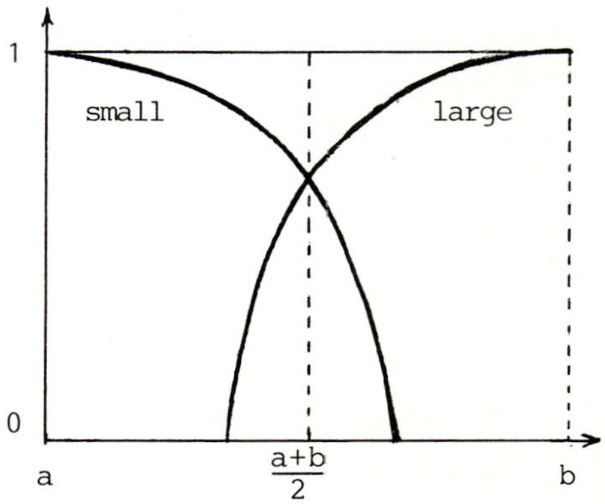

Figure 1.b

cases, the words 'small' and 'large' remain <u>antonyms</u> ; let f_A and f_B be the possibility distributions attached to two perfect antonyms A and B, whose universe of discourse is [a,b], then symmetry of f_A and f_B with respect to the axis $x = \frac{a+b}{2}$ is expressed by the relation

$$\forall\ x \in [a,b],\ f_A(x - \frac{a+b}{2}) = f_B(\frac{a+b}{2} - x) \qquad (1)$$

For a somewhat different view on the treatment of antonyms and synonyms, see [26], [20].

Note that it has been implicity assumed that the endpoints a and b can be precisely determined. This is not always the case, especially when the class of the objects under consideration has a large number of elements ; then a and b are typical minimal and typical maximal values respectively ; in our example, oversized or exceptionally small objects might have their size value beyond b or beneath a respectively. Then the possibility distributions defined on [a,b] would be extended on a larger domain. The possibility distributions do not need to be identified with a great precision, since possibility calculations, which are based on max and min operations, are not very sensitive to slight variations of the possibility degrees. Of importance here is the identification of the subregions of the attribute domain whose values are completely compatible with the considered word (the attached possibility distribution takes the value 1) or are not compatible at all with the word (the attached possibility distribution takes the value 0). Further, in the other subregions gradual transitions between 0 and 1 will be assigned. See [17] for an example of a learning system for identifying the possibility distributions attached to linguistic categories.

Given a collection of labels of linguistic categories $A_1,\ldots,A_n$ defined on a universe [a,b], it can be readily transposed to a new scale represented by the interval [a',b'] (corresponding to a new context) by means of the formula

$$\forall\ i,\ \forall\ x \in [a',b'],\ f_{A_i}^{[a',b']}(x) = f_{A_i}^{[a,b]}(\frac{a(b'-x)+b(x-a')}{b'-a'}) \qquad (2)$$

where $f_{A_i}^{[a',b']}$ and $f_{A_i}^{[a,b]}$ denote possibility distributions on the domains [a',b'] and [a,b] respectively.

In general the linguistic labels which are used to categorize the values of an attribute domain can be hierarchically organized. Figure 2 gives an example of such a hierarchy which includes categories and sub-categories.

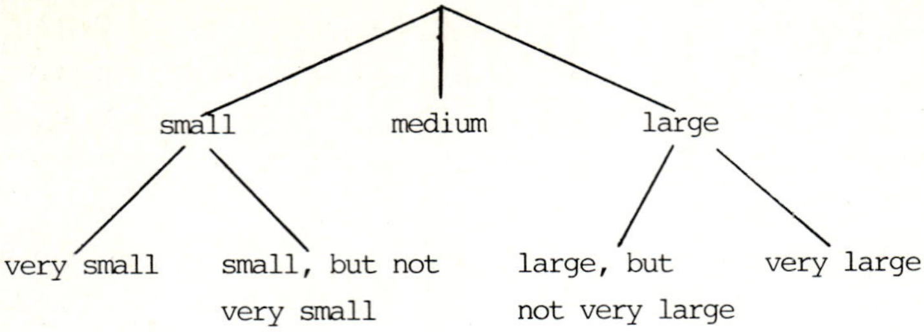

Figure 2

The inclusion of a sub-category A_j in a category A_i translates into the inclusion of the corresponding possibility distributions, defined by

$$\forall x \in [a,b],\ f_{A_j}(x) \leq f_{A_i}(x) \tag{3}$$

The inequality (3) indicates that the compatibility of the attribute value of an object with a category is greater or equal to the compatibility of this value with any sub-category of the considered category. When the attribute domain is linearly ordered (which is the case with measurable attributes), the linguistic categories, which are at a same level in the hierarchy, can then be ordered. Note that in general the possibility distributions which are contiguous in this ordering overlap somewhat (see Figure 1 for an example), since borderline values can be more or less compatible with two contiguous categories.

Until now, we have considered measurable attributes with one-dimensional domains. More complex universes of discourse, however can be handled. For instance, a word like 'dumpy' which means 'short and thick' can be represented in a given context by means of a possibility distribution defined on a Cartesian product of two intervals corresponding to the two dimensions underlying 'dumpy'. In this example the possibility distribution can be approximated by a suitable aggregation of the two possibility distributions defined on one-dimensional domains, which represent 'short' and 'thick' respectively. A choice has then to be made between logical and compensatory conjunctions for modeling the 'and' connective [5]. Shaket [25] represents the meaning of nouns referring to shapes, such as 'cube', 'bar', 'plate' by means of possibility distributions restricting the triples of relative values of the dimensions (width, depth, height), which are more or less compatible which each category.

When the attribute under consideration is no longer measurable, strictly speaking, a representation of linguistic values of this attribute by means of a possibility distribution can still be contemplated, provided that there exists typical values of the attribute, which can be linearly ordered. This finite ordered set of typical values is used as an abstract scale on which the possibility distributions are defined. Let us, for instance, consider the representation of the meaning of words describing the feelings of a person for another person. Five standard levels for instance, may be used for defining the attribute domain : a = hostility, b = antipathy, c = indifference, d = sympathy, e = friendliness. Then the meaning of 'rather friendly' may be represented in a given context by the possibility distribution 0.2/c + 1/d + 0.5/e which means that there is a possibility equal to 0.2 that the feeling corresponds to 'indifference', a possibility equal to 1 that it corresponds to 'sympathy', a possibility equal to 0.5 for 'friendliness' and a possibility equal to zero for any other standard level. Such an approach is used in [23].

For the treatment of linguistic hedges such as 'very', represented by means of operations which modify the shape of possibility distributions, the reader is referred to Zadeh [31], [32], Lakoff [16], Mac Vicar-Whelan [18]. The modeling of aggregation connectives of the 'and'/'or' type in the framework of fuzzy set theory, is surveyed in Dubois, Prade [5].

In this paper we only consider the representation of the meaning of a word by a possibility distribution. The converse problem (i.e. the labeling of a computed possibility distribution by means of a word or expression belonging to a given vocabulary) has been considered by several authors (e.g. [6]).

2) Fuzzy pattern matching

Given two possibility distributions f_A and f_B defined on a same attribute domain U and which represent the meaning of categories A and B, two kinds of measures can be used in order to estimate the similarity of the meaning of A and B ([2], [10]). First, the possibility $\Pi(A;B)$ that A and B refer to common attribute values is defined by

$$\Pi(A;B) = \sup_{x \in U} \min(f_A(x), f_B(x)) \tag{4}$$

which estimates the extent to which the fuzzy sets whose membership functions are f_A and f_B, have a non-empty intersection. Note that $\Pi(A;B) = \Pi(B;A)$. Second, the necessity (or certainty) $N(A;B)$ that B is a sub-category with respect to A is expressed by

$$N(A;B) = \inf_{x \in U} \max(f_A(x), 1-f_B(x)) \tag{5}$$

which is a fuzzy set degree of inclusion. Obviously, $N(A;B) \neq N(B;A)$. By construction, we have :

$$N(A;B) = 1 - \Pi(\bar{A};B) \tag{6}$$

where the complementation is defined by $f_{\bar{A}} = 1-f_A$. The equality (6) expresses that the certainty that A subsumes B, corresponds to the impossibility that B and non A refer to a same attribute value. We always have

$$N(A;B) \leq \Pi(A;B) \tag{7}$$

The equality holds in (7) only if B has a precise meaning with respect to U (i.e. B refers to only one attribute value in U). Obviously, we have $\Pi(A;A) = 1$ since f_A is always supposed to be normalized (i.e. $\exists\, x \in U,\ f_A(x) = 1$, which means that A refers at least to one value of the domain without any restriction). Besides we have only $N(A;A) \geq \frac{1}{2}$; $N(A;A) = 1$ if and only if the meaning of A can be precisely delimited in U (i.e. $\forall\, x \in U$, either $f_A(x) = 1$ or $f_A(x) = 0$). When this is not the case we cannot be completely sure that A always refers exactly to the same set of values in U when A is used several times. This is why we only have $1 > N(A;A) \geq \frac{1}{2}$ when f_A takes values intermediary between 0 and 1 for some x in U.

However the degree of inclusion defined by (5) only satisfies a weak transitivity property. Particularly, it can be proved [3], that $N(A;B) > \alpha$ and $N(B;C) > \alpha$ entails $N(A;C) > \alpha$ for $\alpha > \frac{1}{2}$. However, it is possible to have situations where $N(A;C) < \min(N(A;B),N(B;C))$ with $N(A;B) < \frac{1}{2}$ and $N(B;C) < \frac{1}{2}$. This weak transitivity property indicates that if we are sufficiently sure that a word A has a less specific meaning than a word B and that a word B is more general than a word C, we can consider to the same degree that the word C is more specific than the word A.

When the values of the domain U can be linearly ordered, a fuzzy proximity relation P can be introduced ; $f_P(x,x') \in [0,1]$ estimates the proximity between the values x and x'. It is assumed that $f_P(x,x) = 1$ and $f_P(x,x') = f_P(x',x)$. If U is a part of the real line $\mathbb{R}$, f_P may be defined as $f_P(x,x') = \max(0,\ 1 - \frac{|x-x'|}{\lambda})$; $f_P(x,x') = 0 \Leftrightarrow |x-x'| \geq \lambda$, thus values whose distance between them is less than λ, are regarded as somewhat close in the sense of P. A possibility distribution f_A can be enlarged by means of f_P into a new possibility distribution $f_{A \circ P}$ defined by

$$f_{A \circ P}(x) = \sup_{x' \in U} \min(f_A(x'), f_P(x',x)) \tag{8}$$

We have $\forall\, x,\ f_{A \circ P}(x) \geq f_A(x)$

and $\qquad \Pi(A \circ P;B) \geq \Pi(A;B) \qquad\qquad N(A \circ P;B) \geq N(A;B)$

We may look for the smallest value of the parameter λ (used in the definition of f_P) in order to have $N(A \circ P;B)$ greater or equal to a given threshold ; it will yield an estimate of the distance from A to B. Using the tolerance relation P we are thus able to determine to what extent the fuzzy set of possible interpretations of a linguistic attribute value is not far from the fuzzy set of interpretations of another linguistic value.

The expressions (4), (5) and (8) are easy to compute in practice using a parametered representation of the possibility distributions ; see [2], [4]. A precise identification of the possibility distributions is not required since $\Pi(A;B)$ and $N(A;B)$ are not very sensitive to slight variations of the possibility degrees. This is due to the presence of the max and min operations in their expressions. The degree of matching between two linguistic categories is sometimes computed by means of ad hoc methods. For instance Fieschi et al. [12] estimate this degree only on the basis of the relative positions of the linguistic categories in a tree structure like the one presented in Figure 2. This degree depends only on the level in the tree of the common "ancestor" of the two subcategories which are considered. This heuristic evaluation requires a great homogeneity of the tree structure and does not yield an estimate with a clear meaning.

Given two linguistic attribute values A and B represented by means of possibility distributions, we can compute the degree of overlapping of their respective meaning (by means of $\Pi(A;B)$) and the degree to which one subsumes the other (by means of $N(A;B)$ and $N(B;A)$). Now let us consider a collection of linguistic labels $A_1,\ldots,A_i,\ldots,A_n$ such that $\forall$ i, $\forall$ j, $N(A_i;A_j) = 0$. Moreover, we suppose that they are ordered and that the following condition holds, $\forall$ i, $\forall$ $j \notin \{i-1,i,i+1\}$, $\Pi(A_i;A_j)$ is below a given threshold. In other words, the meaning of A_i may overlap in a significant way only with the meanings of A_{i-1} and of A_{i+1}. However, A_i cannot be considered as more (or as less) specific than A_{i+1}. Under these conditions and provided that the A_i's cover the universe of discourse sufficiently, the A_i's can be used as a basis on which the meaning of a new linguistic label A' can be defined directly. This definition is given in terms of the $\Pi(A';A_i)$'s and of the $N(A';A_i)$'s, rather than by means of a possibility distribution. If there exists only one i such that $\Pi(A';A_i) = 1$ and we have $N(A';A_i) = 0$, it will indicate that A' is subsumed by A_i. In other words, the replacement of A_i by A' would refine the basis. The idea of defining a fuzzy set (in an approximate way) in terms of other fuzzy sets has been already studied by Willaeys [28]. The expression of the approximate synonymy of two words in terms of a degree of overlapping and in terms of a degree of inclusion may be very convenient for representing the related meanings of a family of words on a same universe of discourse. Thus is true even when it is difficult to represent the meaning of each word by means of possibility distributions, as shown in the following section.

3) Nouns and verbs

Let us consider a class of objects having a common name. These objects can be described in terms of attributes pertaining to their appearance or their function (for instance, see [19] for a discussion of this problem from a psychological modelling point of view). The possible values of each attribute for the objects in the class are supposed to be represented by means of a possibility distribution. Then the similarity between two classes of objects described in terms of the same attributes, and thus the synonymy between the names of these classes, can be estimated in terms of possibility and necessity measures. This can be done by aggregation (using min operation ; see [21] for justifications) the possibility measures on the one hand and the necessity measures on the other hand. These are computed for each attribute from the possibility distributions restricting the allowed values of this attribute in each class. When in a given context the similarity of some attributes is considered as more important than the one of others in the matching process, this can be taken into account by weighting the aggregation. This weighting is performed by means of min operation on the elementary possibility (necessity) measures attached to each attribute [4], [5]. Namely, we have

$$r = \min_{i=1,q} \max(r_i, 1-w_i) \qquad (9)$$

with $\max w_i = 1$, where we have q attributes, w_i estimates the importance of attribute i and r_i is the associated rating (possibility or necessity). When all the features have equal importance ($\forall$ i, $w_i = 1$), it reduces to min aggregation.

Even if we are able in theory to estimate the (dis)similarity between the meanings of two names belonging to a same universe of discourse using the approach sketched above, in many practical situations it may be sufficient to have a simpler representation by giving the values of the measures Π and N directly. This point of view is illustrated on the example of Figure 3

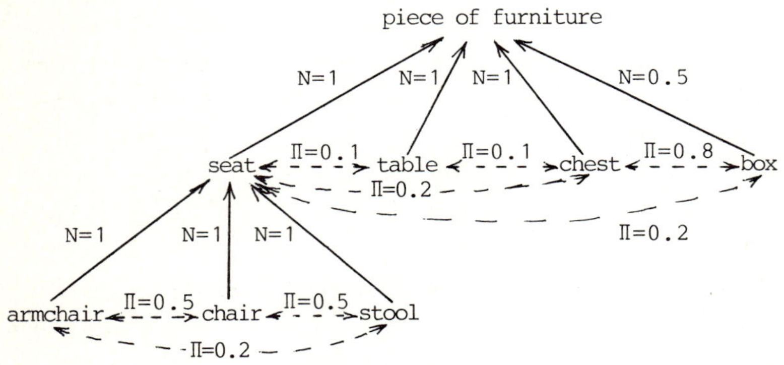

Figure 3

Two kinds of relationship between words are graded in this example :
i) to what degree N a word is more specific than another (i.e. the word is the name of a class of objects which is included in a larger class corresponding to the other word). When N = 1 we have a classical IS-A relation ; when $N < 1$, we are not certain that one word subsumes the other (e.g. box is not strictly speaking a piece of furniture, even if we may consider it is)
ii) to what degree Π the meaning of two words overlap.

Note that Π is not even weakly transitive, contrastedly with N. However the following consistency condition must hold

$$N(A;B) = 1 \Rightarrow \forall\, C,\ \Pi(C;A) \geq \Pi(C;B)$$

The values of N and more especially Π depend on the point of view (i.e. on the attributes which are considered as the more important in the estimation of the "synonymies" ; for instance the attributes pertaining to the function versus the ones pertaining to the appearance).

Other authors (e.g. [15]), have considered that the interchangeability of words or expressions in a given sentence is a matter of degree. In our approach the approximate synonymy is graded according to two different viewpoints. Dealing with the same problem, Sabah [24] similarly points out the difference between "particularization" (a word is a specialization of another) and "analogy" (two words are said to be analogous if they are specializations of a same third word). However this author does not use any such rating. Moreover the proposed definition makes the analogy between words transitive,* which may be dangerous.

What has been said here about the names of classes also applies to verbs for the most part. Indeed the meaning of a verb can be practically represented by means of different facets pertaining to the nature of the underlying action and to the nature of the agents and objects involved in this action. Possibility distributions may be used for expressing the restrictions on the values of each facet. The ideas of specialization (a verb is a specialized form of another) and of "proximity" (two verbs have different but quite close meanings) still make sense. Sabah [24] gives the example of the "analogy" existing between 'to sell' and 'to repair' based on the fact that the agents who sell objects quite often repair them also and conversely.

IV - TAKING INTO ACCOUNT VAGUENESS AND UNCERTAINTY IN NATURAL LANGUAGE INTERFACES

In section II we point out two different kinds of applications of natural language communication in man-machine systems : data base querying and dialogues with robots. In this section we briefly consider some examples of the treatment

(*) provided that the considered words can be viewed as specializations of a same word.

of vagueness in each situation.

1) Knowledge base quering

The possibility and necessity measures introduced in section III.2 play a basic role in the management of relational data bases with incomplete or fuzzy information when vague queries are allowed. The available knowledge (which may be or not linguistically expressed) stored in the data base concerning the possible values of each attribute for each object is represented by means of possibility distributions. Vague queries expressing linguistic restrictions on attribute values and involving fuzzy comparators can be also represented using possibility distributions. Then objects which more or less possibly satisfy the request and the objects which more or less certainly satisfy it can easily be retrieved. If we imagine a data base where linguistic judgements are stored on the levels of students in different courses, we are thus able to deal with queries such as "Find the students who are very good in math and who have approximately the same level in physics" ; see [23].

In some applications the request is progressively transformed in approximately synonymous requests until a request is reached which can be directly answered from the available knowledge ; see [15], [24] for instance. In this process, the words or expressions which are substituted to others may be more specialized, more general or may have only a related meaning ; depending on the nature of these substitutions, the answer which is finally obtained may be more of less appropriate to the request we start with. An estimate of the reliability of the answer may be obtained by combining estimates of the validity of each elementary transformations.

2) Natural language communication with a robot

As explained in section II, the language understanding module has to identify the objects present in the environment to which the sentences refer. In order to understand a phrase like

"... the small box on the dark seat ..."

the system has to look for an object x and object y which can be plausibly designated as a 'small box' and as a 'dark seat' respectively. It then has to check if we can consider that 'x is on y'. Note that we are not necessarily sure that 'on the dark seat' indicates the location of 'the small box' as pointed out in section II ('on the dark seat' may refer to the designation of 'the small box' to move, for instance). From the available knowledge on the objects pertaining to their nature, color, size, ... the system can compute the degree of possibility $\min(\Pi(x;o_i), \Pi(y;o_j))$ and the degree of certainty $\min(N(x;o_i), N(y;o_j))$ that the pair (x,y) corresponds to a pair of objects (o_i,o_j). (In case of complete and precise knowledge, we have $\Pi = N$ as said in section III.2). Then, among the pairs of objects which fit the description in the best way, the system looks for a pair (o_i,o_j) such that $\Pi(on;r(o_i,o_j))$ (and $N(on;r(o_i,o_j))$) are sufficiently large, if such a pair exists, and where $r(o_i,o_j)$ denotes what is known of the spatial relationship between o_i and o_j. If such a pair is found the interpretation is taken on ; if not o_i and o_j must be determined separately, taking into account the objects on which the dialogue possibly focuses. Here we will only sketch how it is possible to accommodate vagueness and uncertainty in a natural language interface system like SYROCO. The reader is referred to [7], [8], [11], for the treatment of the many other aspects of natural language in such a task-oriented dialogue interface. SYROCO includes a language generation module which builds discriminating natural language designations for the objects which are involved in the sentence to be generated, using ordered search methods [9]. If fuzzy adjectives are used when building these designations, the system has to make sure that the description which is built applies much more to the object to be designated than to any other objects, provided that the available knowledge about the environment enables the system to build a sufficiently discriminating designation (i.e. the possibility that the description applies to the other objects must be zero or at least very low).

Generally speaking we may have to use vague words when generating sentence because either the knowledge which is needed to be more precise is not available or it is unnecessary (or even tedious) to be more precise (e.g. we may prefer the information 'it is about 11 h. 30 a.m' to 'it is 11 h. 32 a.m.' in many circumstances). Another situation where a language generation module may have to produce fuzzy descriptions has been simulated by Wahlster [27]. This author considers a room booking system with a natural language interface which is able to describe the room which is proposed to a customer with some exaggeration (e.g. 'the room is quite large'), even if the vague epithets which are used remain always somewhat compatible (in the sense of a possibility measure) with the reality.

V - CONCLUDING REMARKS

Possibility distributions are a particularly adequate tool for representing vague words whose meanings are more or less compatible with different possible values of a domain. Context dependency can be conveniently handled in this framework. In this paper the interest of possibility and necessity measures, which can be built on possibility distributions, has been emphasized. Approximate synonymy can be more precisely analyzed with these two measures than with only one index. We are thus able to distinguish between strict synonymy, specialization and partial synonymy. In general, the values of possibility and necessity measures cannot be determined with great precision in natural language representation. However, these degrees enable one to order the objects to retrieve or to identify, according to the quality of the matching of their descriptions with their known properties. Obviously, although vagueness is a characteristic feature of natural languages, there are many other non-trivial problems which must be handled in order to build natural language interfaces even in limited domain. However it may be important to take into account vagueness in some applications (interfaces with data bases or systems having to understand rough and incomplete descriptions of objects), especially when we want tc capture nuances between close words.

REFERENCES

[1] - T. Ballmer, Fuzzy punctuation or the continuum of grammaticality. Memo ERL-MR90, Univ. of California, Berkeley, 1976.

[2] - M. Cayrol, H. Farreny, H. Prade, Fuzzy pattern matching. Kybernetes, 11, 103-116, 1982.

[3] - D. Dubois, H. Prade, Fuzzy Sets and Systems : Theory and Applications. Academic Press, New York, 1980.

[4] - D. Dubois, H. Prade, Théorie des Possibilités - Applications à la Représentation des Connaissances en Informatique. Masson, Paris, 1985a.

[5] - D. Dubois, H. Prade, A review of fuzzy set aggregation connectives. Information Sciences, 36, 85-121, 1985b.

[6] - F. Eshragh, E.H. Mamdani, A general approach to linguistic approximation. Int. J. Man-Machine Studies, 11, 501-519, 1979.

[7] - F. Evrard, H. Farreny, H. Prade, Non-grammatically-guided system for sentence-analysis in a limited context. Proc. IEEE Int. Conf. on Cybernetics and Society, Atlanta, 284-288, 1981.

[8] - F. Evrard, H. Farreny, H. Prade, A flexible interface for understanding task-oriented unconstrained natural language. Computers and Artificial Intelligence (Bratislava), 2 (6), 497-511, 1983.

[9] - H. Farreny, S. Piquet-Gauthier, H. Prade, Une application des méthodes de recherche ordonnée à la construction de désignations non ambiguës d'objets en génération de phrases. Cybernetica, 26 (4), 315-326, 1983.

[10] - H. Farreny, H. Prade, Flexible matching and its use in analogical reasoning. Proc. Europ. Conf. Artificial Intelligence, Orsay, 43-47, 1982.

[11] - H. Farreny, H. Prade, On the best way of designating objects in sentence generation. Kybernetes, 13 (1), 43-46, 1984.

[12] - M. Fieschi, M. Joubert, D. Fieschi, G. Soula, M. Roux, SPHINX : An interactive system for medical diagnosis aids. In : Approximate Reasoning in Decision Analysis (M.M. Gupta, E. Sanchez, Eds.), North-Holland, 269-275, 1982.

[13] - Y. Gentilhomme, Les ensembles flous en linguistique. Cahiers de Linguistique Théorique et Appliquée (Bucarest) 5, 47-63, 1968.

[14] - J.A. Goguen, Robust programming languages and the principle of maximal meaningfulness. Proc. Milwaukee Symp. Aut. Comput. & Control, 87-90, 1976.

[15] - D. Kayser, An experiment in knowledge representation. Proc. Europ. Conf. Artificial Intelligence, Orsay, 119-124, 1982.

[16] - G. Lakoff, Hedges : A study in meaning-criteria and the logic of fuzzy concepts. J. Philos. Logic, 2, 458-508, 1973.

[17] - R. Lopez de Mantaras, C. Freksa, A learning system for linguistic categorization of "soft" observations. Proc. Colloq. of the Association pour la Recherche Cognitive "Les Modes de Raisonnement", Orsay, 331-343, 1984.

[18] - P.J. Mac Vicar-Whelan, Fuzzy sets, the concept of height and the hedge very. IEEE Trans. on Systems, Man and Cybernetics, 8, 507-511, 1978.
[19] - G.C., Oden, L.L. Lopes, On the internal structure of fuzzy subjective categories. In : Fuzzy Set and Possibility Theory : Recent Developments (R.R. Yager, Ed.), Pergamon Press, 75-89, 1982.
[20] - N. Piera, Morphisms and synonymy. Proc. Fall Inter. Seminar on Applied Logic, Porto Colon, Spain, Sept. 25-30, 1983, 101-109.
[21] - H. Prade, Modal semantics and fuzzy sets theory. In : Fuzzy Set and Possibility Theory : Recent Developments (R.R. Yager, Eds.), Pergamon Press, 232-246, 1982.
[22] - H. Prade, Compréhension de désignations imprécises dans un univers incomplètement connu. T.A. Informations (Paris), 24 (2), 3-12, 1983.
[23] - H. Prade, C. Testemale, Generalizing database relational algebra for the treatment of incomplete/uncertain information and vague queries. Information Sciences, 34, 115-143, 1984.
[24] - G. Sabah, Différentes notions de synonymies liées à la compréhension automatique du langage. Proc. Colloq. of the Association pour la Recherche Cognitive "Les Modes de Raisonnement", Orsay, 477-504, 1984.
[25] - E. Shaket, Fuzzy semantics for a natural-like language defined over a world of blocks. M.S. Thesis, Comput. Science Dept., Univ. Calif., Los Angeles.
[26] - E. Trillas, T. Riera, Towards a representation of "synonyms" and "antonyms" by fuzzy sets. BUSEFAL (L.S.I., Univ. P. Sabatier, Toulouse), n° 5, 42-68, 1981.
[27] - W. Wahlster, Implementing fuzziness in dialogue systems. In : Empirical Semantics. Vol. 1 (B.B. Rieger, Ed.), Brockmeyer Bochum, 259-280, 1981.
[28] - D. Willaeys, Some of the properties of fuzzy discretisation. Proc. IFAC Int. Symp. "Fuzzy Information, Knowledge Representation and Decision Analysis", Marseille, France, July 19-21, Pergamon Press, 61-66, 1983.
[29] - L.A. Zadeh, Fuzzy sets. Information and Control, 8, 338-353, 1965.
[30] - L.A. Zadeh, Quantitative fuzzy semantics. Information Science, 3, 159-176, 1971.
[31] - L.A. Zadeh, A fuzzy set-theoretic interpretation of linguistic hedges. J. of Cybernetics, 2 (3), 4-34, 1972.
[32] - L.A. Zadeh, Fuzzy sets as a basis for a theory of possibility. Fuzzy Sets and Systems, 1 (1), 3-28, 1978a.
[33] - L.A. Zadeh, PRUF-A meaning representation language for natural languages. Int. J. Man-Machine Studies, 10 (4), 395-460, 1978b.
[34] - L.A. Zadeh, Test score semantics for natural languages and meaning representation via PRUF. In : Empirical Semantics, Vol. 1 (B.B. Rieger, Ed.), Brockmeyer Bochum, 281-349, 1981.

Applications of Fuzzy Set Theory in Human Factors, edited by W. Karwowski and A. Mital
Elsevier Science Publishers B.V., Amsterdam, 1986 — Printed in The Netherlands

A NEW DATA-INPUT METHOD BASED ON THE CONCEPT OF EXTENDED FUZZY EXPRESSION & SUBJECTIVE ENTROPY OF PROBABILISTIC SETS

K. HIROTA[1], E. CZOGAŁA[2], and W. PEDRYCZ[2]

[1]Dept. of Instrument & Control Eng., College of Eng., Hosei Univ., Kajino-cho 3-7-2, Koganei-city, Tokyo 184 (Japan)

[2]Dept. of Automatic Control & Comp. Science, Silesian Tech. Univ., Ul Pstrowskiego 16, Gliwice 44-100, (Poland)

ABSTRACT

Human "ambiguity" is one of the most important factors in (conversational) man-machine-systems (MMS's). Two problems in MMS's are commonly dealt with. One is how to express ambiguous data given by human beings, the other what is the optimal number of states in answering questions given by machines/computers in the case that only ambiguous data are available. A decision making model is proposed in this writing by introducing a probability distribution on a [0,1]-interval. An ordinary data-input method is defined as giving only membership information in a "fuzzy" concept. It is shown that a second group of information, called "vagueness", plays an important role in representing ambiguous evaluations of human beings almost completely. It also shows that "three" is the best number of alternatives/states in answering questions. This number is derived from the optimization-process using a concept of subjective entropy. Based on these results, a new MMS data-input method is proposed, where 5-valued membership information is given by human beings with an appropriate use of 3-valued vagueness information. Availability of this method is also confirmed by a simulation expeiment in human character/voice pattern recognition process.

KEY WORDS: man-machine-system, fuzzy, vagueness subjective entropy, pattern recognition

INTRODUCTION

The theory of probabilistic sets has been studied in the past to investigate ambiguous phenomena positively (ref.3). There exist two mutually equivalent expressions of probabilistic sets; one is probabilistic expression, the other is an extended fuzzy expression (ref.5). The latter consists of a countable family of functions called monitors. It has been shown that most information of ambiguous phenomena is expressed by lower monitors, especially by the 1-st monitor "membership" and the 2-nd monitor "vagueness" (ref.5). The former is similar to the one in classical fuzzy theory introduced by L.A.Zadeh (ref.1). The concept of membership and vagueness has been applied to various fields such as decision making, identification of fuzzy systems, pattern recognition, robotics, and MMS's. The main purpose of the present paper is to introduce several recently obtained results which concern MMS's.

One of the most difficult problems in MMS's is man-machine communication (i.e. information exchange between man and machine/computer). In the case of data input from man to computer, the following two problems may arise.

(1) If the man is lacking in exact information, he may input ambiguous information, and

(2) If he selects one answer from several (e.g. 3 - 7) alternatives in such an ambiguous situation, he may have no confidence in choosing the same answer if he is requested to input the same data again in the near future. So the question arises; "What is the suitable number of alternatives/states in answering a question?" We may consider that more precise information will be obtained as the number of states increases, but at the same time we must also consider that the reliability of selecting the answer will be decreased.

In this paper we study these two problems from a viewpoint of the so called theory of ambiguity which includes probability theory, fuzzy set theory by Zadeh (ref.1), information theory by Shannon (ref.2), and probabilistic set theory by Hirota (ref.3). The ordinary data input method to date is characterized as a method which gives only membership information. It is shown concerning the problem (1) that ambiguous information will be expressed correctly if a new measure called "vagueness" is used with the membership information. On problem (2) an optimum number of states is derived by the use of the notion of subjective entropy (ref.4). The results show that the optimum number of states is three followed by four, five, six, two, seven,... (i.e. seven is worse than two). Considering these results we propose a new data input method which consists of a 5-valued membership with a 3-valued vagueness. The additional information "vagueness" index will be useful for man to dissolve the uncertainty of inputting an incorrect data when they have only a superficial understanding about the problem. The availability of this method is confirmed by an experimental result on human performance in character and voice pattern recognition in this study.

SOME PROBLEMS IN MMS DATA INPUT METHOD

The objective of automatic control technique is gradually becoming a total system made up of both men and machines/computers. In such a MMS, human "ambiguity" becomes one of the most important factors. Such a factor should be considered, especially, in man-machine-communication. In this article, we investigate a new data input method from man to machine, where human ambiguity does not affect the accuracy of input data.

Let us consider the following situation: a man is supposed to input his idea to a machine/computer, i.e. he must answer the question given by the computer. If he has a critical mind about the matter, then he may be able to input a yes/no-two valued answer. But sometimes he may face an ambiguous situation. He must input an ambiguous answer other than yes/no when he has few ideas about the question. This ambiguous data input is the first problem in the MMS data input method. In such an ambiguous situation the D.K.(=don't know) answer is usually used in addition to yes and no. However it is also possible to use a five-state-answering-system (where a man is supposed to select a response from yes, almost yes, D.K., almost yes, and no). A seven-state-answering-system may be possible in order to express ambiguous situations more precisely. Finally we will find our way to the fuzzy concept which was introduced by Zadeh (ref.1), where ambiguous states are expressed by a [0,1]-interval (0 and 1 correspond to no and yes, respectively). But it will be a little bit difficult to find any significant differences between slightly different values in a [0,1]-interval, e.g. 0.349 and 0.350. It is natural for man to select one answer from finitely multi-alternatives (not from [0,1]-infinite multi-values). The second problem is concerned with the relation between the number of alternatives and the reliability of input data. It apparently seems that more detailed evaluation is possible as the number of alternatives increases. However too many alternatives may not be desirous, because the man may have no confidence in selecting the same answer if he was asked to answer the same question repeatedly. So there will exist an optimum number of states which maximize a certain information measure of each alternatives. We will investigate these two problems in the following.

CONCEPT OF VAGUENESS

Let us first consider numericalization of input data which are answers given by men to questions presented by computers. Here "affirmative" or "yes" corresponds to 1 and "negative" or "no" to 0.

Other ambiguous, intermediate states are also expressed by suitable values from 0 to 1. This numericalization in a [0,1] interval is the same one as that used in classical fuzzy set theory by Zadeh (ref.1).

Let us also consider the following situation; a man is asked several questions by a computer. If he has only a vague idea about the question, then his answer will be scattered on a [0,1] interval. So the answer may be expressed by the probability distribution $f(\alpha)$ on a [0,1] interval,

$$\int_0^1 f(\alpha)\cdot d\alpha = 1, \qquad f(\alpha) \geq 0. \tag{1}$$

This $f(\alpha)$ will exist in his subconscious and its existence may be guaranteed. However, it will be natural to consider that it is almost impossible to grasp from the outside the precise function-form of $f(\alpha)$. The answer given may be identified as a sample value from the distribution $f(\alpha)$. Of course only one sample value does not reflect correctly all the characteristics of the distribution $f(\alpha)$. In order to get a reliable answer, it is desirable to extract suitably as much information about $f(\alpha)$ as possible. It has already been reported in (ref.5) based on moment analysis that almost all important information about $f(\alpha)$ is concentrated on lower moments such as mean value or variance. A concept of vagueness has been proposed by a reversed idea of moment analysis, which is called an extended fuzzy expression of probabilistic sets. In other words, in an extended fuzzy expression of probabilistic sets a probability distribution on a [0,1]-interval is derived from a countable family of functions

$$\{M, V=M^2, M^3, M^4, \ldots\} \tag{2}$$

under several assumptions. Here, M^n is called n-th monitor of the probabilistic set. In general the number of monitors is countably infinite, but important information exists on lower monitors such as M (called a membership function) and V (a vagueness function), both of which are [0,1]-valued functions on a total space (a universe of discourse).

The value of M indicates the degree of membershipness (belongingness) and has almost the same meaning as the membership function in the classical fuzzy set theory (ref.1), while that of V points out how vague the value of M is.

It could be said from a viewpoint of extended fuzzy expression, that ordinary data-input method in MMS's deals with only membership information. But only membership information is insufficient in general (i.e. inaccuracy of input-data will appear as stated in section 2). Here we would like to propose a new data input method based on membership information combined with its vagueness. The information vagueness expresses the validity of input data (membership value) and is also input by man.

DETERMINATION OF OPTIMUM NUMBER OF STATES

It has been assumed, until now, that values of membership function and vagueness function can take any values in the [0,1]-interval. But it will be a better fit for both man and computer to select one from finitely many alternatives rather than to choose a value freely from a [0,1] interval (infinitely many alternatives). Here the problem stated in section 2 arises (i.e. the number of alternatives (states)). Too many number of alternatives (states) causes a lack of reliability of input-data. So we would like to determine an optimum number of states based on a concept of subjective entropy (ref.4) which is closely related to the well-known entropy referred to in Shannon's information theory (ref.2).

The concept of subjective entropy has been proposed based on the notion of membership with probability as a measure which indicates the structural complexity of probabilistic sets. Let $\{\alpha_i\}_{i=1,2,..,n}$,

$$0 \leqq \alpha_1 < \alpha_2 < \ldots < \alpha_n \leqq 1, \qquad (3)$$

be a finite set in a [0,1] interval, each of which corresponds to an answer given by man to computer. If we restrict the range of answers to this finite set α_i , then the distribution $f(\alpha)$ in eq. (1) may be rewritten in terms of finite sum,

$$\sum_{i=1}^{n} p_i = 1, \quad p_i \geqq 0, \qquad (4)$$

where p_i indicates the probability of inputting an answer α_i by man to computer. Then the subjective entropy H is given by (cf. ref.4)

$$H = -\sum_{i=1}^{n} \{\alpha_i p_i \log_2 \alpha_i p_i + (1-\alpha_i)p_i \log_2 (1-\alpha_i)p_i\} \qquad (5)$$

$$= -\sum_{i=1}^{n} p_i \log_2 p_i + \sum_{i=1}^{n} p_i \cdot h(\alpha_i), \qquad (6)$$

where $h(\alpha)$ stands for Shannon's entropy function (ref.2),

$$h(\alpha) = -\{\alpha \cdot \log_2 \alpha + (1-\alpha)\log_2 (1-\alpha)\}. \qquad (7)$$

This $h(\alpha)$ also indicates a subjective entropy of probabilistic sets in a pure state, where a pure state is characterized as a {0,1} two valued, non-ambiguous state (ref.4). Equation (6) shows that the subjective entropy H is a sum of a p_i-weighted average of entropy $h(\alpha_i)$ of each answer α_i (the second term) and an entropy of the frequency of occurrence (the first term).

Now we can derive an optimum number of states by using subjective entropy measure. Assume that each answer α_i is allocated linearly in a [0,1] interval,

$$\alpha_i = (i-1)/(n-1). \tag{8}$$

That is $\alpha_1=0$, $\alpha_n=1$, and an interval (0,1) is divided into equal parts. It will be natural to consider that an optimal number of states is concerned with the extent to which the information-value goes up when one state is added. Dividing the subjective entropy H (5)(6) by the number of states n, we will obtain an average information value per one state. Since H/n depends upon the probability distribution $\{p_i\}_{i=1,2,\ldots,n}$ in (4), we will denote it by $f(p_1,p_2,\ldots,p_n)$,

$$f(p_1,p_2,\ldots,p_n) = H/n$$
$$= -(1/n)\sum_{i=1}^{n} p_i \log_2 p_i + (1/n)\sum_{i=1}^{n} p_i h((i-1)/(n-1)). \tag{9}$$

An optimum number of states is derived from a maximization-process of $f(p_1,p_2,\ldots,p_n)$. It is shown in (ref.6) that the maximum value f_{max} in (9) is obtained by Lagranges method of undetermined coefficients, and

$$f_{max} = (1/n)\log_2 \sum_{i=1}^{n} 2^{h((j-1)/(n-1))}, \tag{10}$$

when $$p_i = 2^{h((i-1)/(n-1))} / \sum_{j=1}^{n} 2^{h((j-1)/(n-1))}. \tag{11}$$

The relation between f_{max} and n is shown in Fig.1, which tells us that the maximum is attained when n=3 followed by n=4,5,6 in this order. It should also be noted that the f_{max} in n=7 is smaller than that in n=2. So we can conclude that the optimum number of states is 3 and that the validity of three-states-answering method (i.e. "yes", "no" with "D.K.") is confirmed.

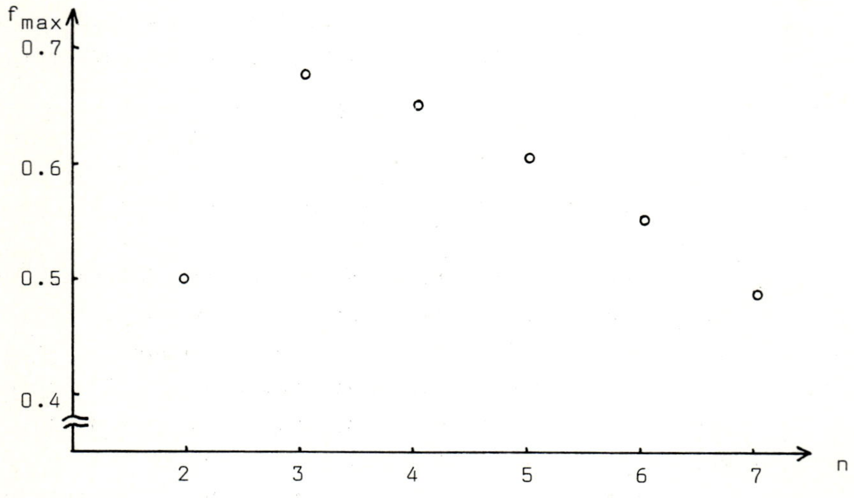

Fig.1 maximum-average-subjective-entropy f_{max} v.s. number of states n

A NEW DATA INPUT METHOD USING MEMBERSHIP & VAGUENESS

Taking into account the concept of vagueness presented in section 3 and the optimum number of states in section 4, we would like to propose a new MMS data input method, an example of which is illustrated in Fig.2. Here the question presented by computer is "Do you have a good appetite?". A man is supposed to input his answer by membership from {"yes", "almost yes", "D.K.", "almost no", "no"}

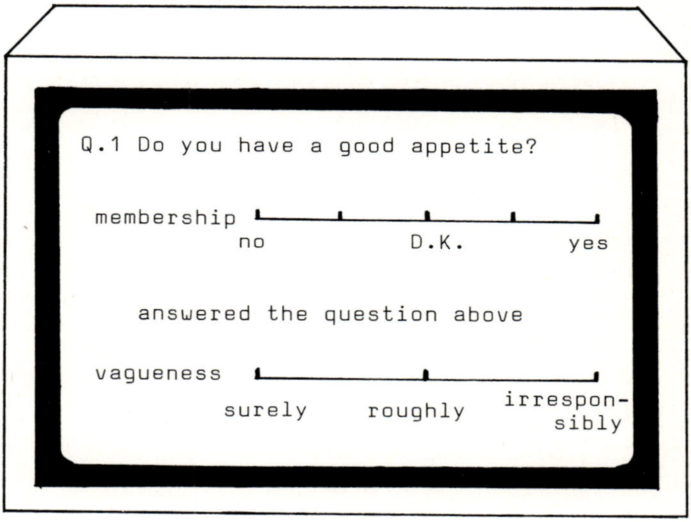

Fig.2 An example of display based on a new MMS data input method

(i.e. five states) with vagueness from {"surely", "roughly", "irresponsibly"} (i.e. three states).

In this method, the second information "vagueness" plays an important role to reduce man's guilty conscience of inputting an unreliable data, if it is really an unreliable one. The repetitive inquiry method is usually used in questionnaires in order to get reliable data from the subjects, but one-time-inquiry method may be enough if both membership and vagueness information are gathered from subjects.

Although the optimum number of states is three from a viewpoint of subjective entropy, the number of states of membership information in Fig.2 is five (that of vagueness is three). This is because we sometimes need to know more precise information for a decision. An example of this need is seen in expert systems such as medical diagnosis. When this is the case reliability of the information decreases a little. Although the number of states in Fig.2 is odd (i.e. 3 or 5), it is, of course, possible to use even (e.g. 4) states according to situation. In this case we use 3 or 5 states since we would like to represent a middle state (i.e. D.K. answer). Taking the optimum number of states into consideration, we should set an appropriate number of states from three, four, and five. It should be noted again, however, that confusion may occur when the number of states is more than or equal to seven.

VALIDITY CHECK BY SIMULATION EXPERIMENTS

Outline of experiments

Two simulation experiments were conducted in order to confirm the validity of the newly presented data input method. The following points for selection of the questions in MMS were taken into consideration:

1. Input answers given by men should preferably be based on an ambiguous decision.
2. Learning effects should not appeared in the case of repetitive answering.
3. Data given by men should not be affected very much by age, sex, education, nationality and so on.

With these in mind, we decided to evaluate human performance in an audio/visual pattern recognition process.

The main purpose of this experiment is to investigate the relationship between ordinary repetitive answering and one-time-answering with membership and vagueness. Two methods were used to accomplish this goal:

1. Repetitive answering; Only membership-value was requested from 26 university students three times at one month intervals.
2. One-time answerng; Both membership and vagueness values were requested from other men (other 25 university students) only once.

Here membership takes five states from "no" to "yes", and vagueness three states from "surely" to "irresponsibly" (cf. Fig.2).

Two kinds of questions were asked; one is concerned with character pattern recognition and the other concerned with voice pattern rocognition. Fundamental patterns were generated as follows:

1. Fundamental character patterns; Twelve handwritten numerics shown in Fig.3 were adopted.
2. Fundamental voice patterns; Twelve voice patterns as shown in Table 1 were generated using a speech synthesis system (COMPUTALKER CT-1 and Synthesis-by-Rule software CSR-1 by D. LLOYD RICE COMPUTALKER CONSULTANTS Corp.).

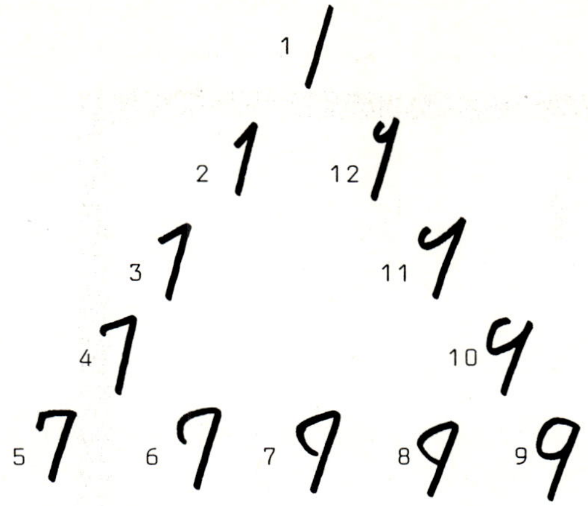

Fig.3 Fundamental 12 character patterns

pattern number	key input symbol	phonetic symbol
1	AAAA	[ɔɔ]
2	AHAA	[ʌɔ]
3	AAER	[ɔə:]
4	AEER	[æə:]
5	UHER	[uə]
6	UHUW	[uu:]
7	UWUW	[u:u:]
8	UHAO	[uɔ:]
9	UHOW	[uou]
10	OWOW	[ouou]
11	AOOW	[ɔ:ou]
12	EROW	[ə:ou]

Tab.1 Fundamental 12 voice patterns

The patterns presented to the subjects consist of every combination of twelve fundamental patterns. In the case of character patterns 144(=12^2) patterns as shown in Fig.4 were presented to subjects. Voice patterns were recorded and edited as shown in Fig.5. Here the order of the presentation was decided randomly using random numbers generated by computers. The order was changed every time in the case of repetitive answering.

Results of experiments

We took a numerical quantification for computer calculation. This is shown in Fig.6. Such numerical data of experiments were fed into the computer (ACOS 77 SYSTEM 650) and sorted. The format of the computer output is shown in Fig.7. Among these results, we found that several replies were made completely irresponsibly. So

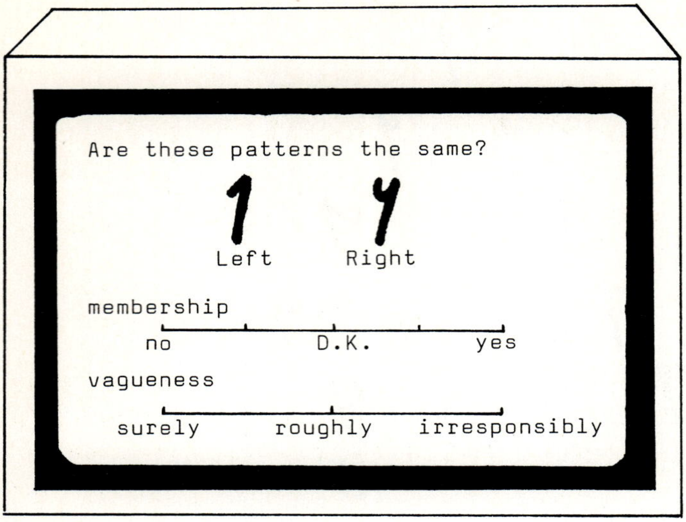

Fig.4 An example of presented character pattern

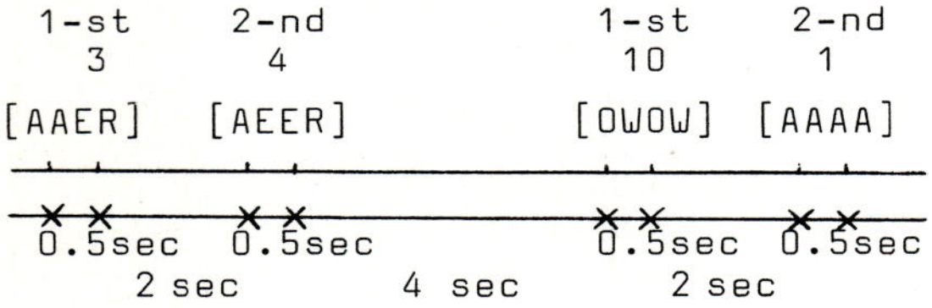

Fig.5 An example of presented voice pattern

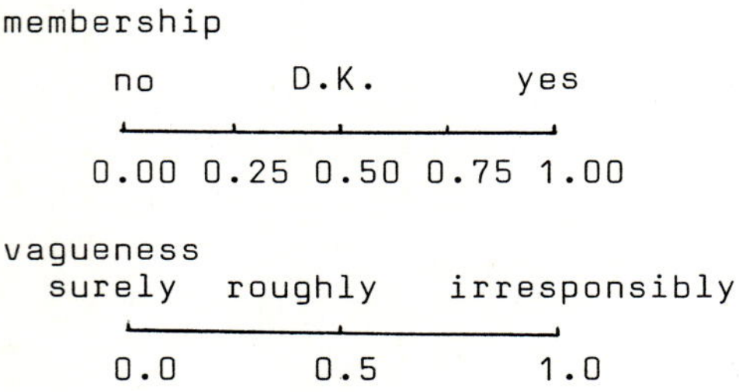

Fig.6 numerical quantification of membership and vagueness

Left/ 1-st pattern \ Right/2-nd pattern	1	2		12
1	1.00	0.75	...	0.50
2	0.75	1.00	...	0.25
.	.	.		.
.	.	.		.
12	0.50	0.25	...	1.00

Fig.7 Sorted data of each subject

we selected appropriate data based on the following criteria:

1. Value of diagonals; Each diagonal component of membership/vagueness should be equal to 1.00/0.00, respectively, because the corresponding presented pattern consists of identical one. Whereas it was possible to find values 0.00 or 0.25 in some of membership matrices. So we count the number of not 1.00 valued diagonal components for each membership matrix. If the number is greater than 3 or the values are less than 0.50, then the result is dismissed.
2. Symmetricity; Both values of (i,j) and (j,i) (i≠j) components should be identical, because the difference between the two corresponding patterns is only the order (i.e. left/right in character patterns or 1-st/2-nd in voice patterns). Therefore we calculate for each matrix the sum of absolute difference of every pair of symmetric components. If the result is too big when compared to the expectated value then the datum is rejected.
3. Lack of filling up; Several questions (especially in the case of voice patterns) were not answered by a certain subject. Such data are rejected.

Taking into account these criteria, we selected appropriate 11/ 10 subjects for repetitive answering/ one-time answering, respectively. Results are arranged in tabular form and presented in Tab.2. The correlation coefficients are also calculated and shown in Tab.3. The numerical relationship will be confirmed from this table. It may not appropriate to evaluate the contentions from these experiments alone, however the availability of vagueness was confirmed through these experiments. Inquiring about both membership and vagueness should be of great assistance to man-machine-communication to facilitate the exactness of information input.

Repetitive answering

MEAN VALUE

1.000	.758	.371	.023	.015	.023	.008	.000	.000	.015	.008	.167
.667	.995	.738	.121	.023	.045	.008	.015	.000	.030	.083	.364
.333	.605	1.000	.720	.136	.258	.091	.091	.000	.061	.136	.348
.023	.197	.629	1.000	.788	.705	.182	.083	.061	.053	.144	.144
.000	.045	.205	.795	1.000	.826	.174	.083	.000	.045	.098	.091
.015	.045	.174	.639	.917	1.000	.424	.174	.030	.083	.129	.121
.003	.033	.061	.205	.159	.508	.970	.780	.591	.333	.311	.091
.000	.008	.076	.136	.106	.159	.856	.985	.818	.539	.220	.174
.000	.000	.045	.038	.023	.038	.515	.856	1.000	.659	.447	.114
.033	.030	.114	.250	.045	.053	.364	.545	.379	.992	.773	.576
.033	.106	.220	.174	.136	.061	.189	.258	.371	.712	1.000	.636
.250	.439	.265	.144	.121	.098	.106	.144	.182	.402	.614	.995

VARIANCE

.000	.073	.160	.005	.007	.009	.002	.000	.000	.004	.002	.093
.131	.004	.072	.065	.005	.021	.002	.007	.000	.014	.033	.140
.151	.159	.000	.105	.046	.104	.041	.049	.000	.015	.057	.125
.009	.079	.077	.000	.080	.104	.073	.048	.023	.022	.053	.038
.000	.021	.073	.051	.000	.044	.093	.059	.000	.009	.041	.026
.004	.024	.081	.129	.018	.000	.138	.085	.014	.033	.057	.046
.002	.012	.023	.104	.067	.142	.007	.067	.121	.105	.114	.033
.000	.002	.047	.084	.065	.086	.042	.004	.073	.114	.098	.070
.000	.000	.021	.008	.009	.012	.144	.053	.000	.094	.139	.042
.040	.007	.053	.117	.021	.026	.118	.134	.163	.002	.054	.134
.040	.042	.090	.062	.057	.015	.076	.104	.137	.076	.000	.133
.129	.167	.125	.039	.046	.041	.034	.065	.069	.117	.140	.004

One-time-answering

MEAN VALUE OF MEMBERSHIP

1.000	.900	.400	.075	.000	.000	.000	.000	.000	.025	.150	.475
.900	1.000	.725	.150	.000	.000	.025	.000	.000	.050	.225	.525
.500	.700	1.000	.450	.000	.150	.075	.075	.000	.075	.300	.450
.025	.125	.575	1.000	.375	.725	.125	.025	.000	.050	.150	.275
.000	.000	.100	.875	1.000	.900	.075	.000	.000	.025	.150	.150
.000	.000	.175	.750	.300	1.000	.425	.075	.000	.025	.100	.200
.000	.025	.050	.125	.025	.400	1.000	.725	.725	.575	.425	.200
.000	.000	.000	.050	.025	.075	.925	1.000	.900	.600	.275	.225
.000	.000	.000	.025	.000	.175	.500	.750	1.000	.825	.350	.225
.000	.050	.025	.100	.050	.050	.475	.675	.900	.975	.725	.475
.075	.125	.450	.275	.175	.350	.450	.400	.400	.650	.950	.675
.475	.475	.500	.225	.100	.250	.300	.200	.250	.425	.575	1.000

VARIANCE OF MEMBERSHIP

.000	.015	.140	.026	.000	.000	.000	.000	.000	.006	.090	.106
.015	.000	.106	.065	.000	.000	.006	.000	.000	.010	.043	.081
.112	.047	.000	.047	.000	.040	.026	.013	.000	.013	.097	.060
.005	.028	.076	.000	.016	.106	.053	.006	.000	.022	.052	.043
.000	.000	.027	.016	.000	.015	.026	.000	.000	.006	.027	.040
.000	.000	.038	.037	.077	.000	.076	.026	.000	.006	.027	.035
.000	.006	.010	.023	.006	.077	.000	.106	.081	.113	.076	.060
.000	.000	.000	.010	.006	.026	.013	.000	.027	.127	.056	.073
.000	.000	.000	.006	.000	.101	.050	.112	.000	.013	.077	.118
.000	.010	.006	.027	.010	.022	.068	.051	.015	.006	.031	.068
.013	.016	.060	.068	.051	.077	.097	.102	.065	.027	.010	.026
.055	.081	.050	.031	.027	.075	.072	.085	.075	.138	.051	.000

MEANVALUE OF VAGUENESS

.000	.150	.200	.150	.000	.000	.050	.050	.000	.000	.050	.300
.150	.000	.100	.050	.000	.000	.050	.000	.050	.000	.200	.250
.200	.350	.000	.200	.000	.200	.050	.000	.000	.050	.100	.300
.100	.050	.350	.000	.200	.100	.050	.000	.000	.100	.150	.200
.000	.000	.100	.150	.000	.100	.050	.000	.000	.000	.150	.250
.000	.000	.050	.300	.150	.050	.250	.150	.000	.050	.050	.150
.000	.000	.050	.100	.000	.400	.000	.150	.200	.450	.350	.200
.000	.000	.050	.000	.000	.050	.050	.000	.150	.050	.200	.050
.000	.000	.000	.000	.000	.050	.300	.150	.100	.400	.150	.200
.000	.050	.050	.000	.050	.050	.400	.100	.150	.000	.350	.450
.200	.150	.350	.250	.150	.200	.350	.300	.150	.300	.150	.350
.250	.350	.250	.250	.150	.150	.150	.050	.150	.200	.500	.000

VARIANCE OF VAGUENESS

.000	.102	.060	.052	.000	.000	.022	.022	.000	.000	.022	.060
.102	.000	.040	.022	.000	.000	.022	.000	.022	.000	.060	.112
.050	.102	.000	.060	.000	.110	.022	.000	.000	.022	.040	.160
.040	.022	.102	.000	.110	.040	.022	.000	.000	.040	.052	.060
.000	.000	.040	.052	.000	.090	.022	.000	.000	.000	.052	.063
.000	.000	.022	.110	.102	.022	.112	.102	.000	.022	.022	.052
.000	.000	.022	.040	.000	.090	.000	.052	.110	.122	.102	.110
.000	.000	.022	.000	.000	.022	.022	.000	.052	.022	.060	.022
.000	.000	.000	.000	.000	.022	.060	.052	.090	.090	.052	.060
.000	.022	.022	.000	.022	.022	.140	.040	.052	.000	.102	.072
.050	.102	.052	.112	.052	.110	.102	.060	.052	.110	.052	.102
.063	.102	.063	.112	.052	.052	.052	.022	.052	.110	.150	.000

(a) character pattern recognition

Tab.2 Sample mean and sample variance

Repetive answering

MEAN VALUE

.752	.752	.233	.227	.212	.068	.068	.348	.106	.083	.129	.076
.970	.777	.364	.273	.144	.068	.076	.333	.121	.083	.106	.068
.333	.295	.762	.848	.364	.265	.129	.295	.220	.083	.220	.197
.230	.275	.732	.985	.567	.303	.159	.227	.136	.076	.250	.265
.167	.177	.303	.568	.792	.341	.288	.318	.242	.197	.167	.174
.105	.053	.242	.071	.311	.770	.864	.167	.697	.288	.591	.826
.053	.015	.197	.106	.227	.303	.962	.182	.689	.644	.727	.705
.424	.275	.288	.205	.379	.288	.250	.939	.432	.212	.439	.432
.106	.063	.220	.061	.182	.750	.727	.364	.932	.606	.939	.977
.038	.033	.186	.220	.144	.348	.561	.258	.364	.970	.576	.667
.076	.136	.152	.114	.174	.561	.515	.318	.932	.477	.962	.879
.053	.098	.220	.106	.182	.561	.545	.333	.932	.432	.811	.970

VARIANCE

.033	.019	.118	.100	.072	.016	.024	.102	.046	.029	.046	.021
.007	.005	.125	.107	.047	.012	.025	.109	.038	.025	.034	.012
.124	.081	.012	.094	.079	.079	.031	.100	.072	.014	.048	.060
.074	.104	.035	.004	.128	.086	.052	.081	.042	.028	.087	.083
.073	.032	.101	.126	.002	.094	.095	.092	.035	.079	.059	.055
.038	.014	.035	.022	.075	.007	.027	.056	.060	.067	.124	.036
.015	.004	.055	.053	.065	.071	.008	.046	.095	.083	.085	.093
.127	.108	.107	.062	.122	.076	.033	.011	.077	.080	.129	.156
.034	.020	.075	.034	.052	.083	.073	.084	.016	.091	.023	.009
.012	.025	.050	.120	.047	.117	.110	.081	.163	.007	.119	.067
.035	.050	.053	.042	.073	.140	.110	.084	.020	.142	.012	.061
.022	.034	.079	.030	.065	.125	.077	.086	.039	.134	.080	.010

One-time-answering

MEAN VALUE OF MEMBERSHIP

1.000	1.000	.250	.333	.083	.056	.111	.389	.083	.083	.139	.056
.972	1.000	.139	.083	.111	.056	.056	.528	.139	.111	.194	.056
.250	.056	1.000	.972	.694	.444	.139	.194	.361	.222	.361	.583
.194	.139	1.000	1.000	.778	.111	.139	.417	.194	.056	.278	.278
.111	.023	.722	.528	1.000	.333	.528	.361	.361	.111	.194	.361
.055	.083	.278	.139	.389	1.000	.889	.250	.528	.444	.528	.806
.023	.023	.111	.056	.222	.944	1.000	.306	.722	.667	.750	.611
.500	.417	.222	.222	.444	.306	.361	.972	.417	.278	.389	.500
.023	.111	.250	.028	.194	.806	.639	.417	.972	.778	.944	.972
.055	.111	.167	.167	.167	.583	.528	.222	.583	1.000	.500	.556
.023	.083	.361	.194	.222	.722	.583	.500	1.000	.639	1.000	1.000
.056	.023	.167	.139	.222	.333	.750	.389	.889	.417	.861	1.000

VARIANCE OF MEMBERSHIP

.000	.007	.059	.125	.023	.011	.057	.154	.028	.014	.057	.011
.035	.000	.037	.028	.015	.011	.011	.117	.057	.015	.094	.011
.111	.011	.000	.006	.164	.108	.043	.025	.099	.075	.113	.125
.135	.037	.000	.000	.103	.029	.057	.093	.094	.011	.090	.131
.043	.075	.157	.157	.000	.111	.090	.071	.099	.015	.066	.099
.011	.055	.159	.027	.127	.000	.099	.083	.137	.136	.159	.094
.005	.005	.027	.025	.043	.011	.000	.122	.131	.139	.125	.127
.111	.167	.131	.103	.080	.074	.113	.005	.153	.090	.099	.125
.005	.057	.077	.005	.052	.094	.127	.083	.005	.062	.025	.006
.011	.057	.056	.097	.028	.139	.117	.076	.153	.000	.139	.122
.005	.023	.158	.108	.070	.117	.153	.097	.000	.099	.000	.000
.011	.005	.069	.027	.076	.097	.097	.097	.099	.111	.057	.000

MEAN VALUE OF VAGUENESS

.000	.055	.333	.278	.000	.056	.056	.056	.167	.056	.000	.056
.157	.111	.111	.111	.056	.056	.000	.056	.056	.167	.056	.000
.157	.055	.000	.000	.111	.222	.111	.222	.222	.111	.222	.333
.222	.055	.056	.111	.278	.111	.056	.389	.167	.111	.222	.111
.000	.000	.056	.055	.111	.222	.444	.167	.222	.056	.111	.111
.055	.055	.167	.111	.167	.000	.056	.222	.111	.167	.222	.167
.000	.000	.222	.056	.111	.111	.000	.222	.222	.111	.111	.222
.157	.167	.278	.222	.222	.222	.167	.056	.111	.056	.111	.389
.157	.000	.222	.000	.111	.222	.389	.333	.222	.222	.111	.111
.055	.000	.056	.056	.111	.333	.167	.167	.167	.056	.167	.167
.000	.111	.278	.111	.167	.278	.056	.278	.000	.278	.278	.000
.000	.056	.157	.222	.167	.000	.500	.222	.111	.056	.278	.167

VARIANCE OF VAGUENESS

.000	.025	.111	.173	.000	.025	.025	.025	.056	.025	.000	.025
.055	.043	.043	.099	.025	.025	.000	.025	.025	.111	.025	.000
.111	.025	.000	.000	.079	.062	.043	.117	.117	.043	.062	.167
.117	.025	.025	.097	.117	.043	.025	.154	.111	.043	.117	.099
.000	.000	.025	.025	.043	.117	.136	.056	.117	.025	.099	.043
.025	.025	.056	.043	.111	.000	.025	.117	.099	.056	.117	.056
.000	.000	.117	.025	.043	.043	.000	.062	.173	.043	.043	.117
.111	.056	.117	.117	.117	.117	.056	.025	.043	.025	.099	.099
.056	.000	.117	.000	.043	.062	.099	.111	.117	.062	.099	.043
.025	.000	.025	.025	.043	.111	.111	.111	.111	.025	.111	.056
.000	.043	.117	.099	.056	.173	.025	.117	.000	.117	.062	.000
.000	.025	.111	.117	.056	.000	.167	.173	.099	.025	.117	.056

(b) voice pattern recognition

Tab.2 Continued

Character pattern recognition

Rep.mean : One-time mean of memb.	0.902
Rep.var. : One-time mean of vag.	0.673
Rep.var. : One-time var. of memb.	0.583
One-time var. of memb. : One-time mean of vag.	0.450

Voice pattern recognition

Rep.mean : One-time mean of memb.	0.953
Rep.var. : One-time mean of vag.	0.412
Rep.var. : One-time var. of memb.	0.674
One-time var. of memb. : One-time mean of vag.	0.558

Tab.3 Correlation coefficients

Conclusion

Two problems in MMS data input method were investigated. The first is "innaccuracy of input data based on ambiguous information". The other is "an optimum number of states necessary in order to get exact data". These problems were solved by introducing a concept of vagueness and an optimization process of subjective entropy. As a result a new MMS data input method was proposed using a five valued membership with a three valued vagueness. Validity of proposed method was confirmed through two experiments on man's character/ voice pattern recognition process. It was shown that inquiring both membership and vagueness indices will be of great assistance to man-machine-communication in order to get exact information even if it is an ambiguous question. We are now studying industrial applicability of the proposed method to expert systems and machine intelligence.

REFERENCES

1 L.A.Zadeh, Fuzzy Sets, Information & Control, Vol.8, 1965, pp. 338-353

2 C.E.Shannon, Mathematical Theory of Communication, Bell Syst. Tech. J., Vol.27, 1948, pp.370-423

3 K.Hirota, Concepts of Probabilistic Sets, Fuzzy sets & Systems, Vol.5, 1981, pp.31-46

4 K.Hirota, Ambiguity based on the Concept of Subjective Entropy, in Fuzzy Information and Decision Processes (M.M.Gupta, E.Sanchez eds.), North-Holland, Amsterdam, 1982, pp.29-40

5 K.Hirota, Extended Fuzzy Expression of Probabilistic Sets, in Advances in Fuzzy Set Theory and Applications (M.M.Gupta, R.K. Ragade, R.R.Yager eds.),North-Holland, Amsterdam,1979,pp.201-214

6 K.Hirota, Concept of Vagueness and Entropy Analysis in Questionnaire (in Japanese Language), Operations Research Japan, Vol.26-1, 1981, pp.38-44

Applications of Fuzzy Set Theory in Human Factors, edited by W. Karwowski and A. Mital
Elsevier Science Publishers B.V., Amsterdam, 1986 — Printed in The Netherlands

TOWARDS AN ALGORITHMIC/PROCEDURAL "HUMAN CONSISTENCY" OF DECISION SUPPORT SYSTEMS : A FUZZY LOGIC APPROACH

J. KACPRZYK

Systems Research Institute, Polish Academy of Sciences,
ul. Newelska 6, 01-447 Warsaw (Poland)

ABSTRACT

To enhance human acceptance and hence implementability of guideline solutions provided by a decision support system, we propose some means to make decision making models to be used by the system more "human-consistent". Basically, we try to introduce commonsense knowledge into some multicriteria and multistage (control) models. Commonsense knowledge is represented and manipulated in terms of Zadeh's fuzzy-logic-based approach, i.e. as a collection of dispositions (propositions with implicit fuzzy linguistic quantifiers).

INTRODUCTION

One of the currently most important and fastest growing disciplines is what is termed ergonomics or human factor (engineering). Its basic research area is the design and use of machinery, work-places, etc. taking into account human features and characteristics. In principle, for practical reasons, the idea of "fitting the task to the man" is mostly pursued. Needless to say, ergonomics is a multidisciplinary science.

Ergonomics is an outgrowth of the so-called "aviation psychology" developed in the USA during World War II. Thereafter it went beyond military settings to the factory, office, etc. wherever some human - machine relation occurred.

In recent years, as a consequence of an unprecedented growth in the use of computers in virtually all human activities in business, technology, commerce, science, etc., the analysis of man-computer systems is becoming one of the more important issues.

Among the man-computer systems, a particular role is played by those which are meant to support the human being in performing functions involving some inherently human capabilities as, e.g., reasoning or decision making. We will concentrate here on the latter, i.e. decision making.

Although experience accumulated by mankind throughout the centuries does make it possible to effectively and efficiently cope with a multitude of decision situations, current decisions are made in difficult, complex, competitive and ill-structured settings. These settings are full of uncertainty, subjectivity, imprecision, etc. in data, relations or value systems.

This, as well as unusually high potential gains or losses due to a proper or improper decision, suggests that human decision making processes should be assisted by some (computerized) decision support systems. Recent developments in computing technology do justify this idea.

At present, and presumably in the foreseeable future, it seems that the most efficient use of decision support systems will certainty be to assist and help the decision maker arrive at a proper decision but by no means to fully replace him or her. The system should therefore carry out some of the tasks it is better suited for, such as routine processing of relatively well defined and structured data, and then provide the user with some "good" solution guidelines. The final choice involving a delicate analysis of preferences, tradeoffs, etc. should then be left to the user whose human skills are far superior in this respect.

To efficiently arrive at a proper and well timed decision within the above basic framework of decision support, some synergy between man and machine should exist. This includes trying to use the best capabilities and features of both parties involved, and attempting to make one party's conduct consistent with that of the other. The former aspect has already been mentioned. The latter, which is more important in our context, is somehow unidirectional. Namely, it is true that humans usually change their conduct while operating in a computerized environment. Unfortunately, the change of human nature is difficult to obtain. The other direction, to make the machine more "human-consistent", i.e. "to fit the task to the man" as mentioned before, seems more promising and will also be adopted here. Paranthetically, let us notice that the rationale behind the 5th Generation computing technology parallels this reasoning.

The human consistency of decision support systems has two aspects. The first is related to communication (interface) between the user and system and involves, among others, input of data and commands and output of results in a user-friendly way, preferably in a natural language which is the only fully natural means of human communication. The second aspect is related to algorithms, procedures. etc. employed by the system to obtain a solution. They are normally built upon some technical mathematical concept as, e.g., optimum, pure rationality, clear-cut constraints, etc. which need not necessarily reflect their human perception. This inconsistency may often inhibit the human acceptance of the results provided by the system and hence make doubful their usefulness.

The above two types of human consistency of decision support systems are of utmost practical importance, and both should be taken into account. The first, which might be called the input/output consistency, is more often dealt with. The second, which might be termed the algorithmic/procedural consistency, unfortunately, is not often considered in the field of decision support systems (a related need of "soft" models and approaches in systems analysis seems to be more strongly emphasized - see, e.g., Rapoport, 1970 or Checkland, 1972).

In this paper we deal with the algorithmic/procedural consistency. Our basic philosophy is that, from a pragmatic point of view, the "quality criterion" of a decision support system is its usefulness, i.e. ability to provide the user with implementable solution guidelines. And only those guidelines which do not depart to much from the user's experience, perception or commonsense may fulfill their purpose. The algorithms and procedures to be employed should therefore somehow parallel the way the human user perceives their essence and intention.

Among attempts to attain that, an important one is to use models which might be called "knowledge-based" as opposed to the conventional "data-based" ones. Human reasoning is certainly much more "knowledge-based", in the sense that it uses many non-numeric data and production-rule-like dependences, than data-based, i.e. based on numeric data, mathematical equations, etc. Paranthetically, let us notice that expert systems, which will hopefully be the most powerful means for dealing with diverse real world problems, are knowledge-based too.

One of the most important types of knowledge is commonsense knowledge. It is extensively used by humans making it possible to find a solution even in situations with almost no information. Clearly, such commonsense solutions may not be ideal but they are rarely really bad, and never absurd. Commonsense is a formidable human feature which is unfortunately not possessed by the computer - with all of the negative implications as, e.g., a danger of absurd results in case of incomplete or unreliable data. Introduction of commonsense knowledge into decision support systems would therefore greatly improve their human consistency, and hence facilitate their practical use. Unfortunately, a formal representation and manipulation of commonsense knowledge is conceptually difficult and far from being solved.

For practical purposes, Zadeh's (1984) approach to commonsense knowledge is presumably the most promising. It views commonsense knowledge as a collection of dispositions, i.e. propositions involving implicit linguistic quantifiers. For instance, a disposition "winter days are cold" is in fact meant as, say, "most winter days are cold", where "most" is a linguistic quantifier. Manipulation of dispositions is done by some fuzzy-logic-based calculus. The approach is simple and elegant.

We will show in this paper how the use of Zadeh's approach to commonsense knowledge leads to a new class of more human-consistent multicriteria and multistage (control) decision making models. These models are chosen because virtually all decisions made in reality involve multiple aspects and some dynamics. Basically, in the multicriteria case the models allow one to find an optimal solution which best satisfies, say, most (almost all, much more than 50%, etc.) of the important criteria. Notice that in conventional models we seek an optimal solution to best satisfy all of the criteria. In the multistage (control) case, the models allow one to find an optimal sequence of controls to best satisfy the goals and constraints at, say, most (almost all, etc.) of the earlier control stages. Let us also notice that a similar approach leads to a new class of group decision making and consensus formation models (see, e.g.,Kacprzyk, 1984a, 1985a, 1985b, 1985d) which will not be presented here.

First, we sketch the idea of Zadeh's approach to commonsense knowledge. Then, we consecutively apply it to derive new multicriteria and multistage (control) decision making models. Mathematics will be kept to a minimum and technicalities will be avoided to assure readability. Finally, we give some concluding remarks and an extensive bibliography.

For convenience to the reader let us briefly review some of the basic fuzzy-sets-related elements and notation which will be employed.

A fuzzy set A in X, written $A \subseteq X$, say A="large"={0,1,..., 10} to be meant as a fuzzy set A labelled "large (number)", is represented by - and often equated with - its membership function $f_A : X \to [0,1]$ which states to what degree x belongs to A : from 0 for full nonbelongingness to 1 for full belongingness, through all intermediate values. For a finite $X=\{x_1,\ldots,x_n\}$, we write $A = f_A(x_1)/x_1+\ldots+f_A(x_n)/x_n$ where "+" is set-theoretic and "$f_A(x_i)/x_i$" means the pair $(x_i, f_A(x_i))$.

Very important for our purposes is a general framework for decision making under fuzziness according to Bellman and Zadeh (1970). Its basic elements are: a fuzzy goal $G \subseteq X$, a fuzzy constraint $C \subseteq X$, and a fuzzy decision $D \subseteq X$. To show the essence of this approach, let us use Fig.1.

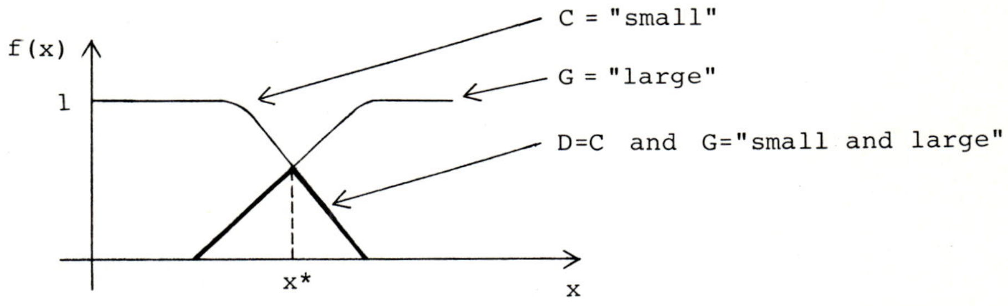

Fig.1. Basic elements of Bellman and Zadeh's approach to decision making under fuzziness.

Let us assume that our fuzzy constraint is C="small (number)" and our goal is G="large (number)"whose membership functions, $f_C(x)$ and $f_G(x)$, are as given.

We wish to

"satisfy C <u>and</u> attain G"

which corresponds to the fuzzy decision $D \subseteq X$ whose membership function is

$$f_D(x) = f_C(x) \wedge f_G(x) \quad \text{for each} \quad x \in X$$

where "$\wedge$" is "minimum", i.e. $a \wedge b = \min(a,b)$, and represents the connective "and".

The fuzzy decision gives the "goodness" of each x as a solution of the considered problem. Thus, x* is the best (optimal) solution because $f_D(x)$ takes on its maximum value for x*; we will write $x^* = \arg\max_{x \in X} f_D(x)$, which means that x* is an x which maximizes $f_D(x)$ with respect to x.

This general framework may easily be extended to cover the cases of multiple fuzzy constraints and goals, and of fuzzy goals and constraints defined in different spaces which are relevant for our purposes.

For more information on fuzzy sets, see, e.g., Kacprzyk (1983b) or an introductory section in this volume.

BASIC ELEMENTS OF ZADEH'S APPROACH TO THE REPRESENTATION AND MANIPULATION OF COMMONSENSE KNOWLEDGE

In Zadeh's (1984) approach, commonsense knowledge is viewed as a collection of dispositions, i.e. propositions with implicit fuzzy linguistic quantifiers. For instance, a disposition "winter days are cold" does implicitly involve some fuzzy quantifier, say "most", "almost all", etc., i.e. should in fact be read as "most winter days are cold". Since the traditional logical systems provide no means for handling propositions with fuzzy quantifiers, Zadeh (1983, 1984) develops the following fuzzy-logic-based calculus of linguistically quantified propositions.

A linguistically quantified proposition, exemplified by "most experts are convinced", may be generically written as

$$QY\text{'s are } F \tag{1}$$

where Q is a linguistic quantifier (most), $Y=\{y\}$ is a set of objects (experts), and F is a property (convinced).

We can also add importance, assumed to be a fuzzy set $B \underset{\sim}{\subseteq} Y$, to (1) obtaining

$$QBY\text{'s are } F \tag{2}$$

i.e. "most of the important experts are convinced".

The main problem now is to determine truth (QY's are F), or truth (QBY's are F), knowing truth (y_i is F) for each $y_i \in Y$. In the classic approach proposed by Zadeh (1983, 1984), a linguistic quantifier Q is assumed to be a fuzzy set in $[0,1]$, $Q \underset{\sim}{\subseteq} [0,1]$, characterized by its membership function $f_Q(r)$, $r \in [0,1]$. Notice that this is the case for a proportional quantifier, say "most", while for an absolute quantifier, say "about 5", $Q \underset{\sim}{\subseteq} R$, i.e. is a fuzzy set in the real line. Throughout this paper we will use the proportional quantifiers which seem to be better suited for our purposes. Analogous properties also hold for the absolute quantifiers.

Property F is defined as a fuzzy set in $Y, F \subseteq Y$, whose membership function $f_F(y_i)$ gives to what degree $y_i \in Y$ possesses property F. If $Y=\{y_i,\ldots,y_p\}$, then it is assumed that truth (y_i is F) $=f_F(y_i)$, $i=1,\ldots,p$.

The calculation of truth (QY's are F) is based on the nonfuzzy cardinalities (the so-called Σ Counts, see Zadeh, 1983) of the respective fuzzy sets and proceeds as follows:

1. Calculate

$$r = \Sigma\text{Count}(F) / \Sigma\text{Count}(Y) = \frac{1}{p} \sum_{i=1}^{p} f_F(y_i) \tag{3}$$

2. Calculate

$$\text{truth (QY's are F)} = f_Q(r) \tag{4}$$

Importance may be introduced into the above as follows. B = "important" is defined as a fuzzy set in $Y, B \subseteq Y$, such that $f_B(y_i) \in [0,1]$ is a degree of importance of y_i: the higher its value, the more important y_i.

We first rewrite "QBY's are F" (e.g., "most of the important experts are convinced") as "Q(B and F) Y's are B" (e.g., "most of the (important and convinced) experts are important") which leads to the following counterparts of (3) and (4):

1. Calculate

$$r' = \sum_{i=1}^{p} (f_B(y_i) \wedge f_F(y_i)) / \sum_{i=1}^{p} f_B(y_i) \tag{5}$$

2. Calculate

$$\text{truth (QBY's are F)} = f_Q(r') \tag{6}$$

Example 1. Let us have 3 experts, X,V and Z, i.e. Y="experts" = {X,V,Z}. Let F="convinced"=0.1/x + 0.6/V + 0.8/Z, that is X is convinced (as to an issue in question) to degree 0.1, i.e. practically not at all, V - to degree 0.6, i.e. moderately, and Z - to degree 0.8, i.e. quite strongly. Let B="important"=0.2/X + 0.5/V + 0.6/Z, that is the importance of X is 0.2, that of V is 0.5 and that of Z is 0.6; notice that none of the experts is considered very important (e.g., competent). Let Q="most" be given as

$$f_{\text{"most"}}(x) = \begin{cases} 1 & \text{for } x \geqslant 0.8 \\ 2x-0.6 & \text{for } 0.3 < x < 0.8 \\ 0 & \text{for } x \leqslant 0.3 \end{cases} \tag{7}$$

Then, on the one hand, r=0.5 and

$$\text{truth ("most experts are convinced")}=2\cdot 0.5-0.6=0.4$$

On the other hand, r=0.9 and

$$\text{truth ("most of the important experts are convinced")}=1$$

For our particular purposes the above basic means for the representation and manipulation of commonsense knowledge in Zadeh's setting are sufficient, although in the sequel we introduce some other, more specific notions. Evidently, it is easy to imagine that the basic problem associated with the use of that knowledge in most applications is inference based on dispositions. Some details may be found in Zadeh (1984) but the issue is far from a real solution.

Let us remark that in the procedure outlined above, we can also use fuzzy cardinalities of the respective fuzzy sets, the so-called FCounts, instead of the nonfuzzy ones (ΣCounts). This does not change the essence of the approach but may complicate the calculations to a considerable extent. Some details may be found in Kacprzyk (1985c) and Zadeh (1983).

Let us also note that the presented method may be viewed as yielding a consensory-like aggregation of the pieces of evidence "y_i is F", i=1,...,p. For details on this important issue, refer to Kacprzyk (1983a), Kacprzyk and Yager (1984a, 1984b) or Yager (1983).

The presented approach to dealing with linguistically quantified propositions is not the only one. An alternative procedure, yielding a competitive-like aggregation, has been proposed by Yager (1983) and used in Kacprzyk (1983a), Kacprzyk and Yager (1984a, 1984b), Yager (1983), etc. We will not discuss it here.

Finally, let us notice that Zadeh's approach to commonsense knowledge is not the only one. Among some other approaches, which may be useful for our purposes as they are constructive and relatively simple conceptually and computationally, hence operational, some default-logic-based approaches (Reiter, 1980; Reiter and Criscuolo, 1983) should be mentioned. For some details, see also Ferreny and Prade (1984).

We will now present the application of Zadeh's approach to derive some new, more human-consistent multicriteria and multistage (control) decision making models.

COMMONSENSE-KNOWLEDGE-BASED MULTICRITERIA DECISION MAKING MODELS

Virtually all decisions to be made in non-trivial real situations must take into account the existence of multiple, often conflicting, objectives or criteria. This has triggered much research on multicriteria decision making.

For our purposes multicriteria decision making under fuzziness may be meant as follows. $A=\{a\}=\{a_1,\ldots,a_q\}$ is a set of possible options (alternatives, decisions...) and $0=\{0_1,\ldots,0_p\}$ is a set of fuzzy objectives (fuzzy constraints and/or fuzzy goals). The degree to which option $a \in A$ satisfies objective $0_i \in 0$ is given by

$$\text{truth } (0_i \text{ is satisfied (by a))} = f_{0_i}(a), i=1,\ldots,p. \tag{8}$$

Traditionally, it is postulated (e.g., Bellman and Zadeh, 1970; Kacprzyk, 1983b) that $a \in A$ satisfy "0_1 <u>and</u> ... <u>and</u> 0_p" i.e. <u>all</u> the fuzzy objectives, and hence the degree of that satisfaction is given by the fuzzy decision

$$f_D(a|\text{"all"}) = \text{truth } (0_1 \underline{\text{ and}} \ldots \underline{\text{and }} 0_p \text{ are satisfied}) =$$

$$= \text{truth } (\text{"all" } 0_i'\text{s are satisfied}) = \text{truth } (0_1 \text{ is satisfied}) \wedge$$

$$\ldots \wedge \text{ truth } (0_p \text{ is satisfied}) = f_{0_1}(a) \wedge \ldots \wedge f_{0_p}(a) \tag{9}$$

The problem is to find an optimal decision $a^* \in A$, such that

$$a^* = \arg \max_{a \in A} f_D(a|\text{"all"}) \tag{10}$$

The requirement to satisfy "all" the fuzzy objectives may be viewed too rigid and restrictive for practical purposes. An idea for replacing "all" by some milder requirement spacified by a linguistic quantifier Q, say "most", appeared in Yager (1983), Kacprzyk and Yager (1984a, 1984b), etc. Basically, it consists of seeking an optimal decision that best satisfies Q (e.g., "most") fuzzy objectives. We will now discuss this type of problem.

Following Section 2, we introduce first the fuzzy set S = "satisfied" $\underset{\sim}{\subset} 0$, such that $f_S(0_i)=\text{truth}(0_i$ is satisfied (by a)) = $f_{0_i}(a)$, $i=1,\ldots,p$, is to what degree objective 0_i is satisfied (by option a), and a fuzzy set B="important" $\underset{\sim}{\subset} 0$, such that $f_B(0_i) \in [0,1]$ is the degree of importance of objective 0_i: from 0 denoting "unimportant at all" to 1 denoting "definitely important", through all intermediate values. $Q \underset{\sim}{\subseteq} [0,1]$ is a fuzzy linguistic quantifier.

The fuzzy decision is now written as

$$f_D(a|Q,B) = f_D(a|Q,\text{"important"}) = \text{truth}(QB\ O_i's \text{ are satisfied (by a)}) = \text{truth}(Q \text{ "important" } O_i's \text{ are satisfied}) \quad (11)$$

and gives the degree to which Q of the important (B) fuzzy objectives are satisfied (by a).

Employing (6), we obtain

$$f_D(a|Q,B) = f_Q(\sum_{i=1}^{p} (f_B(O_i) \wedge f_{O_i}(a)) / \sum_{i=1}^{p} f_B(O_i)) \quad (12)$$

and the problem is to find an optimal option $a^* \in A$, such that

$$a^* = \arg\max_{a \in A} f_D(a|Q,B) \quad (13)$$

i.e. an optimal option which best satisfies Q of the important (B) fuzzy objectives.

Notice that if we do not wish to account for importance, we set $f_B(O_i)=1$ for each $O_i \in O$, i.e. we assume that all $O_i's$ are equally important, and (12) and (13) become, respectively:

$$f_D(a|Q) = f_Q(\frac{1}{p} \sum_{i=1}^{p} f_{O_i}(a)) \quad (14)$$

and

$$a^* = \arg\max_{a \in A} f_D(a|Q) \quad (15)$$

It is easily seen that it is difficult to say something about the solution of (15) for an arbitrary linguistic quantifier Q, i.e. for an arbitrary function $f_Q(r)$. Fortunately enough, there are some particularly relevant quantifiers in our context, the so-called nondecreasing quantifiers whose essence may be subsumed as "the more objectives that are satisfied the better". For such quantifiers, the solution of (15) is relatively easy; for details, see, e.g., Yager, 1983 or Kacprzyk and Yager, 1984a, 1984b).

Example 2. Let us have 3 options, i.e. $A=\{a_1,a_2,a_3\}$, and 3 fuzzy objectives O_1, O_2 and O_3. Let $O_1=1/a_1+0.7/a_2+0.2/a_3$ which may be read as: option a_1 is definitely the best one, option a_2 may be chosen although it is not a definitely preferred choice (only to degree 0.7), and option a_3 is a relatively bad (to degree

0.2) choice although still possible. Let $0_2=0.2/a_1+1/a_2+0.5/a_3$ and $0_3=0.2/a_1+0.3/a_2+1/a_3$ to be meant analogously. Moreover, let Q="most" be given by (7), and B="important"$=0.3/0_1+0.8/0_2+0.1/0_3$ which expresses importances of the particular objectives as in Example 1.

Our problem is to find "an optimal option which best satisfies most of the important objectives". For the particular options we obtain the following values of the fuzzy decision

$$f_D(a_1|\text{"most", "important"}) = 0$$

$$f_D(a_2|\text{"most", "important"}) = 1$$

$$f_D(a_3|\text{"most", "important"}) = 0.9 \qquad (16)$$

that is the optimal option sought is a_2.

COMMONSENSE-KNOWLEDGE-BASED MULTISTAGE DECISION MAKING (CONTROL) MODELS

Most decision situations involve some dynamics, i.e. decisions that are currently made influence not only outcomes at the following time (stage) but also those in the more distant future. To account for that, multistage decision (control) models are developed.

For our purposes, multistage decision making (control) under fuzziness may be formalized as follows. At each time (control stage) t, the control $u_t \in U=\{c_1,\ldots,c_m\}$ is subjected to a fuzzy constraint $f_{C^t}(u_t)$, and on the state attained $x_{t+1} \in X=\{s_1,\ldots,s_n\}$ a fuzzy goal $f_{G^{t+1}}(x_{t+1})$ is imposed: the state transitions are governed by $x_{t+1}=g(x_t,u_t)$; $x_t, x_{t+1} \in X$, $u_t \in U$, $t=0,1,\ldots,N$; N is some termination time.

It is commonly postulated (e.g., Bellman and Zadeh, 1970; or Kacprzyk, 1982, 1983b) that at each t, the control u_t satisfy the fuzzy constraint C^t and the fuzzy goal G^{t+1} (in fact the fuzzy goal is satisfied not by u_t but by the resulting new state x_{t+1}), to be written as P_{t+1}: "C^t and G^{t+1} are satisfied (by u_t)". This satisfaction is evidently to the degree equal to

$$\text{truth } P_{t+1} = \text{truth } (C^t \text{ and } G^{t+1} \text{ are satisfied"}) =$$

$$= f_{C^t}(u_t) \wedge f_{G^{t+1}}(x_{t+1}) \qquad (17)$$

Traditionally, we require a sequence of controls to satisfy the fuzzy constraints and fuzzy goals at all the subsequent control stages, hence the fuzzy decision expressing the degree of that satisfaction is

$$f_D(u_0,\ldots,u_{N-1}|x_0,\text{"all"}) = \text{truth}(P_1 \text{ and}\ldots\text{and } P_N \,|\text{"all"}) =$$

$$= \bigwedge_{t=0}^{N-1} \text{truth } P_{t+1} = \bigwedge_{t=0}^{N-1} (f_{C^t}(u_t) \wedge f_{G^{t+1}}(x_{t+1})) \tag{18}$$

where $\bigwedge_{t=0}^{N-1} a_t = a_0 \wedge \ldots \wedge a_{N-1}$.

The problem is to find an optimal sequence of controls $u_0^*,\ldots,u_{N-1}^*$, such that

$$u_0^*,\ldots,u_{N-1}^* = \arg \max_{u_0,\ldots,u_{N-1}} f_D(u_0,\ldots,u_{N-1}|x_0,\text{"all"}) \tag{19}$$

For details and some extensions of the above basic formulation, see Kacprzyk (1982, 1983b).

As in the case of multicriteria decision making, "all" may be viewed too restrictive, and its replacement by a milder linguistic quantifier Q, say "most", has been proposed by Kacprzyk (1983a). Therefore an optimal sequence of controls is sought, $u_0^*,\ldots,u_{N-1}^*$, which best satisfies the fuzzy constraints and goals at Q control stages, i.e.

$$u_0^*,\ldots,u_{N-1}^* = \arg \max_{u_0,\ldots,u_{N-1}} f_D(u_0,\ldots,u_{N-1}|x_0,Q) =$$

$$= \arg \max_{u_0,\ldots,u_{N-1}} (\bigwedge_{t=0}^{N-1} | Q)(f_{C^t}(u_t) \wedge f_{G^{t+1}}(x_{t+1})) \tag{20}$$

where $(\bigwedge_{t=0}^{N-1}|Q)a_t$ means that in $a_0 \wedge \ldots \wedge a_{N-1}$ only Q, say "most", a_i's are included.

For simplicity, we will assume that importance is not accounted for, i.e. the importance of each control stage is the same. Let us notice that the introduction of importance, i.e. derivation of a model which allows one to seek an optimal sequence of controls best satisfying the fuzzy goals and constraints at, say most of the earlier control stages, may be viewed as the introduction of discounting which has a long tradition in conventional approaches. This important issue is however beyond the scope of this paper.

For the solution of (20) we use the two steps of Zadeh's procedure, and consecutively obtain:

$$r(u_0,\ldots,u_{N-1}|x_0) = \frac{1}{N}\sum_{t=0}^{N-1}(f_{C^t}(x_t)\wedge f_{G^{t+1}}(x_{t+1})) \tag{21}$$

$$f_D(u_0,\ldots,u_{N-1}|x_0,Q) = f_Q(r(u_0,\ldots,u_{N-1}|x_0)) =$$

$$= f_Q(\frac{1}{N}\sum_{t=0}^{N-1}(f_{C^t}(u_t)\wedge f_{G^{t+1}}(x_{t+1}))) \tag{22}$$

Therefore we seek an optimal sequence of controls, such that

$$u_0^*,\ldots,u_{N-1}^* = \arg\max_{u_0,\ldots,u_{N-1}} f_D(u_0,\ldots,u_{N-1}|x_0,Q) \tag{23}$$

that is, the one which best satisfies the fuzzy constraints and fuzzy goals at Q, say "most", control stages.

Similarly as in the case of multicriteria decision making, it is also difficult to say something about the solution of (23) for an arbitrary Q. Fortunately enough, for the so-called nondecreasing quantifiers whose essence is now "the more control stages at which the fuzzy constraints and goals are satisfied the better", the solution of (23) may be obtained by dynamic programming. For details, see Kacprzyk (1983a).

Example 3. Let us have 3 control stages, i.e. N=3. Let the fuzzy constraints and goals at the particular control stages be

$$C^o = 0.5/c_1 + 1/c_2 \qquad G^1 = 0.1/s_1 + 0.6/s_2 + 1/s_3$$

$$C^1 = 1/c_1 + 0.7/c_2 \qquad G^2 = 0.6/s_1 + 1/s_2 + 0.5/s_3$$

$$C^2 = 1/c_1 + 0.6/c_2 \qquad G^3 = 1/s_1 + 0.8/s_2 + 0.3/s_3$$

to be understood as in Example 2.

Let the state transitions be governed by a state transition equation represented by the table

$x_{t+1} =$ u_t

	x_t: s_1	s_2	s_3
c_1	s_3	s_3	s_3
c_2	s_2	s_2	s_2

i.e. if, for example, the current state is $x_t=s_2$ and the current control is $u_t=c_1$, the next state becomes $x_{t+1}=s_3$.

Using dynamic programming we solve the problem "backwards", i.e. first for the last control stage t=N-1, then for t=N-2,..., and finally for t=0, determining the so-called policy functions $a_t^*(x_t)$ which give for each state x_t an optimal control at a particular control stage t.

Thus, for t=N-1=2, we obtain the policy functions

$$a_2^*(s_1) = c_2, \quad a_2^*(s_2) = c_2, \quad a_2^*(s_3) = c_2;$$

i.e. if at t=2 we are in state s_1 then the optimal control is c_2, for $a_2^*(s_1)=c_2$, etc. Next, for t=1 and t=0, respectively

$$a_1^*(s_1) = c_2, \; a_1^*(s_2) = c_2, \; a_1^*(s_2) = c_2;$$

$$a_0^*(s_1) = c_1 \text{ or } c_2, \; a_0^*(s_2)=c_1, \; a_0^*(s_3)=c_1 \text{ or } c_2.$$

SOME REMARKS ON COMMONSENSE KNOWLEDGE IN GROUP DECISION MAKING AND CONSENSUS FORMATION MODELS

Zadeh's approach to commonsense knowledge employed here has also proved to be an efficient means for making some other types of decision making models, more human-consistent.

More specifically, it is easy to notice that a fuzzy linguistic quantifier, say, "most" or "almost all", is a natural representation of a (fuzzy) majority as perceived by humans. This is a point of departure in Kacprzyk (1984a, 1985a) where some new solution concepts for group decisions are proposed as, e.g., some fuzzy cores, consensus winners, etc., using the approach employed here. Moreover, in Kacprzyk (1985b) the approach is used to define the notion of a "soft" consensus and its degree to model and monitor real, not ideal, consensus reaching processes.

CONCLUDING REMARKS

In this paper we show the introduction of some elements of commonsense knowledge, using one of its representation which seems to be particularly promising due to its simplicity, elegance and constructiveness, into some multicriteria and multistage (control) decision making models.

The very purpose of the above is to make the models, which are to be used in decision support systems, more human-consistent.

This should greatly facilitate interaction between the user and the system because the human user will presumably be more willing to adopt solution guidelines provided by the system if the system's "reasoning" and conduct parallels to some extent those of his or her. The implementability of the system's solutions, hence the system's usefulness, should therefore be enhanced.

This paper is an attempt towards developing a research direction, to be eventually called the algorithmic/procedural "human-consistency" of decision support systems, which we feel is extremely important and should be further pursued.

REFERENCES

R.E. Bellman and L.A. Zadeh (1970), Decision-making in a fuzzy environment, Management Science 17, 151-169.

P.B. Checkland (1972), Towards a system-based methodology of real world problem solving. Journal of Systems Engineering 3.

H. Ferreny and H. Prade (1984), Mechanization of default reasoning in terms of posibilities (in French) Proc. 5th WOGSC Congress, AFCET, Paris.

J. Kacprzyk (1982), Multistage decision processes in a fuzzy environment: a survey. In M.M. Gupta and E. Sanchez (Eds.): Fuzzy Information and Decision Processes, North-Holland, Amsterdam, 251-265.

J. Kacprzyk (1983a), A generalization of fuzzy multistage decision -making and control via linguistic quantifiers. International Journal of Control 38, 1249-1270.

J. Kacprzyk (1983b), Multistage Decision-Making under Fuzziness: Theory and Applications, Verlag TÜV Rheinland, Cologne.

J. Kacprzyk (1984a), Collective decision making with a fuzzy majority. Proc. 5th WOGSC Congress, AFCET, Paris.

J. Kacprzyk, Guest Ed. (1984b), Special Issue on Fuzzy Sets and Possibility Theory in Optimization Models, Control and Cybernetics 13, 4.

J. Kacprzyk (1984c), A generalized formulation of fuzzy multistage decision making and control under fuzziness, In Sanchez E. and M.M. Gupta (Eds.), Proc. IFAC Symposium on Fuzzy Information, Knowledge Representation and Decision Analysis (Marseille , 1983) Pergamon, Oxford.

J. Kacprzyk (1985a), Some "commonsense" solution concepts in group decision making using fuzzy linguistic quantifiers. In Kacprzyk and Yager (1985).

J. Kacprzyk (1985b), On some fuzzy cores and "soft" consensus measures in group decision making. In Bezdek J. (Ed.) The Analysis of Fuzzy Information. CRC Press (forthcoming).

J. Kacprzyk (1985c), Zadeh's commonsense knowledge and its use in multicriteria, multistage and multiperson decision making. In M.M. Gupta et al. (Eds.): Approximate Reasoning in Expert Systems. North-Holland, Amsterdam , 105-121.

J. Kacprzyk (1985d), Group decision making with a fuzzy linguistic majority. Fuzzy Sets and Systems (forthcoming).

J. Kacprzyk (1985e), Towards "human-consistent"multistage decision making and control models using fuzzy sets and fuzzy logic. In A.O. Esogbue (Guest Ed.), Bellman Special Issue, Fuzzy Sets

and Systems (forthcoming).
J. Kacprzyk and R.R. Yager (1984a), "Softer" optimization and control models via fuzzy linguistic quantifiers. Information Sciences 34, 157-178.
J. Kacprzyk and R.R. Yager (1984b), Linguistic quantifiers and belief quantification in fuzzy multicriteria and multistage decision making. In Kacprzyk (1984b), 155-174.
J. Kacprzyk and R.R. Yager (1985), Management Decision Support Systems Using Fuzzy Sets and Possibility Theory. Verlag TÜV Rheinland, Cologne.
L. Lesmo, L. Saitta and P. Torasso (1984), Evaluation of quantified expressions in expert systems. Proc. 5th WOGSC Congress, AFCET, Paris.
A. Rapoport (1970), Modern systems theory - an outlook for coping with change. In General Systems Yearbook, Society for General Systems, XV.
R. Reiter (1980), A logic for default reasoning. Artifical Intelligence 13, 81-132.
R. Reiter and G. Criscuolo (1983), Some representational issues in default reasoning. Computers and Mathematics with Applications 9, 15-27.
R.R. Yager (1983), Quantifiers in the formulation of multiple objective decision functions. Information Sciences 31, 107-139.
L.A. Zadeh (1983), A computational approach to fuzzy quantifiers in natural languages, Computers and Mathematics with Applications 9, 149-184.
L.A. Zadeh (1984), A theory of commonsense knowledge. In H.J. Skala S. Termini and E. Trillas (Eds.): Aspects of Vagueness, Reidel, Dordrecht, 257-295.

Applications of Fuzzy Set Theory in Human Factors, edited by W. Karwowski and A. Mital
Elsevier Science Publishers B.V., Amsterdam, 1986 — Printed in The Netherlands

A FUZZY METHOD FOR THE MODELING OF HUMAN-COMPUTER INTERACTIONS IN INFORMATION RETRIEVAL TASKS

Guy A. Boy[1] and Pierre M. Kuss
ONERA-CERT-DERA[2] 2, avenue Edouard Belin 31055 Toulouse Cedex France

ABSTRACT

The task of document retrieval in a library requires a procedure for categorizing the documents and an information processing technique for retrieving them. The data provided by the user are often incomplete, imprecise and/or uncretain and the librarian has to perform a fuzzy problem solving task to retrieve the requested document. The theory of fuzzy sets can thus provide a theoretical framework to describe the information processing done by the librarian. Fuzzy techniques also enable the development of meta-knowledge, which captures important aspects of human experience.

Taking this into account, a computer-based retrieval aid (BIBLIO) has been developed and a protype implemented in an Artificial Intelligence language (Lisp). In BIBLIO, document characateristics are represented by fuzzy models. User propositions can be imprecise with a particular level of uncertainty. A thesaurus (set of keywords) has been implemented as a network relating key-words. Each piece of knowledge (e.g. author names, dates) is called a pattern and represented by a triplet (attribute, object, value). The "value" of an "attribute" can be qualitative or quantitative. It is represented by a possibility distribution in the attribute domain. The uncertainty attached to a value is taken into account by a scalar (subjective probability). Fuzzy connectives (e.g. conjunction, disjunction) have been defined and allow the user to send complex requests. Each user request is fuzzy-matched with the characteristic patterns of each reference in the data base. In order to know the system response relevance, and classify the selected documents, optimistic and pessimistic correlation measures between request and data are calculated and presented to the user.

While the example chosen here is the library retrieval task, the approach could be extended easily to other retrieval applications, and more generally to computer-aided information seeking.

Keywords. Fuzzy pattern matching; Fuzzy sets; Human engineering; Human-machine systems; Models; Retrieval system;

1. INTRODUCTION

Provision of computer software as an aid to the user in any task, alters the user's role in performing that task. Human factors studies aimed at understanding human-computer interactions. Such studies in turn provide guidelines for designing more easy-to-use computer aided systems. Two different approaches can be considered (ref. 10): a *normative* analysis which idealizes information processing by the human operator and gives a rational model of "an optimum" human-machine interaction system, using relevant mathematical methods and techniques; and a *descriptive* analysis which try to extract human operator's belief and preference as they are, not as they should be. Human information processing is a subtle mix of rational and irrational

[1] At the present moment, Research Associate, NASA-Ames Research Center, Aero-Space Human Factors Research Division, Mail Stop 239-3, Moffett Field, CA 94035, U.S.A..

[2] ONERA: Office National d'Etudes et de Recherches Aerospatiales ; CERT: Centre d'Etudes et de Recherches de Toulouse ; DERA: Departement d'Etudes et de Recherches en Automatique.

features, and the objective of the present paper is to propose a framework (essentially normative) which also takes into account descriptive features of human behavior. In human engineering, both normative and descriptive analyses are performed. They have been applied to the design of physical devices such as cockpits and computer terminals, and corresponding abstract products such as procedures and computer languages.

A task analysis of document retrieval reveales a number of fuzzy patterns and processes. The information on the documents is generally incomplete, and the users' requests are often imprecise, incomplete, and uncertain. The librarian must process this information and give a result about his own investigation. This investigation is based on librarian's personal experience. It would thus be interesting to feed the potential aid system with heuristics (ref. 2) taking into account such experience.

Moreover, if the user does not have the precise reference of the document he is searching for, he knows the significance of this document. This significance is subjective in nature, and correlated to his knowledge and needs (ref. 3).

There are three key considerations in the design of a retrieval aid system.

1. The knowledge base is crucial. The ability to infer relationships among data is largely determined by the type of knowledge representation chosen. Inference is based on possible and necessary propositions.

2. The judgement of the user is a "subjective" criterion which can be assessed with subjective scales. These scales can be personalized to account for differences among the users. This personalization can be carried out automatically if the system is provided with criteria for different "classes" of user.

3. The interface between the problem solver and the user is a limiting factor. Often, an increase in precision results in a decrease in certainty in such a process. For example, precise observable descriptors are seldom adequate to codify the significance of the requested document to the user.

In classical retrieval systems, all the descriptors are crisp and precise. Generally, the most clever combinations of such descriptors are not sufficient to retrieve the pertinent documents. The real world proves that when a reasonable level of imprecision is kept, then the significance is also kept and a more certain retrieval is generally assured.

2. RETRIEVAL TASK ANALYSIS

The document retrieval task, like other information retrieval tasks, can be modelled by an Information Processing System (IPS) (see Fig. 1) (ref. 14). At the most general level, the retrieval task can be split into two parts: information representation (how a reference is coded or registered from a basic knowledge of its content and librarian experience), and information retrieval (how a reference is retrieved from given external data).

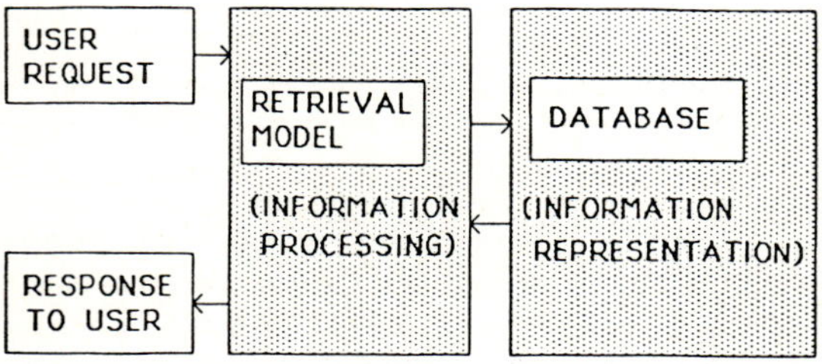

Figure 1. General Mechanism of a Retrieval System

2.1. Information Representation

First, the librarian must *analyze* the document to store in the library. Generally the result of this analysis is card-indexed. Each card is stored in a directory, and includes the main characteristics describing the document.

The representation process begins by an abstraction, i.e. the strictly-necessary-information is extracted from the initial information (see Fig. 2). In the present application the question is : what is strictly-necessary-information for retrieving a document ? Once this minimal information has been identified, there are several choices for the "final" representation. When the support is hardcopy (generally cards) or computer software, minimal information is generally written (author name, title, date, codes, keywords). Three methods are often used to represent documents:

- natural language abstract,
- key word list,
- mixed.

The natural language abstract representation is the method which best preserves the significance of the document. It is easy to implement during the representation phase, but many ambiguities are possible in the retrieval phase. The key word list representation is the most commonly used in practice. It demands more attention during the representation phase, but is not ambiguous. The mixed representation is a combination of both of the previous ones. For all these representations, the main problem remains their in completeness and imprecision. In this study, the key word list representation has been chosen.

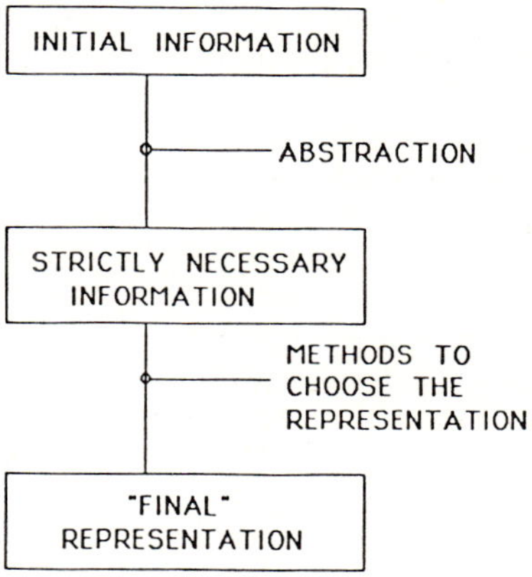

Figure 2. Representation Process

2.2. Information Retrieval

Information retrieval operations can be modelled as follows:

- the user formulates the request,
- the system analyzes the user's request,
- the system displays the results.

The request is formulated following a retrieval equation. This equation includes the characteristics of the document to be sought out (e.g. author name(s), date, key word(s), etc.), and boolean operators connecting these characteristics, e.g. :

- logical (and, or, exept),
- extension (word prefix search),
- comparison (less than, etc.),
- weighting (importance of certain characteristics in the equation).

[3] For example, the word "aircraft" can be extended to "airplane", "jet", "aviation", "flying", etc.

The request is an input to the retrieval system. The analysis of the request can be divided into three phases:

- taking into account extension operators[3],
- consulting the thesaurus to identify the "complete" retrieval key word set,
- pattern matching of the characteristics (authors name(s), date, key words), taking into account logical, comparison and weighting operators.

2.3. Objectives of the BIBLIO system

In order to maximize its usefulness, BIBLIO includes aids for both information representation and information retrieval.

Aid for representing information

Generally, it is difficult, and sometimes impossible, to find the "right" key words to describe a document. Most of the time, the choice of such key words is intuitive either when the librarian makes a key word list, i.e. the reference of the document in the library, or when the user tries to formulate a request. In all the cases, the representation becomes imprecise, and a credibility and/or possibility factor should be attached to each key word describing the document.

Aid for retrieving information

Generally, the user doesn't know the document characteristics very well. Rather his data is imprecise (e.g. incorrect spelling of the author(s) name(s), fuzzy date), or uncertain (e.g. confidence degree on a characteristic).

Imprecision : If the user doesn't know the correct spelling of an author's name, this imprecision can be taken into account by the system. Such an imprecision may be explained by either a phonetic or an incorrect spelling of the name. Thus, two types of pattern matching models can be designed, one based on phonetics and another on incorrectness. Another example is the date of a reference; it may be 1976, between 1975 and 1978, after 1975, around 1976.

Uncertainty : The user might be able to associate a confidence rating to an author name or a date, for instance. This confidence rating (certainty) will be chosen on a subjective scale.

3. A WAY TO REPRESENT UNCERTAINTY AND IMPRECISION

3.1. Quantification of the Uncertainty

A piece of information declared as certain is considered to be true. A fact, formulized as a proposition, is uncertain as long as its truth is not proved. This comes from the lack of confidence about the information source.

Probabilities are the oldest and best known model of uncertainty. In the probability theory, if the truth or falsity of a proposition "p" is not known a priori (ignorance), it is natural to choose:

$$P(p) = P(\neg p) = 0.5 \text{ , where } \neg p \text{ is the negation of p.}$$

But, if we do one thousand experiments where the proposition "p" is statistically confirmed as true in five hundred of them and disconfirmed in the other five hundred cases, then we are going to know that the process is uniform and the probability of "p" is equal to 0.5. Thus, complete ignorance is represented exactly the same as a uniform process, i.e. lack of knowledge is represented by a very well known piece of knowledge.

Moreover, if more than two alternatives are present, "complete ignorance" is difficult to represent. Consider the paradox described by the following example :

p1 = "the author of the book is American"
p2 = "the author of the book is not American"

If nothing is known about the book, using the probability theory, the a-priori ignorance is represented by the distribution :

$$P\ (p1) = P\ (p2) = 0.5$$

Now if two other propositions are added, saying :

p3 = "the author of the book is Californian"
p4 = "the author of the book is not Californian"

The representation of the a-priori ignorance becomes a really tricky problem ! Because (p3 $\rightarrow$ p1), the following relation can be written :

$$P\ (p3) = \alpha * P\ (p1) \text{ with } 0 < \alpha < 1$$

Because p4 is the complement of p3, the axiom $\Sigma_{i=1,4}$ P (pi) = 1 is difficult to satisfy. The a-priori ignorance is too complex to find satisfactory probability distributions and Bayesian reasoning is generally impossible when the number of propositions becomes important.

In the "Mathematical Theory of Evidence" (ref. 4, 11, 17, 22, 23) the belief in a proposition "p" is represented by an interval [Cr(p), Pl(p)]. The lower bound Cr(p) is the ***credibility*** degree associated with the proposition "p". Credibility is defined as the sum of all the amounts of belief m(q), also called "mass", assigned to each proposition "q" implying "p":

$$Cr\ (p) = \Sigma_{q \rightarrow p} \{ m(q) \}$$

The upper bound Pl(p) is the ***plausibility*** degree defined as the sum of all the amounts of belief m(q) assigned to each proposition "q" not implying "p". The following relation can be written:

$$Cr(p) + Pl(\neg p) = 1$$

Thus, the more a proposition is credible, more its countrary is not plausible.

The four previous propositions (p1, p2, p3, p4) are considered, and several additional propositions are set :

p5 = "the author of the book is French"
p6 = "the author was an undergraduate and Ph.D student at MIT"
p7 = "the author is married to a famous American woman"
p8 = "the author of the book is employed by NASA-Ames Research Center"
p9 = "the author is a professor at Stanford University"

1.00	VERY HIGH SIGNIFICANCE
0.75	IMPORTANT SIGNIFICANCE
0.50	MEDIOCRE SIGNIFICANCE
0.25	LOW SIGNIFICANCE
0.00	INSIGNIFICANCE

Figure 3. Subjective Scale

For each of the nine propositions, generally called focal propositions or focal elements, a subjective rating can be performed using a [0,1] scale and the user's belief. A five level subjective scale has been chosen to help the user in rating (see Fig. 3). A result of such a rating might be :

rating (p1) = 0.8	rating (p2) = 0.2	rating (p3) = 0.4
rating (p4) = 0.6	rating (p5) = 0.3	rating (p6) = 0.7
rating (p7) = 0.9	rating (p8) = 1.0	rating (p9) = 0.8

These ratings must be transformed into belief amounts following the normalization :

for all pi $\in$ {p1, ..., p9}

$$m(pi) = rating(pi) \; / \; [\; rating(p1) + \ldots + rating(p9) \;]$$

The resulting belief amounts are :

m(p1) = 0.140	m(p2) = 0.035	m(p3) = 0.070
m(p4) = 0.105	m(p5) = 0.053	m(p6) = 0.123
m(p7) = 0.158	m(p8) = 0.176	m(p9) = 0.140

Then if the following implications are considered :

$p3 \rightarrow p1$	$p6 \rightarrow p1$	$p7 \rightarrow p1$	$p8 \rightarrow p1$	$p9 \rightarrow p1$
$p5 \rightarrow p2$				
$p8 \rightarrow p3$	$p9 \rightarrow p3$			
$p2 \rightarrow p4$	$p5 \rightarrow p4$	$p6 \rightarrow p4$		

Credibilities for the initial propositions p1, p2, p3, and p4 are :

Cr(p1) = 0.684 = m(p1) + m(p3) + m(p7) + m(p8) + m(p9)
Cr(p2) = 0.088 = m(p2) + m(p5)
Cr(p3) = 0.386 = m(p3) + m(p8) + m(p9)
Cr(p4) = 0.316 = m(p2) + m(p4) + m(p5) + m(p6)

Associated plausibilities are :

Pl(p1) = 0.912 = 1 - Cr($\neg$p1) = 1 - Cr(p2)
Pl(p2) = 0.316 = 1 - Cr($\neg$p2) = 1 - Cr(p1)
Pl(p3) = 0.684 = 1 - Cr($\neg$p3) = 1 - Cr(p4)
Pl(p4) = 0.614 = 1 - Cr($\neg$p4) = 1 - Cr(p3)

These results can be interpreted as follows. It is reasonably credible (0.684) that the author is American with a residual ignorance of 22.8 % (i.e.: Pl(p1) - Cr(p1) = 0.228). It is not very much plausible (0.316) that the author is not American with a residual ignorance of 22.8%. It is difficult to decide if the author is Californian,

[Cr(p3)=0.386, Pl(p3)=0.684],

or not,

[Cr(p4)=0.316, Pl(p4)=0.614],

but we know this with a reasonably residual ignorance of 29.8%. This example shows that we are able to discriminate that the author is American, but with this level of knowledge represented by the nine focal propositions it is impossible to decide if he is Californian or not.

"The primary advantage of this approach is that each knowledge source can express itself at a level of detail of its own choosing. When there is no clear reason to prefer one proposition to another, that jugment can be suspended" (ref. 11). Thus, a library user can express some belief that a document is characterized by a given author name without having to speculate as to that document's title. "A Bayesian approach would require that a precise probability be assigned to each type (attribute), no matter how noisy the origin of information, and no matter how little statistical data are available from which to make justifiable estimates" (ref. 11).

Note that ***probability*** is a particular case of credibility and plausibility measures. A probability measure corresponds to precise information, i.e. the focal elements are singletons, but information is contradictory, disonant, i.e. the focal elements are disjoints. In this case the following relations are satisfied : Cr(p) = Pl(p) = P(p). *Necessity* and *possibility* (ref. 21) are another particular case of credibility and plausibility. They correspond to consistent information, i.e. the focal elements are consonant (e.g., p1→p2, p2→p3, ..., p(n-1)→p(n)), but information is imprecise. For all "p" and "q" the following relations are verified :

$$Cr\ (p\ \&\ q) = \min\ (\ Cr(p), Cr(q)\ )\ \rightarrow\ Cr = N\ \ (\text{necessity})^4$$
$$Pl\ (p\ V\ q) = \max\ (\ Pl(p), Pl(q)\ )\ \rightarrow\ Pl = \Pi\ \ (\text{possibility})^5$$

A proposition must be possible in order to be probable, and probable in order to be necessary (ref 5, 6) :

$$N\ (p)\ \leqslant\ P\ (p)\ \leqslant\ \Pi\ (p)\ \text{, see Fig. 4.}$$

A possibility measure Π (A) can be built from a possibility distribution π (x) as follows :

$$\Pi\ (A) = \sup_{x \in A}\ \{\ \pi\ (x)\ \}$$

Example :

A = "Publication dates around 1977"

This expression is similar to the expression of a probability measure P from its probabilty distribution p :

$$P\ (A) = \Sigma_{x \in A}\ \{\ p\ (x)\ \}$$

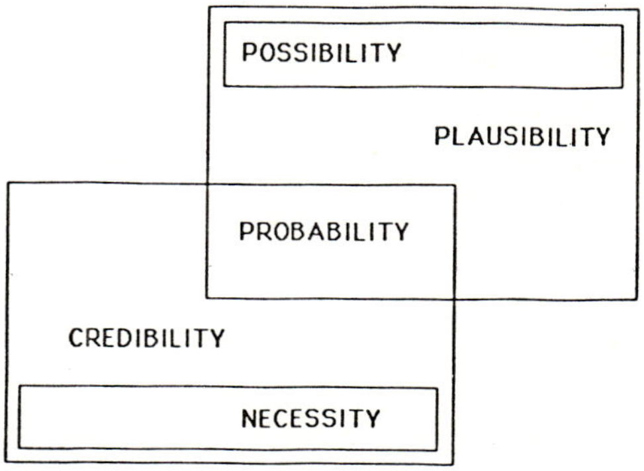

Figure 4. Lower and Upper Probabilities (Dubois, 1983)

[4] & represents the intersection operator (disjunction).

[5] V represents the union operator (conjunction).

The possibility of an event is calculated from the best case (in the possibilty distribution), and not from an accumulation of more or less good cases as in probability theory. Thus the possibility of an event tries to evaluate its feasability. If the possibility of a proposition "p" is 1, the possibility of "$\neg$p" may be 1 also. If the necessity of a proposition "p" is 1 ("p" is certain), then the possibility and necessity of "$\neg$p" is 0.

3.2. Quantification of the Imprecision

Fuzzy Sets

The following propositions are imprecise :

"This book has been recently published",
"This book has been ordered before 1978".

The latter above proposition shows a simple imprecise specification of a value, e.g. "before 1978" suggests that one should consider the set of dates corresponding to the interval bounded by the oldest date in the library and 1978. The former shows an imprecise predicate ("recently"). If the predicate "recently" is defined on a set of dates, any crisp interval cannot represent it successfully. The set of dates which defines "recently" is a fuzzy set (ref. 6, 18). A

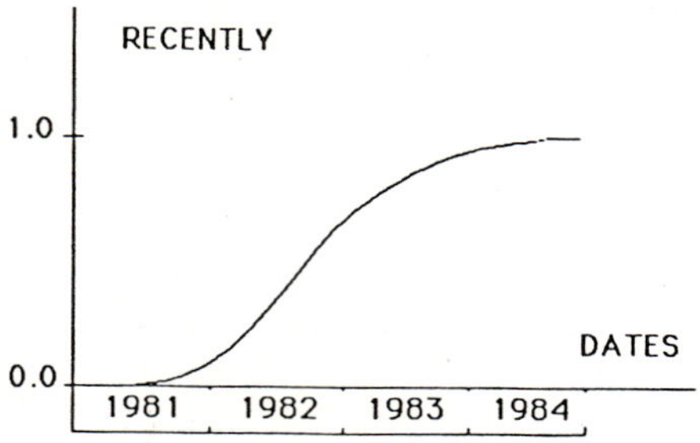

Figure 5. 'recently' membership function

fuzzy predicate (e.g. recently) is defined as a function from a given domain (e.g. dates) into the real interval [0,1]. This function is called membership function. In the figure 5, each point represents the membership degree (ordinate) according to the predicate "recently" for a particular date (abscisse).

Fuzzy Set vs. Possibility

A fuzzy set can be used to represent a soft constraint (i.e ill-defined). Such a constraint bounds a set of more or less acceptable values. A possibility distribution represents a fuzzy set of the more or less possible values of a variable. Then the membership function is interpreted as a degree of possibility (ref. 21). Consequently, a fuzzy set allows one to define the set of the more or less possible values of a variable limited by a soft constraint.

3.3. Truth of a Proposition

The truth of a proposition is its conformity with facts or reality as percieved by the user.

Representing an Imprecise Proposition

Considering the proposition : R = "d is A" (e.g. this document is recently published), where d is a given sub-set of a set D (e.g. documents), and A a predicate, sometimes vague (e.g. appeared recently), which limits the possible values of d. The content of a proposition can be represented by a possibility distribution $\pi_{a(d)}$ attached to the function a(d) (e.g. publication date of the documents d). $\pi_{a(d)}$ is an application from a given set $X \equiv \{x_i\}$ (e.g. publication dates) onto the interval [0,1], and is the membership function f_A attached to the predicate A:

$$\text{for all } x \in X, \quad \pi_{a(d)}(x) = f_A(x).$$

Fuzzy Pattern Matching

Let two propositions, a request R and a data base pattern a(d), represented by possibility distributions, resp. π_R and $\pi_{a(d)}$. Let consider the interval [N(a(d)/R), Π(a(d)/R)], where Π(a(d)/R) and N(a(d)/R) the possibility and necessity that a(d) would be true given R :

$$\Pi\ (a(d)/R) = \sup_{x \in X} \{ \min \{\pi_R(x), \pi_{a(d)}(x)\} \}$$

$$N\ (a(d)/R) = \inf_{x \in X} \{ \max \{1 - \pi_R(x), \pi_{a(d)}(x)\} \}$$

Π(a(d)/R) evaluates the degree to which the intersection of the fuzzy sets characterized by π_R and $\pi_{a(d)}$ is not empty. N(a(d)/R) evaluates the degree to which the fuzzy set characterized by $\pi_{a(d)}$ is included in the fuzzy set characterized by π_R.

If the data is precise, then $\pi_{a(d)}$ is characterized by a possibility distribution null for all the points except for one value. In that case : Π(a(d)/R) = N(a(d)/R). This degree can be interpreted as a degree of truth.

4. THE DESIGN OF BIBLIO

Each data stored in the data base and each user's request may be imprecise. Precision is considered as a specific case of imprecision. Then each piece of information is represented by a possibility distribution.

4.1. Knowledge Representation

An elementary piece of information can be represented by the triplet (attribute, object, value), where "value", qualitative or quantitative, is an element member of the attribute domain. The following examples are given :

(author, book-X, Miller),
(publication-date, book-Y, 1978).

The value of an attribute may be precise or imprecise. The possible values of an attribute can be represented by a possibility distribution on the attibute domain. In such a context, a precise value is represented by a possibility distribution null for all points except for the point corresponding to the precise value, at this particular point the possibility is 1. An unknown value is represented by a possibility distribution equal to 1 for all the points of the attribute domain.

The value of an attribute is often uncertain. This uncertainty comes from the lack of confidence of the user about the information source. To take into account such uncertainty, a five degree subjective scale has been built :

Degree # 5 : I am sure that the information is good
Degree # 4 : I think that the information is good
Degree # 3 : I do not know
Degree # 2 : I think that the information is not good
Degree # 1 : I am sure that the information is not good

Each degree can be related to a subjective probability "u", for example :

Degree # 5 --> 1.0
Degree # 4 --> 0.7
Degree # 3 --> 0.5
Degree # 2 --> 0.3
Degree # 1 --> 0.0

Quantitative Attributes

The possibility distribution attached to a quantitative attribute is modelled by a trapezium (see Fig. 6). This representation is very simple to create and manipulate. It is a good approximation of the common sense associated with the imprecision concept (ref. 1). It can be expressed by the quintuplet (z, a, b, c, d). It includes a prefered domain [b, c], an allowed domain [a, b] ∪ [c, d], and an unacceptable domain which is the complement of [a, d]. "z" is an index equal to 1 for a regular configuration, 0 for a possibility distribution identical to 1 for any point, and -1 for an inverse configuration (see Fig. 7).

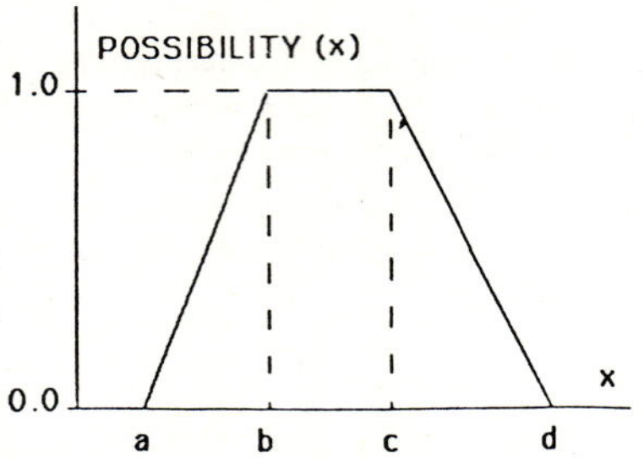

Figure 6. Possibility Distribution Representation of a Quantitative Variable

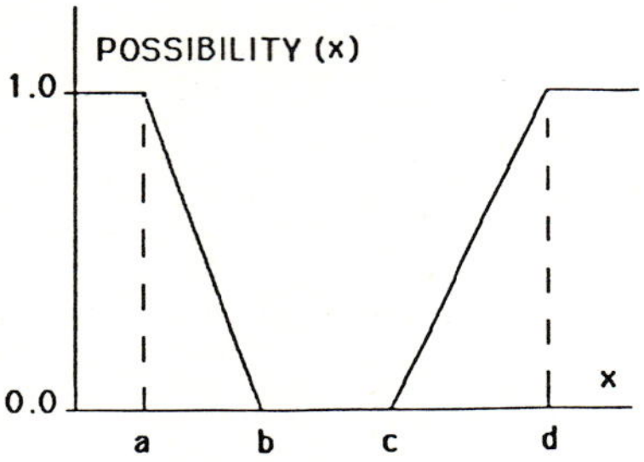

Figure 7. Inverse Configuration

The uncertainty is assumed to modify the imprecision, i.e. the possibility distribution, taking into account three observations :

1. when the uncertainty increases, the possibility distribution spreads out,
2. when the proposition is totally uncertain (total ignorance), the possibility distribution is equal to 1 for all the points of the attribute domain,
3. when the certainty of the opposite proposition increases, the possibility distribution will be complemented to 1 (negation of the proposition).

Let "u" be the subjective probability attached to a given proposition p and (z, a, b, c, d) the possibility distribution representing the imprecision of p. Then the resulting possibility distribution (z', a', b', c', d') can be such that the preferred domain [b, c] and the acceptable domain [a, d] are expended by a coefficient equal to 1/(2u-1). Moreover :

if $0.5 < u \leqslant 1$ then $z' = z$
if $u = 0.5$ then $z' = 0$
if $0 \leqslant u < 0.5$ then $z' = -z$

In BIBLIO, 6 options have been implemented to enter a date: "in x", "after x", "before x", "between x and y", "around x", and "recently". Five of these possibility distributions are presented (see Fig. 8). "before x" and "after x" are symetrical. Only the degrees # 5 and # 4 are represented because the possibility distribution relative to degrees 2 and 1 are their complements to 1. Degree 3 corresponds to a distribution equal to 1 in all the cases.

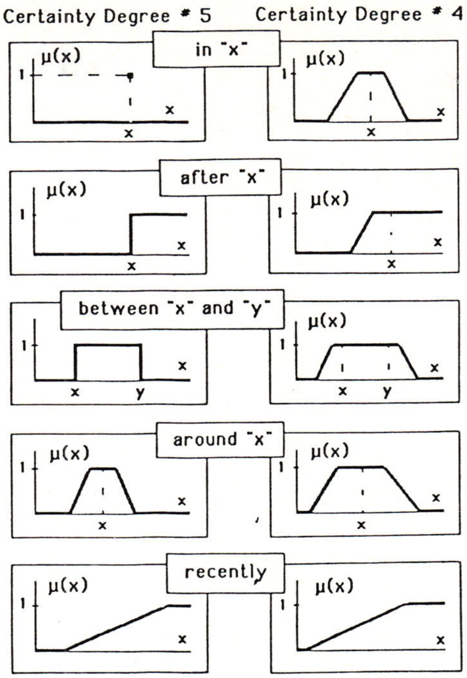

Figure 8. Examples of Possibility Distribution Represented in BIBLIO

Qualitative Attributes

The possibility distribution attached to a qualitative attribute is modelled by a set of couples {({x1},t1), ..., ({xn},tn)} (see Fig. 9). Where {x1}, ..., {xn} are sub-sets characterized respectively by the discrete possibility distribution (t1, ..., tn). This representation is very simple to create and manipulate. In the present example, all the qualitative attributes are author-names or key-words. A request from the user can be the following :

> "x is A1 or A2 ... or An" (e.g. "I think the author can be Miller (1), but he can be also Brooks (0.6), or Smith (0.4), or anybody else (0.1)") represented by the possibility distribution :

{ ({A1}, 1) , ({A2}, 0.6}) , ({A3}, 0.4}) , (X-{A}, 0.1) }

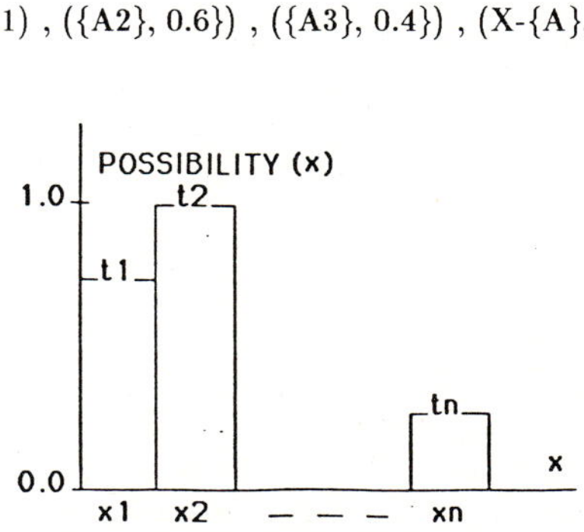

Figure 9. Possibility Distribution Representation of a Qualitative Variable

It is assumed that these attributes can have either a well defined value, or a unknown value. Let X be the attribute domain (e.g. a author-name list). Let "x" be an attribute and "A" a value of this attribute. In the data base, two kinds of propositions are possibles :

"x is A" (e.g. "the author name is Miller") represented by the possibility distribution :

$$\{ (\{A\}, 1) , (X\text{-}\{A\}, 0) \}$$

"x unknown" represented by the possibility distribution :

$$\{ (X, 1) \}.$$

Both user's request "x is A" and "x includes A" are represented by the possibility distribution : $\{(\{A\},1), (X\text{-}\{A\},0)\}$.

User uncertainty is taken into account (at the request level) as follows : if "u" is the subjective probability attached to a given proposition p and $\{ (\{A\}, 1) , (X\text{-}\{A\}, 0) \}$ the possibility distribution representing the imprecision of p, the resulting possibility distribution can be the following :

if $0.5 < u \leqslant 1$	$\{ (\{A\}, 1) , (X\text{-}\{A\}, 1\text{-}u) \}$
if $u = 0.5$	$\{ (X, 1) \}$
if $0 \leqslant u < 0.5$	$\{ (\{A\}, u) , (X\text{-}\{A\}, 1) \}$

Examples :

- Let the proposition "editor is Elsvier" have a certainty degree # 4. The subjective probability attached to degree # 4 is 0.6. The possibility distribution attached to the proposition is :

$$\{ (\{Elsvier\}, 1) , (X\text{-}\{Elsvier\}, 0.4) \}$$

where X is the editor attribute domain.

- Let the proposition "author is Brown" have a certainty degree # 1 (the user is sure that author's name is not "Brown"). The subjective probability attached to degree # 1 is 0.0. The possibility distribution attached to the proposition is :

$$\{ (\{Brown\}, 0) , (X\text{-}\{Brown\}, 1) \}$$

where X is the author attribute domain.

4.2. Database

In BIBLIO, the basic entity is the *document*. A document is characterized by various attributes, such as : Reference, Author(s), Title, Title Complement, Editor, Publication Date, Order Date, Order Origin, Words List, Comments List. The entity *document* is modelled as a frame including four main fields :

- Field 1 : (Reference, Author(s), Title, Editor, Publication Date)
- Field 2 : (Reference, Title Complement, Order Origin, Order Date)
- Field 3 : (Reference, Key-Words List)
- Field 4 : (Reference, Comments List)

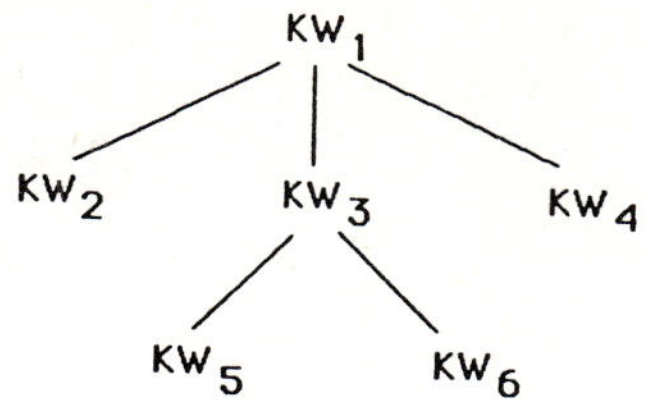

Figure 10. Thesaurus of Keywords

In parallel with the management of these data, the system is able to update a thesaurus of key-words. This thesaurus includes trees of key-words such as (see Fig. 10) :

$$((KW_1) ((KW_2) (KW_3) (KW_4)))$$

$$((KW_3) ((KW_5) (KW_6)))$$

4.3. Evaluation of the requests

In terms of relational algebra, the evaluation of a request R gives an operation of "generalized selection" $\sigma(R)$ characterized by a basic condition R. The result of a selection $\sigma(R)$ is represented by two sets :

- The set of documents which satisfy R "possibly",
- The set of documents which satisfy R "necessarily".

These selected documents are characterized by :

$$\sigma\ (R) = \{ \sigma_\Pi\ (R), \sigma_N\ (R) \}$$

Atomic Request Evaluation

In the data base, a document "d" is characterized by a possibility distribution $\pi_{a(d)}\ (x)$ according to the attribute A (e.g. publication date), and "x" member of the attribute domain (e.g. 1978).

An atomic request is characterized by a condition R and is represented by a possibility distribution $\pi_R\ (x)$. The possibility that R would be satified given the possibility distribution of A for the document d is :

$$\Pi\ (\ a(d)\ /\ R\ ) = \sup_{x \in X} \{ \min\ (\ \pi_R\ (x)\ ,\ \pi_{a(d)}\ (x)\ )\ \}$$

The necessity degree is the following :

$$N\ (\ a(d)\ /\ R\ ) = \inf_{x \in X} \{ \max\ (\ 1 - \pi_R\ (x)\ ,\ \pi_{a(d)}\ (x)\ )\ \}$$

The possibility [resp. necessity] degree $\Pi(a(d)/R)$ [resp. $N(a(d)/R)$] defines the membership grade of the document "d" to a set of documents which possibly [necessarily] satisfies the condition R. These two numbers give us a measure of the relevance of the document. $\Pi(a(d)/R)$ [resp. $N(a(d)/R)$] gives an "optimistic" [resp. "pessimistic"] correlation measure. The difference $[\Pi(a(d)/R) - N(a(d)/R)]$ gives a measure of the residual ignorance.

Composed Request Evaluation

A composed request includes atomic requests and logical connectives (e.g. and, or, except). The composed request evaluation is a combination of atomic request evaluations. Let us consider two request conditions R1 and R2, for example. In a retrieval problem, these conditions are "non-interactive" (ref. 20), i.e. the elements of the first condition do not depend on the elements of the second condition, and if the union and intersection of fuzzy sets are defined by the max and min operators, then the following formulas are applicable :

$$\sigma\,(\,R1 \vee R2\,) = \{\,[\sigma_{\Pi}(R1) \cup \sigma_{\Pi}(R2)]\,,\,[\sigma_{N}(R1) \cup \sigma_{N}(R2)]\,\}$$
$$\sigma\,(\,R1\ \&\ R2\,) = \{\,[\sigma_{\Pi}(R1) \cap \sigma_{\Pi}(R2)]\,,\,[\sigma_{N}(R1) \cap \sigma_{N}(R2)]\,\}$$

Key-Word Request Evaluation

A task analysis of a search by key-words shows that the user proposes a list of notions. He attaches to each notion a more or less important significance according to the objective of the search. This is not uncertainty, but a significance weight . A possibility distribution is not appropriate for representing this kind of knowledge. The request is then simply represented by a list of notions weighted by a coefficient between 0 and 1, a significance weight (or rating, see Fig. 3) of that notion according to the user objective.

Let R be the set of key-words which defines the request. Each key-word is attached to a rating which can be interpreted as a membership function $f_{R(x)}$ according to the set R, i.e. the given key-word belongs to R at the degree $f_{R(x)}$. In order to find the relevance of each document of the data base according to the key-word request, two heuristic measures can be built, J1 optimistic and J2 pessimistic.

R is represented as follows :

$$R = \{\,(KW_1, t1)\,,\,(KW_2, t2)\,,\,\dots\,,\,(KW_n, tn)\,\}$$

Each reference of the data base is characterized by a key-word list a(d) (pattern) :

$$a(d) = \{\,(KW_{\alpha}, k_{\alpha})\,,\,(KW_{\beta}, k_{\beta})\,,\,\dots\,,\,(KW_{\omega}, k_{\omega})\,\}$$

J1 quantifies the non-emptiness of the set $R \cap a(d)$. R and a(d) are approximations of possibility distributions, then the operator :

$$\sup_{x \in X}\,(\,f_{R \cap a(d)}(x)\,)$$

is too restrictive while $f_{R \cap a(d)}(x)$ is not a possibility distribution. A less restrictive operator has been chosen (ref. 5) :

$$\min\,[\,1\,,\,\Sigma_{x \in X}\,(\,f_{R \cap a(d)}(x)\,)\,]$$

Moreover, this measure must be over-estimated if the number $n_{R \cap a(d)}$ of elements of $R \cap a(d)$ is large, then if n_R is the number of elements of R :

$$J1\,(d) = \max\,\{\,n_{R \cap a(d)}\,/\,n_R\,,\,\min\,[\,1\,,\,\Sigma_{x \in X}\,(\,f_{R \cap a(d)}(x)\,)\,]\,\}$$

To build J2, the membership grade $f_{R \cap \neg a(d)}$ of the keywords which belong to R and do not belong to a(d) is taken into account.
The global membership grade can be evaluated by :

$$\sup_{x \in X}\,(\,f_{R \cap \neg a(d)}(x)\,)$$

and the global not-membership grade is then :

$$1 - \sup_{x \in X} \left(f_{R \cap \neg a(d)}(x) \right)$$

moreover, this measure must be under-estimated if the number of elements of $R \cap a(d)$ is small, then J2 is defined as :

$$J2\ (d) = \min \left\{ n_{R \cap a(d)} / n_R \, , \, 1 - \sup_{x \in X} \left(f_{R \cap \neg a(d)}(x) \right) \right\}$$

Example :

Let the request :
R ≡ { (human factors, 0.9), (man-machine interactions, 0.8), (computer science, 0.1) }

Let { d1, ..., d8 } be documents from the database characterized by :

a (d1) ≡ { (human factors, 1), OTHERS }
a (d2) ≡ { (man-machine interactions, 0.9), OTHERS }
a (d3) ≡ { (computer science, 0.8), OTHERS }
a (d4) ≡ { (human factors, 0.2), (man-machine interactions, 0.6), OTHERS }
a (d5) ≡ { (human factors, 0.4), (computer science, 1), OTHERS }
a (d6) ≡ { (computer science, 1), (man-machine interactions, 0.7), OTHERS }
a (d7) ≡ { (human factors, 1), (computer science, 1), (man-machine interactions, 1), OTHERS }
a (d8) ≡ { OTHERS }

where "OTHERS" are other keywords not present in the request.

Taking into account the calculations for d4, for instance :

$R \cap a(d4)$ ≡ { (human factors, 0.2), (man-machine interactions, 0.6) }
$R \cap \neg a(d4)$ ≡ { (human factors, 0.8), (man-machine interactions, 0.4), (computer science, 0.1) }

$$n_{R \cap a(d4)} / n_R = 0.66$$

J1 (d4) = max (0.66 , min (1 , 0.2 + 0.6)) = 0.8
J2 (d4) = min (0.66 , 1 - sup (0.8, 0.4, 0.1)) = 0.2

The following results are found :

documents	J1	J2
d1	0.90	0.20
d2	0.80	0.10
d3	0.33	0.10
d4	0.80	0.20
d5	0.66	0.20
d6	0.80	0.10
d7	1.00	1.00
d8	0.00	0.00

As shown in this retrieval example, fuzzy sets and fuzzy logic (ref. 12, 13, 16, 19, 20) allow the designer to fashion pattern matching operators more or less restrictive, more or less soft, etc., according to the application. A list of such operators is given in (ref. 5).

5. DISCUSSION AND CONCLUSION

The task of information retrieval presents a user with two potentially conflicting demands : keeping the significance of the semantics of his request, and expressing this request in more or less precise and certain language. To express the request the user uses a model which is a discrete set of descriptors. If the user knows the document he is looking for very well, he can either express his request in a quasi infinite set of descriptors (everything is precise, certain, and complete), or he can find an aggregated sub-set of descriptors which summerizes the document perfectly. In this case the request is precise and certain. Similarly, the same considerations apply to the indexation of the document in the library.

If the user does not know precisely and certainly the document(s) he is looking for, to interact with the librarian he also needs to formulate a discrete set of descriptors. In this case, the problem is different. If the user gives descriptors that are too precise and too certain, then the librarian is likely to attempt a precise and certain retrieval task. If the user gives descriptors that are too imprecise and too uncertain, then everything is possible. However, a reasonable amount of "fuzziness" can keep the significance of the request and help direct the librarian in his search. Thus, people are assumed to "fuzzify" qualitative patterns to give significance to the whole. This "out-of-focus" image of a discretized world is a kind of "making-sense cement" assembling components which have no significance when they are taken out of context.

To assess and extract the above significance, experimental analyses can be carried out. Pairs of subjects (user and librarian) can be tested in a set of particular retrieval tasks using conventional retrieval aids. The subjects can be asked to "think aloud" during the retrieval process which can be videotaped. Protocols (the chronological history of all the user's interactions) can also be transcribed for later analysis (ref. 8, 15). Such protocol analysis constitutes the first step in identifying the sources of imprecision and uncertainty on both sides, i.e. user and library-librarian-system. A second type of experiment is necessary to implement a "good" knowledge base in the BIBLIO system. This experiment is based on an analysis of the interaction of subjects with the system. Subjects can be experts, i.e. librarians, or non-experts, i.e. library users. Then an iterative process is carried out using protocol analyses with different versions of the BIBLIO system.

The representation of fuzzy information is an important matter for further investigation in the field of human factors, especially in cognitive psychology. Improved understanding of the fuzziness of human information should provide a very good tool to aid design of human-machine interfaces. Natural language modelling is a particularly interesting application of fuzzy sets in human-computer interactions. For example, extracting key-words from a complex request is not easy in practice. A human prefers to keep the meaning of his request using an imprecise representation rather than selecting a set of precise key-words and thus losing an important part of the significance of what he wants to retrieve. This implies that it is necessary to include flexible models of the possible imprecision and uncertainty of the user in computerized retrieval aid systems (e.g. BIBLIO). Moreover, it is reasonable to think that these models could be adapted on the basis of user interactions with the system. Classes of users could then be identified and more personalized retrieval aids could be provided.

REFERENCES

1 Boy, G.A., & Tessier, C, (1985), "Cockpit Analysis and Assessment By the MESSAGE Methodology" *2nd IFAC/IFIP/IFORS/IEA Conference on Analysis, Design, and Evaluation of Man-Machine Systems,* Villa Ponty, Varese, Italy, September 10-12.

2 Buchanan, B.G., (1981), *Research on Expert Systems,* Heuristic Programming Project, Report no. HPP 81-1, Feb. 1981, Computer Science Department, School of Humanities and Sciences, Stanford University.

3 Danilowicz, C., (1984), "Users and Experts in the Document Retrieval System Model", *Int. J. Man-Machine Studies,* 21, pp. 245-252.

4 Dempster, A.P., (1968), "A Generalization of Bayesian Inference", *Journal of the Royal Satistical Society,* Series B, Vol. 30, pp. 205-247.

5 Dubois, D., (1983), *Modeles Mathematiques de l'Imprecis et de l'Incertain en vue d'Applications aux Techniques d'Aide a la Decision,* Thesis, Universite Scientifique et Medicale de Grenoble, France.

6 Dubois, D., & Prade, H., (1980), *Fuzzy Sets and Systems : Theory and Applications,* Academic Press, USA.

7 Dubois, D., & Prade, H., (1982), "On Several Representations of an Uncertain Body of Evidence", *In Fuzzy Information and Decision Processes (M.M. Gupta, E. Sanchez, eds.),* Noth-Holland, pp. 167-181.

8 Ericsson, K., & Simon, H., (1980), "Verbal Reports as Data", *Psychological Review,* 87, n.3, pp. 215-251.

9 Gordon, J., & Shortliffe, E.H., (1985), "A Method for Managing Evidential Reasoning in a Hierarchical Hypothesis Space" *Artificial Intelligence,* to appear in no. 424, vol. 26(3), July.

10 Kahneman, D., & Tversky, A., (1984), "The Psychology of Decision Making", *Naval Research Reviews,* Office of Naval Research, One/1984, Vol XXXVI.

11 Lowrance, J.D., & Garvey, T.D., (1983), *Evidential Reasoning: An Implementation for Multisensor Integration,* Technical Note 307, Artificial Intelligence Center, 333 Ravenswood Ave., Menlo Park, CA.

12 Mandani, E.H., (1977), "Application of Fuzzy Logic to Approximate Reasoning Using Linguistic Systems", *IEEE Trans. on Computers,* C-26, pp. 1182-1191.

13 Negoita, C.V., (1985), *Expert Systems and Fuzzy Systems,* The Benjamin/Cumming's Publishing Company Inc., 2727 Sand Hill Road, Menlo Park, California 94025, USA.

14 Newell, A., & Simon, H.A., (1972), *Human Problem Solving,* Englewood Cliffs, Prentice Hall.

15 Nisbett, R., & Wilson, T., (1977), "Telling More than We Can Know: Verbal Reports on Mental Processes", *Psychological Review,* 84, n.3, pp. 231-259.

16 Prade, H., (1985), "A Computational Approach to Approximate and Plausible Reasoning with Applications to Expert Systems", *IEEE Transactions on Pattern Analysis and Machine Intelligence,* Vol. PAMI-7, no. 3, May, pp. 260-283.

17 Shafer, G, (1976), *A Mathematical Theory of Evidence,* Princeton University Press, Princeton, NJ.

18 Zadeh, L.A., (1965), "Fuzzy Sets", *Information and Control,* no. 8, pp. 338,353.

19 Zadeh, L.A., (1975), "Fuzzy Logic and Approximate Reasoning", *Synthese,* no. 30, pp. 407-428.

20 Zadeh, L.A., (1975b), "The Concept of a Linguistic Variable and its Application to Approximate Reasoning", *Information Sciences,* Part 1, vol. 8, pp. 199-249, Part 2, vol. 8, pp. 301-357, Part 3, vol. 9, pp. 43-80.

21 Zadeh, L.A., (1978), "Fuzzy Sets as a Basis for a Theory of Possibility", *Fuzzy Sets and Systems,* Vol. 1, no. 1, pp. 3-28.

22 Zadeh, L.A., (1979), "On the Validiy of Dempster's Rule of Combination of Evidence", *Memo. UCB/ERL/M79/24, University of California, Berkeley.*

23 Zadeh, L.A., (1984), "A Simple View of the Dempster-Shafer Theory of Evidence" *Memo. Berkeley Cognitive Science Report no. 27,* University of California, Berkeley.

ACKNOWLEDGEMENTS

We are thankful to Dr. Philippa Gander for insightful comments about this paper.

Part III.

HUMAN OPERATOR: JUDGEMENT AND RELIABILITY

Applications of Fuzzy Set Theory in Human Factors, edited by W. Karwowski and A. Mital
Elsevier Science Publishers B.V., Amsterdam, 1986 — Printed in The Netherlands

TECHNIQUES FOR FITTING FUZZY CONNECTIVE AND LOGICAL OPERATORS TO HUMAN JUDGMENT DATA IN DESIGN AND EVALUATION OF MAN-MACHINE SYSTEMS

M.J. SMITHSON

Behavioural Sciences Department, James Cook University, Queensland (Australia)

ABSTRACT

This chapter presents techniques for assessing the degree of fit between fuzzy set information processing models and judgment data. Solutions for two kinds of problem are proposed. The first utilizes a modified least-squares technique for fitting fuzzy aggregation models to situations requiring a human judge to aggregate several attributes or properties of a stimulus to form an overall rating of that stimulus on yet another attribute. The second kind employs bivariate prediction methods to assess the extent to which human logical or predictive judgments correspond to fuzzy logical models.

INTRODUCTION

Modern human factors research and design requires accurate models of human information processing and decision making, both in designing semiautomatic systems which respond effectively to human judgments and simulating those judgment processes in automatic control systems. Two types of judgment tasks are basic to human information processing. One might be called an 'aggregation' task, in which the judge must combine two or more attribute judgements to form a third more global judgement about a stimulus. In natural language, such combinations often are expressed by conjunctions (e.g., 'and', 'or'). They frequently take the form 'A is B and C.' The second type of task may be termed an 'implicative' judgment, which requires the judge to infer one attribute-state from another. These judgments often take the form 'If X, then Y, otherwise Z.'

Fuzzy set theory provides an intuitively plausible mathematics for modeling both these tasks, and a number of automata and control systems have utilized fuzzy aggregation and logical operators to simulate or respond to input from human judgment. Several kinds of operators have been proposed, however, and that has generated considerable discussion about their relative merits. Most of this discussion has been conceptual. The few empirical studies that have been conducted on the correspondence between certain operators and actual human judgments all contain some methodological problems. One reason for this state of affairs is the lack of adequate tools for measuring the correspondence between fuzzy aggregation or logical models and judgment data.

The approach taken in this chapter is to provide methods that enable the designer or researcher to optimize the fit between fuzzy set theory and judgment data, by selecting an appropriate operator and solving for certain data-driven parameters. Not only is this approach useful in design problems where the simulation of judgment is required, but it is well-suited for product evaluation and marketing research. For instance, an evaluation researcher will be able to determine the extent to which a typist's overall rating of a chair's performance is a function of the degree to which s/he finds it 'comfortable *but not* soporific.'

A REVIEW OF FUZZY AGGREGATION OPERATORS

Much has been written on fuzzy aggregation operators and their properties (see, e.g., refs. 1 & 2). This review performs two functions: One is to summarize the familiar operators and their properties, and the other is to lay the foundations for modeling aggregation and implication by introducing some extensions of those operators. For notational convenience, the membership function $f_k(x_i)$, which denotes 'the membership of the ith element of some collection X in fuzzy set A_k', usually will be abbreviated f_{ki} or in cases where no specific element needs be referred to, f_k. Also for convenience, 'f_1 and f_2' is denoted $A(f_1,f_2)$, 'f_1 or f_2' by $O(f_1,f_2)$ (for the conjunctive or), and '$f_1$xor f_2' by $XO(f_1,f_2)$ (for the disjunctive or).

The list below includes the most widely mentioned desirable properties for fuzzy aggregation operators:

1. $A(1,f_k) = f_k$,
2. Nondecreasingness,
3. Commutativity,
4. Associativity,
5. Distributivity: $A(f_1,O(f_2,f_3)) = O(A(f_1,f_2),A(f_1,f_3))$ and $O(f_1,A(f_2,f_3)) = A(O(f_1,f_2),O(f_1,f_3))$,
6. Archimedeanity: $A(f_k,f_k) \leqq f_k$ and $O(f_k,f_k) \geqq f_k$,
7. De Morgan Laws: $O(f_1,f_2) = 1 - A(1-f_1,1-f_2)$ and $XO(f_1,f_2) = A(O(f_1,f_2),1-A(f_1,f_2))$,
8. $A(f_1,f_2) + O(f_1,f_2) = f_1+ f_2$.

Several investigators have proven that the only operator satisfying all of these properties is the min-max pair initially proposed by Zadeh (refs. 3 & 4):

$$A(f_1,f_2) = \min(f_1,f_2) \text{ and}$$
$$O(f_1,f_2) = \max(f_1,f_2). \qquad (1)$$

Several other operators exist which satisfy all properties except distributivity. They have been widely discussed in the literature, although the first two are by far more popular than the third:

Product operators:

$$A(f_1,f_2) = f_1 f_2 \text{ and}$$
$$O(f_1,f_2) = f_1 + f_2 - f_1 f_2 \qquad (2)$$

Bounded sum operators:

$$A(f_1,f_2) = \max(0, f_1 + f_2 - 1) \text{ and}$$
$$O(f_1,f_2) = \min(1, f_1 + f_2) \qquad (3)$$

Log operators:

$$A(f_1,f_2) = \log(((s^{f_1} - 1)(s^{f_2} - 1)/(s-1)) + 1) \text{ and}$$
$$O(f_1,f_2) = 1 - \log(((s^{1-f_1} - 1)(s^{1-f_2} - 1)/(s-1)) + 1) \qquad (4)$$

where $s>0$ and $\neq 1$.

Frank (ref.5) has shown that these are the only operators satisfying all properties except distributivity, and that other popular operators fail to have property 8. It is worth noting, however, that some other operators may very nearly fulfill 8. One example is what might be called a 'Lorenztian' operator, in which $O(f_1,f_2)$ may be interpreted as the Lorentzian sum of two velocities in the same direction:

$$A(f_1,f_2) = f_1 f_2/(2 - f_1 - f_2 + f_1 f_2) \text{ and}$$
$$O(f_1,f_2) = (f_1 + f_2)/(1 + f_1 f_2) \qquad (5)$$

The maximum difference between $A(f_1,f_2)+O(f_1,f_2)$ and f_1+f_2 is 0.011.

The extent to which these operators fit judgment data depends largely on how 'severe' or 'generous' they are, which in turn is determined by their Archimedeanity (i.e., how far below f_k the value of $A(f_k,f_k)$ is). It would be useful to have a family of operators whose Archimedeanity varies according to a parameter which may then be optimally chosen. The s parameter in the Log operators and the λ parameter in Yager's (ref. 6) family both vary inversely with Archimedeanity. However, neither parameter lends itself to straightforward optimization on the basis of best fit to real data. If we consider the convex sum of any pair of operators from the group (1)-(4):

$$A(f_1,f_2) = qA_1(f_1,f_2) + (1-q)A_2(f_1,f_2)$$
$$O(f_1,f_2) = qO_1(f_1,f_2) + (1-q)O_2(f_1,f_2) \qquad (6)$$

for $0 \leq q \leq 1$, then these pseudo-operators preserve all properties except associativity, and (as will be shown in the next section) we may obtain least-squares solutions for q. The q parameter varies linearly with Archimedeanity if $A_2(f_1,f_2) \geq A_1(f_1,f_2)$. For judgments involving single aggregations, the loss of associativity is not problematic.

Some researchers (ref. 7) have indicated a need for operators which lie in between a pure 'and' and 'or'. The most obvious kind, again, is a pseudo-operator which is the convex sum of an 'and' and 'or'. Likewise, a variable 'or' may be formed by a convex sum of the conjunctive and disjunctive 'or':

$$\&_{12} = qO(f_1,f_2) + (1-q)A(f_1,f_2) \text{ and}$$
$$@_{12} = qO(f_1,f_2) + (1-q)XO(f_1,f_2) \quad (7)$$

For those operators which satisfy property 8, we may express $\&_{12}$ in the form $\frac{1}{2}(f_1+f_2) + (2q-1)H(f_1,f_2)$ where $H(f_1,f_2) = \frac{1}{2}(f_1+f_2) - A(f_1,f_2)$, whence we see that linearity is attained when $q= 1/2$. Zimmermann and Zysno (ref. 8) have proposed a pseudo-connective similar to $\&_{12}$ which is linear in the logarithmic scale:

$$\S_{12} = (A(f_1,f_2))^{1-q}(O(f_1,f_2))^q \quad (8)$$

Again, these pseudo-operators lend themselves to least-squares solutions for the parameter q.

SINGLE AGGREGATION OPERATOR MODELS

Given judgments by a finite number of judges on three attributes for a finite number of stimuli, the problem is to find the operator that best models f_3, say, as a fuzzy aggregate of f_1 and f_2. Those models which specify a conventional fuzzy set operator in advance may be tested in a straightforward fashion using ordinary least-squares, as long as only f_3 is permitted a linear transformation (e.g., $kf_3+ c = A(f_1,f_2)$, where k and c are to be determined by the regression procedure).

Models which solve only for one kind of pseudo-operator may be solved via the predictive model

$$kf_{3i} + c = B_i + hD_i \quad (9)$$

where k, c, and h are parameters to be estimated, B_i and D_i are specific functions of f_1 and f_2, and $i= 1,2,\ldots,N$ which indexes the judgment data-points. The resulting system of equations for the three parameters is:

$$\hat{c} = \overline{B} - \hat{k}\overline{f}_3 + \hat{h}\overline{D} \quad (10)$$

$$\hat{k} = \sum_i(\hat{h}(D_i - \overline{D}) + (B_i - \overline{B}))(f_{3i} - \overline{f}_3)/ \sum_i(f_{3i} - \overline{f}_3)^2 \quad (11)$$

$$\hat{h} = \sum_i(\hat{k}(f_{3i} - \overline{f}_3) - (B_i - \overline{B}))(D_i - \overline{D})/ \sum_i(f_{3i} - \overline{f}_3)^2 \quad (12)$$

where the ^ sign denotes an estimate of the parameter.

The nature of h, B_i and D_i depends on the aggregation model being used, and they are tabulated below. Further details on the derivation of these equations and alternative models for aggregation are provided in Smithson (ref. 9).

& aggregator model:

$$B_i = A(f_{1i}, f_{2i}),$$
$$D_i = O(f_{1i}, f_{2i}) - A(f_{1i}, f_{2i}) \text{ and}$$
$$h = q. \qquad (13)$$

@ aggregator model:

$$B_i = XO(f_{1i}, f_{2i}),$$
$$D_i = O(f_{1i}, f_{2i}) - XO(f_{1i}, f_{2i}) \text{ and}$$
$$h = q. \qquad (14)$$

Pure 'and' model:

$$B_i = A_2(f_{1i}, f_{2i}),$$
$$D_i = A_2(f_{1i}, f_{2i}) - A_1(f_{1i}, f_{2i}) \text{ and}$$
$$h = -q. \qquad (15)$$

Pure 'or' model:

$$B_i = O_2(f_{1i}, f_{2i}),$$
$$D_i = O_2(f_{1i}, f_{2i}) - O_1(f_{1i}, f_{2i}) \text{ and}$$
$$h = q. \qquad (16)$$

Note that for the & and @ models, we require connectives from the group in (1)-(4) for the results to be interpretable along the lines of this discussion. The general q-type solution for a 'xor' model is not quite as tractable as the others. However, for the case where the operator pair consists of the min-max and bounded sum versions of 'xor', we have

$$B_i = (1+|f_{1i} - f_{2i}| - |1-(f_{1i}+f_{2i})|)/2,$$
$$D_i = (1-|f_{1i} - f_{2i}| - |1-(f_{1i}+f_{2i})|)/2, \text{ and}$$
$$h = q^2. \qquad (17)$$

It is possible to solve for the parameters in both kinds of pseudo-operator models simultaneously. Let q_1 refer to the parameter in (15) and (16), and q_2 refer to the parameter in (13) and (14). Then the & model, for instance, yields a system of four equations in c, k, q_1, and q_2. The first three equations are identical to (10)-(12), with $B_i = A^*(f_{1i}, f_{2i})$ and $D_i = O^*(f_{1i}, f_{2i}) - A^*(f_{1i}, f_{2i})$ where A^* and O^* are q-type operators defined as in (6) with q_1 as their parameter. The third equation is quadratic in q_1, and so the system must be solved via the usual modified Newton-Raphson methods. Since the Jacobians are directly evaluatable, there is some computing economy. The remaining (fourth) equation is:

$$\hat{q}_1 = \frac{\sum_i (\hat{k}(f_{3i} - \bar{f}_3) - \hat{q}_2(D_{2i} - \bar{D}_2) - (A_{2i} - \bar{A}_2))(G_i - \bar{G})}{(2\hat{q}_2 - 1)\sum_i (G_i - \bar{G})} \tag{18}$$

where $A_{2i} = A_2(f_{1i}, f_{2i})$, $D_{2i} = O_2(f_{1i}, f_{2i}) - A_2(f_{1i}, f_{2i})$, and $G_i = A_2(f_{1i}, f_{2i}) - A_1(f_{1i}, f_{2i})$. For many practical situations, however, it may suffice to solve for only one parameter while varying the other between 0 and 1 in small steps.

I shall conclude this section with some examples. Thole et. al. (ref. 10) asked subjects to rate the degree to which 20 stimulus items were 'metallic', 'containers', and 'metallic containers.' The objective here was to determine what aggregator subjects were using to combine the attributes in the phrase 'metallic containers.' Thole concluded that $\min(f_1, f_2)$ worked well if f_3 were multiplied by a constant. We may evaluate a variety of operators using the & model since the authors implied in this and a later article (ref. 8) that operators between 'and' and 'or' might capture such aggregation judgments.

The top half of Table 1 shows the estimates for c, k, and q using (10)-(13). For all operators other than the bounded sum, the values of q are near 0, indicating a strong tendency toward 'and'. The values of c and k for the min-max operator do not agree with the values given by Thole et. al., but this is due to the nature of the estimation methods they used. The multiple correlation coefficient (R^2) indicates the proportion of variance in f_3 explained by the operators. It is evident that the more Archimedean the operator, the worse the degree of fit with the data. The small negative value of q for the Log operator when s=0.5 points up a minor practical problem with least-squares estimation: the estimates are unconstrained and may exceed the range-limits of q, especially when the true value is very close to 0 or 1. Finally, it is worth noting that virtually all operators fit the data better than a simple linear model (conventional regression of f_3 on f_1 and f_2 yields R^2 = 0.639).

TABLE 1

Least squares estimates for Thole et. al. data

Operator-type	q	k	c	R^2	
min-max	0.091	0.742	0.073	0.976	
Eq. (7), q=0.5	0.056	0.702	0.000	0.879	(using min-max, bounded sum)
Bounded sum	0.395	0.564	0.215	0.668	
Product	0.022	0.749	-0.058	0.904	
Lorentzian	0.052	0.702	-0.021	0.848	
Log (s=0.2)	0.000	0.734	-0.051	0.882	
Log (s=0.5)	-0.032	0.760	-0.058	0.923	

TABLE 1 (Continued)

Operator-type	w_1	w_2	q	k	c	R^2
min-max	1.00	1.00	0.091	0.742	0.073	0.976
Eq. (7), =0.5	0.75	0.65	0.020	0.830	0.026	0.955
Bounded sum	0.55	0.55	0.302	0.360	0.643	0.774
Product	0.75	0.65	0.047	0.793	0.053	0.943
Lorentzian	0.65	0.65	0.181	0.724	0.152	0.910
Log (=0.2)	0.75	0.65	0.064	0.785	0.056	0.932
Log (=0.5)	0.75	0.75	0.020	0.802	0.028	0.951

The second half of Table 1 contains estimates for models in which f_1 and f_2 were permitted exponential weights w_1 and w_2. Since they cannot be analytically solved for, these weights were found iteratively in the usual manner. As might be expected, permitting such weights improves the performance of most operators but for this data-set the min-max 'and' proves most successful.

Reanalysis of similar data using these methods reveals no consistent basis for preferring one operator over another. Zimmermann and Zysno (ref. 8) made much of the poor fit between min or max and their data, and ultimately concluded that the product operator (geometric mean, in fact) is superior. But a reanalysis of their data with these methods reveals an excellent fit for an & operator using the min-max pair, with $R^2 = 0.943$. The product operator does not substantially outperform min-max, even permitting exponential weights; the best fit is $R^2 = 0.964$. An improvement of only 2% in explained variance hardly warrants a rejection of the min-max operator. I have also analyzed data-sets for which the bounded sum and Lorentzian have provided the best fit. A FORTRAN program is available (described in Smithson, Ref. 11) which performs the analysis described in this section.

MULTIPLE AGGREGATION MODELS

Judgment tasks involving the aggregation of more than two fuzzy attributes pose considerable problems for model-building. Owing to their nested (hierarchical) nature, most multiple operator models require direct-search algorithms to find optimal values for the parameters and weights used in the previous section. The sole exception is the product operator for models involving aggregators lying in between 'and' and 'or'. The only straightforwardly general result is obtained for the & model, and so it is presented here (derivations are given in ref. 9).

Let q, B, and D have the meanings associated with them in (7) and (13), and assume that B and D are defined in terms of the product operator. Let a double-bar over any variable denote the difference between its value and the mean value for that variable (e.g., $\bar{\bar{B}}_i = B_i - \bar{B}$). We also require a system of subscripts.

Let upper-case characters index attributes and lower-case characters index data-points. The subscripting system for attributes will obey the following rules:

1. $B_{KLi} = A(f_{Ki}, f_{Li})$ and D_{KLi} likewise,
2. $\&_{KLi} = B_{KLi} + q_K D_{KLi}$;
3. $B_{K(LM)i} = A(f_{Ki}, \&_{KLi})$ and $D_{K(LM)i}$ likewise,
4. $\&_{K(LM)i} = B_{K(LM)i} + q_K D_{K(LM)i}$;
5. $B_{K(L(MN))i} = A(f_{Ki}, \&_{L(MN)i})$ and $D_{K(L(MN))i}$ likewise,

and so forth.

6. In any of the variables listed above, let $K(.M)$ replace the extended general notation for subscripts $K(K+1(....(M-1\ M))....)$.
7. Let $Q_{Ji} = O(q_J, f_{Ji}) - A(q_J, f_{Ji})$.

Now, we define the general multiple & operator model by

$$kf_i + c = \&_{1(.M)i} \tag{19}$$

where f_i denotes the aggregate judgment scale which the model predicts. Denote the partial derivative of $\&_{1(.M)i}$ with respect to q_K by $\&^{(K)}_{1(.M)i}$. Also, let

$$T_{Ki} = \prod_{J=1}^{K-1} Q_{Ji} B_{K(.M)i}\ , \tag{20}$$

where by definition $\prod_{J=1}^{1-1} Q_{Ji} = 1$; and

$$S_{Ki} = \sum_{J=1}^{K-1} q_J f_{Ji} \prod_{L=1}^{J-1} Q_{Li}\ . \tag{21}$$

Then the system of equations in q_K are:

$$\hat{c} = \overline{\&}_{1(.M)} - \hat{k}\overline{f} \tag{22}$$

$$\hat{k} = \sum_i \overline{\overline{f}}_i \overline{\overline{\&}}_{1(.M)i} / \sum_i \overline{\overline{f}}_i^2 \tag{23}$$

$$\hat{q}_1 = \sum_i (\hat{k}\overline{\overline{f}}_i - \overline{\overline{B}}_{1(.M)i})\overline{\overline{D}}_{1(.M)i} / \sum_i \overline{\overline{D}}^2_{1(.M)i} \quad \text{and for } K \neq 1, \tag{24}$$

$$\hat{q}_K = \sum_i (\hat{k}\overline{\overline{f}}_i - \overline{\overline{S}}_{Ki} - \overline{\overline{T}}_{Ki})(\overline{\overline{\&}}^{(K)}_{1(.M)i}) / \sum_i (\overline{\overline{\&}}^{(K)}_{1(.M)i})^2 \tag{25}$$

All equations in q_K are quadratic in q_L for all $L \neq K$. Thus, all equations have directly evaluatable Jacobians and the system may be solved with Newton-Raphson methods. Models which involve variables or attributes grouped at the same hierarchical level may be treated in an obvious way by considering the grouped attributes as composite attributes. For instance, the model

$$kf_i + c = \&_{(12)(34)i} \tag{26}$$

would have three parameters requiring estimation in addition to k and c. H_1 would be associated with the connective linking the pair of terms $\&_{12}$ and $\&_{34}$, while H_2 and H_3 would link f_1 with f_2 and f_3 with f_4, respectively. Further details of such models are presented in ref. 9.

FUZZY IMPLICATION MODELS

Least-squares approaches and the multiple correlation coefficient provide an adequate basis for assessing 1-1 predictive models for judgment data, including the aggregation models reviewed so far. However, they will not do for 'if-then-else' or other 1-many predictions. The basic framework for testing such predictions is presented in this section, with discussion restricted to simple 'if-then' models.

Let $f_{1\to 2i}$ denote the degree of truth in the statement 'if x_i belongs to S_1 then x_i belongs to S_2, where S_1 and S_2 are fuzzy sets. Then this truth-value has two definitions which are widely used in the literature (ref.12). The first is the Arithmetic Rule:

$$f_{1\to 2i} = O(1-f_{1i}, f_{2i}). \tag{27}$$

The second is the Maxmin Rule:

$$f_{1\to 2i} = O(A(f_{1i}, f_{2i}), 1-f_{1i}). \tag{28}$$

As if the embarrassment of riches in having two definitions for fuzzy implication were not enough, any aggregation operators may be used in these definitions. Furthermore, there are several alternative multiple-valued logics with their own definitions for implication (ref.13). Two of the most popular are the Godelian and Ratio Rules:

$$\begin{aligned} f_{1\to 2i} &= 1 \text{ when } f_{1i} \leq f_{2i} \\ &= f_{2i} \text{ otherwise;} \end{aligned} \tag{29}$$

$$\begin{aligned} f_{1\to 2i} &= 1 \text{ when } f_{1i} \leq f_{2i} \\ &= f_{2i}/f_{1i} \text{ otherwise.} \end{aligned} \tag{30}$$

Finally, there is a well-known class of Minkowskian 'distance' models which assign truth-values to observations that decrease in proportion to their distance from a prediction region. Smithson and Knibb (ref. 14) have shown that these models may be subsumed under the Arithmetic Rule by using a pair of pseudo-operators (which are nonassociative):

$$\begin{aligned} A(f_1, f_2) &= 1 - (\max(0, 1-f_1+f_2))^n \text{ and} \\ O(f_1, f_2) &= (\max(0, f_1+f_2-1))^n \text{ for } n \geq 1. \end{aligned} \tag{31}$$

The relevant model for implication is:

$$\begin{aligned} f_{1\to 2i} &= 1 \text{ when } f_{1i} \leq f_{2i} \\ &= 1 - (f_{1i}-f_{2i})^n \text{ otherwise.} \end{aligned} \tag{32}$$

For n=1, the model reduces to the Arithmetic Rule using the bounded sum operator.

The problem before us is to measure the extent to which fuzzy implication models successfully predict judgment data of the form (f_1,f_2). The problem with ordinary measures of predictive success such as the multiple correlation coefficient is that they test only propositions of the form 'x_i is in S_2 iff x_i is in S_1.' A proposition such as 'if x_i is heavy then x_i will be big' is not a 1-1 prediction because it does not predict anything when x_i is not heavy.

However, a general 'proportional reduction of error' framework exists for constructing suitable measures of predictive success for 1-many propositions. Hildebrand et. al. (ref. 15) outlined a paradigm for the development of such measures:

1. Specify a rule 'K' for predicting the state of a dependent variable when the state of the independent variable is known;
2. Specify a rule 'U' for predicting the state of a dependent variable when the state of the independent variable is unknown;
3. Define what constitutes predictive error and how it is to be counted or measured; and
4. Compute a measure of predictive success which has the form $\nabla = (U-K)/U$, where U = predictive error under rule U, and K = predictive error under K.

Given the material presented thus far, the measure of predictive error in step 3 is apparent. It must be the extent to which the proposition 'if x_i is in S_1 then x_i is in S_2' is falsified by the observation (f_{1i},f_{2i}), which is defined by $1 - f_{1\to 2i}$. Henceforth this quantity will be denoted by $e_{1\to 2i}$. Several examples of $e_{1\to 2i}$ are given below.

Arithmetic Rule:

$$e_{1\to 2i} = \min(f_{1i},1-f_{2i}) \quad \text{(min-max)} \tag{33}$$
$$e_{1\to 2i} = f_{1i} - f_{1i}f_{2i} \quad \text{(product)} \tag{34}$$
$$e_{1\to 2i} = (\max(0,f_{1i} - f_{2i}))^n \quad \text{(distance)} \tag{35}$$

Maxmin Rule:

$$e_{1\to 2i} = \min(f_{1i},\max(1-f_{1i},1-f_{2i})) \quad \text{(min-max)} \tag{36}$$
$$e_{1\to 2i} = f_{1i} - f_{2i}f_{1i}^2 \quad \text{(product)} \tag{37}$$

Now, the benchmark prediction Rule U (step 2) conventionally is constructed on the basis of 'best guesses' or chance performance. Here, following ref. 15, we predict the value of f_{2i} on the basis of the marginal probability density of f_{2i}, denoted by p_{2i}. Thus, the joint distribution of (f_{1i},f_{2i}) under the condition of statistical independence would be determined by the product of the marginal probability density functions p_{1i} and p_{2i}, and so the amount of predictive error under Rule U (statistical independence) is measured by

$$U = \int_0^1 \int_0^1 e_{1\to 2i} p_{1i} p_{2i} \mathrm{d}f_1 \mathrm{d}f_2 \tag{38}$$

Rule K predictive success, on the other hand, is determined by the observed joint probability density function p_{12i}, and so the amount of error under Rule K is measured by

$$K = \int_0^1 \int_0^1 e_{1\to 2i} p_{12i} \mathrm{d}f_1 \mathrm{d}f_2 \tag{39}$$

Obviously, the integrals are replaced by summation when the probability density functions are discrete. The proportional reduction of error measure, ∇, may be interpreted as the proportional reduction in error when predicting S_2 by knowing S_1.

That the various ways of measuring error in (33)-(37) as well as in alternative logics yield different values for ∇ may be demonstrated in a simple example. Table 2 displays data from a study of attitudes toward political violence (cf. Muller, ref. 16). Hildebrand et. al. (ref. 15, pp.108-110) analyzed these data using an unfuzzy logic for measuring predictive error. Muller's hypothesis translates approximately into the proposition that 'if a person intends to engage in political violence (IPV), he or she will approve of the use of political violence (APV).' Muller measured both APV and IPV on a Likert-type agreement scale, so we may let f_2 and f_1 measure the extent to which a person is a member of the set of people who agree with APV and IPV, respectively. It is worth noting that two other appropriate translations of Muller's hypothesis are 'IPV is sufficient (but not necessary) to produce APV' or 'APV is necessary (but not sufficient) to produce IPV', depending on which way the researcher wishes to make the prediction.

TABLE 2
Muller's Data: Frequency Crosstabulation

f_2 (APV)	f_1 (IPV) 0.0	0.2	0.4	0.6	0.8	1.0
1.0	2	2	2	4	4	4
0.8	2	4	11	8	3	2
0.6	9	17	8	13	1	0
0.4	43	45	39	3	1	1
0.2	89	75	5	0	0	0
0.0	97	5	0	0	0	0

Table 3 shows the ∇ values obtained for different operators and implication rules. The differences for predictive success as measured by ∇ are wide-ranging, due mainly to the extent that various operators or implication rules penalize observations falling in the upper left-hand triangular region of Table 2, where most of them lie. The bounded sum, distance, Godelian, and Ratio versions of $e_{1\to 2i}$ do not penalize observations in that region at all, whereas the min-max and product operators do under both the Arithmetic and Maxmin Rules.

TABLE 3

Prediction Success Measures for Muller's Data

Operator type	U	K	∇	Π	Γ
Unfuzzy (ref. 15)	102.82	17.00	0.835	0.271	0.227
AR min-max	78.19	62.80	0.197	0.426	0.084
AR product	59.21	39.88	0.326	0.325	0.106
AR bounded sum	33.67	3.60	0.893	0.243	0.217
AR distance (=2)	14.51	0.80	0.945	0.215	0.203
MM min-max	81.97	70.20	0.144	0.447	0.064
MM product	74.06	60.98	0.177	0.370	0.065
Godelian Rule	89.69	12.20	0.863	0.260	0.225
Ratio Rule	79.60	9.60	0.879	0.236	0.208

Note: AR = Arithmetic Rule
MM = Maxmin Rule

Given that ∇ ranges from 0.144 to 0.945 when its range is in the interval from 0 to 1, the choice of logic and operator is nontrivial. Hildebrand et. al. point out that predictive success is not the sole criterion for assessing a 1-many predictive model. Such predictions can vary in their precision or scope, and so predictive precision requires a measure as well. In a modification and extension of the Hildebrand framework, Smithson and Knibb (ref. 14) defined a measure, Π, which is the ratio of U in a 1-many prediction to U^* in its 1-1 maximally stringent counterpart. Thus, U for any 'if-then' prediction is compared with $U^*$for its 'iff' counterpart. One way of combining ∇ and Π in comparing the overall performance of various operators is to define $\Gamma = (\nabla)(\Pi)$, or $(U-K)/U^*$. The right-most columns of Table 3 show Π and Γ, respectively.

Inspection of Table 3 reveals that the high-precision operators yield low values for predictive success, and vice-versa. It is interesting to note that the unfuzzy version compares favorably with all others in its Γ value, although the Godelian Rule and bounded sum models are close. The discussion in ref. 14 reviews four sets of criteria for selecting predictive logics: Substantive, formal, statistical, and predictive. Substantive criteria refer to the match between an operator or logical rule and what the human judge means in a prediction statement. Formal criteria refer to purely logical properties, while statistical criteria refer to bias in estimates of population values for ∇ and Π from sample data. The conclusions in ref. 14 are based on a large Monte-Carlo study. Unfortunately, no operator or logic emerges as clearly superior to others under all conditions, although the bounded sum for the Arithmetic Rule is arguably the best compromise candidate.

GENERAL PREDICTIVE MODELS AND OPTIMIZATION

An extension of the basic predictive framework for assessing the fit between judgment data and more complex predictive judgment models is handled in a fairly obvious fashion via fuzzy logical and possibility theory. Generalized versions of the Arithmetic and Maxmin Rules which model 'if-then-else' propositions with specified alternatives have been extensively studied in the literature (e.g., refs. 12 and 17). Denote truth of 'if x_i is S_1 then S_2, else S_3' by $f_{1\to2\downarrow3i}$

Arithmetic Rule:

$$f_{1\to2\downarrow3i} = A(O(1-f_{1i},f_{2i}),O(f_{1i}\ f_{3i})) \qquad (40)$$

Maxmin Rule:

$$f_{1\to2\downarrow3i} = O(A(f_{1i},f_{2i}),A(1-f_{1i}\ f_{3i})) \qquad (41)$$

Likewise, Mizumoto has provided a generalized version of the Godelian Rule using the Arithmetic Rule format in which Godelian implication replaces O in (40):

$$f_{1\to2\downarrow3i} = A(f_{12i},f_{13i}), \qquad (42)$$

where

f_{12i} = 1 when $f_{2i} \geqq f_{1i}$
= f_{2i} otherwise; and
f_{13i} = 1 when $f_{3i} \geqq 1-f_{1i}$
= f_{3i} otherwise.

The Ratio Rule may be generalized in a similar fashion. In many applications, S_2 and S_3 refer to the same underlying attributes but are different fuzzy sets nonetheless. If that is the case, then the required framework for assessing predictive success uses the same marginal and bivariate probability density functions as outlined in the previous section. Otherwise, U and K are based on the three marginal density functions p_{1i}, p_{2i}, p_{3i}, and the joint density function p_{123i}; and (38) and (39) are extended in the obvious way.

It is not difficult to construct procedures for optimization which permit a linear (or other) transformation of one of the membership functions (f_1, say) while maximizing ∇ or Π by selecting values for the transformation parameters. In the simplest case, of fuzzy implication with no specified alternative ('if-then') and a linear transformation $af_{1i}+ b$, we have a (non)linear programming problem requiring requiring the constraints

$$a > 0,$$
$$a\min(f_{1i}) + b \geqq 0, \text{ and} \qquad (43)$$
$$a\max(f_{1i}) + b \leqq 1.$$

One may also wish to impose the constraint $b \leq c$, for some $c > 0$, to avoid those solutions where the value of a comes arbitrarily close to 0. It is easy to show that this region always is convex. It is noteworthy that the min-max, bounded sum, and distance operators require only the constraints $b \leq c$ and $a \leq d$, where c and d are in the interval between 0 and 1.

Since U-K is proportional to Γ in these optimization problems (because U^* is held constant if only f_1 is transformed), it may serve as the function to be maximized. The Arithmetic Rule product operator model is the only case which yields a linear function in a and b:

$$U\text{-}K = a\int_0^1\int_0^1 p_{1i}p_{2i}V_{12i}f_{1i}(1-f_{2i})\,df_{1i}\,df_{2i} + b\int_0^1\int_0^1 p_{1i}p_{2i}V_{12i}(1-f_{2i})\,df_{1i}\,df_{2i} \quad (44)$$

where $V_{12i} = 1 - (p_{12i}/(p_{1i}p_{2i}))$.

All other models require nonlinear (direct-search) approaches.

FURTHER POSSIBILITIES

This paper has presented approaches for solving two common kinds of problem in fitting fuzzy logical models to human judgment data. These techniques should find application in the design and evaluation of any man-machine system whose performance relies on a model of human judgment and information processing. Empirical evaluations of these systems and of how well fuzzy set theory models human reasoning and knowledge still are rather rare, and I hope this work enables such evaluations to be more routinely carried out.

A major objective in constructing the techniques presented here has been to render them as independent of specific aggregation operators as possible, so as to maximize user and designer choice. In the single-aggregation and simple implication models, this aim has largely been realized. However, further work remains to be done on multiple-aggregation and complex implication models. Both are rather intractable, and may require 'brute-force' numerical methods for a generalizable approach. Likewise, the dilemmas posed by multiple fuzzy set operators and logics are still very much with us, although comparative empirical evaluations will go a considerable way towards empowering designers and researchers to make informed choices.

REFERENCES

1 E. Trillas, C. Alsina, and L. Valverde, Do we need max, min, and 1-j in fuzzy set theory? in R.R. Yager (Ed.) Fuzzy Set and Possibility Theory: Recent Advances, Pergamon, New York, 1982, pp.275-297.
2 S. Weber, Fuzzy Sets and Systems, 11 (1983) 115-134.
3 R. Bellman and M. Giertz, Information Sciences, 5 (1973) 149-156.

4 H. Hamacher, On logical connectives of fuzzy statements and their affiliated truth functions, Proceedings of the Third European Meeting on Cybernetics and Systems Research, Vienna, 1976.
5 M.J. Frank, Aequationes Mathematicae, 19 (1979) 194-226.
6 R.R. Yager, Fuzzy Sets and Systems,4 (1980) 235-242.
7 D. Dubois and H. Prade, Fuzzy Sets and Systems: Theory and Applications, Wiley, New York, 1980.
8 H.J. Zimmermann and P. Zysno, Fuzzy Sets and Systems,4 (1980) 37-51.
9 M. Smithson, Mathematical Social Sciences,7 (1984) 231-251.
10 U. Thole, H.J. Zimmermann, and P. Zysno, Fuzzy Sets and Systems,2 (1979) 167-180.
11 M. Smithson, Behavior Research Methods, Instruments, and Computers, 16 (1984) 553-556.
12 L. A. Zadeh, in L.A. Zadeh et. al. (Eds.) Fuzzy Sets and their Applications to Cognitive and Decision Processes, Academic Press, New York, 1975, pp.1-39.
13 N. Rescher, Many-Valued Logic, McGraw-Hill, New York, 1969.
14 M. Smithson and K. Knibb, Mathematical Social Sciences (in press).
15 D. K. Hildebrand, J.D. Laing, and H. Rosenthal, Prediction Analysis of Cross Classifications, Wiley, New York, 1977.
16 E.N. Muller, American Political Science Review, 66 (1972) 928-959.
17 M. Mizumoto, in R.R. Yager (Ed.) Fuzzy Set and Possibility Theory: Recent Advances, Pergamon, New York, 1982, pp.211-223.

Applications of Fuzzy Set Theory in Human Factors, edited by W. Karwowski and A. Mital
Elsevier Science Publishers B.V., Amsterdam, 1986 — Printed in The Netherlands

A Fuzzy Model for the Accumulation of Judgments by Human Experts

Alf C. ZIMMER
University of Regensburg, Universitätsstraße 31, 8400 Regensburg, W-Germany

ABSTRACT

A formal model for the justification and checking of arguments is proposed for the development of expert systems. The two major parts of the model consist in the detection of violations of the rules governing rational discourse and in the evaluation of conclusions in fuzzy syllogistic reasoning.

The application of the model consists in the interaction between the expert ('arguer') and the formal model. In the beginning the train of arguments is checked for rule violations and the expert ('arguer') is asked to mend his/her argument accordingly. Then the acceptable train of arguments, that is, a syntactically valid chain of propositions is evaluated according to its fuzziness. Upper and lower bounds for the acceptability of the train of arguments are calculated and fed back to the expert ('arguer') in either numerical or verbal form. After that the expert decides either to choose a level of acceptability in the given bounds or suggests an expansion of the justification. In the latter case the evaluation in both steps starts anew.

First experimental results in predictions for everyday events indicate that the model is viable, fits into the subjects' cognitions, and helps the subjects to arrive at trains of arguments which are more acceptable for them than their initial justifications.

INTRODUCTION

During the last 20 years the role of the human operator in technical settings has changed from immediate manipulation towards a more and more remote monitoring of the equipment and its behavior. Due to this fact, the task of the operator has become more abstract and the required skills pertain mostly to correct decision making, especially in critical situations for which no standard procedures exist (the TMI incident is a typical example for failures in decision making in such a control situation, for other examples see Reason & Mycielska, 1982).

In order to enable the operator to decide correctly even in situations not encountered before or not trained for, it is necessary that he or she has an internal representation, or more specifically, a model (see Johnson-Laird, 1980) of the equipment and process to supervise. There a two major requirement for such a model: (i) it has to be designed in a way consistent with the information-processing capacity and style of the operator, and (ii) it has to contain all states of the equipment and the relations between them that are essential for the operator to plan the necessary interventions ahead.

The results of cognitive psychology suggest that such a model has to be qualitative rather than quantitative not only in regard to the states but also in regard to the relations, that is, instead of a physically correct description in terms of differential equations descriptions of the following kind are to be embodied in a mental model: "a rapid increase of X in A normally causes a drop of pressure in B unless C is in state Y."

Forbus & Stevens (1981) have demonstrated that even for comparatively simple equipments and processes (e.g. steam engines for USN ships) a model in qualitative terms is not only sufficient for the operator to generate a complete control structure for the process, but also that it leads to less errors than an instruction outline based on the set of differential equations underlying the process. This and other results have fostered the development of 'qualitative physics' (for an overview see Gentner & Stevens, 1983) as a means for the development of tutoring systems tailored to the control structure of a given task. In experiments on probability assessment and prediction Zimmer (1983, 1984) has shown that for most people qualitative expressions for quantifiable events have not only a comparatively stable meaning but are also less prone to interference than numerical expressions. Furthermore, for people (even experts) it is easier to combine qualitative information from different sources together with their relations than to integrate quantitative data about components of a system which are functionally related.

There are three main conclusions to be drawn from the indicated evidence about the human operator in control and supervision tasks: (i) the desription of the system in question and the tutoring of the controller have to concentrate on the states of the system and their relations which are relevant for making decisions, (ii) these descriptions ought to be qualitative because in most cases the crucial decisions are about intervening or not intervening; and even the decisions about how to intervene are normally qualitative, too, (iii) the control panels have to be designed in such a way that the location of the instruments and controls are in accordance with the qualitative model.

PRINCIPLES FOR THE DEVELOPMENT OF INTERNAL MODELS

Unfortunately, there is no simple algorithm to translate quantitative into qualitative descriptions and even if this translation has been made it cannot be guaranteed that it fits the information-processing capacity of the human operator and enables him or her to develop a control structure that is simple and at the same time comprehensive enough to plan an intervention in a novel situation. For these reasons the way to come to a functional qualitative description is to find out what kind of internal models experts have and how they derive control decisions from this kind of representation. On the first glance, simply asking experts what they do, for instance, in landing a plane smoothely and safely, seems to be the appropriate approach. However, there are a couple of problems connected with this method. The main problem is the pre-knowledge (correct or incorrect) of the interviewer about the equipment and process in question; not only that it influences the kind of questions posed, furthermore the sequence and the content of the questions can lead to a biased report by the expert, that is, he or she conforms unwittingly to the implicit presuppositions contained in the questions. Of equal importance is that a human interviewer processes the obtained information only partially, usually in dependence on his or her implicit model, and is therefore not able to assess the consitency of a report adequately or to contrast the reports by different experts.

These shortcomings of the traditional approach even if it is centered on the 'critical incidences' of a process (see Flanagan, 1954) have been the motivation to devise a computer-aided interactive system for the elicitation of expert knowledge. This system is aimed at two major and interdependent goals:

(i) to help the expert to arrive at a description that is consistent and complete (as viewed by the expert)

(ii) to render different reports comparable.

If the second goal is attained, it is possible to use already obtained reports for the derivation of novel questions and to decide if there is one unique representation for the process or if there are multiple and possibly disparate representations.

In order to devise such an interactive system it is necessary to make a-priori assumptions about the format of the experts' knowledge. The gist of the assumptions made is that technical and scientific knowledge can be described as consisting of chains of arguments which are tied together by the rules of proposing, questioning, and defending these arguments. Furthermore it is assumed that such arguments differ according to the degree of belief attached to them and according to the importance they have for the overall goal of argumentation (e.g. are they necessary preconditions for further arguments or do they only together with other arguments support a certain chain of argumentation;). There is sufficient empirical support for these assumptions (see Zimmer, 1985, 1986) which underly Toulmin, Rieke & Janik's (1979) treatise on arguments, to envisage them as a viable starting point for a human-computer interaction in the elicitation of expert knowledge.

Toulmin, Rieke, and Janik (1979) assume that the statements of experts (e.g. claims, diagnoses, predictions) together with the given justifications (usually reasons in a given context) can be viewed as arguments in a rational discourse. The analysis of arguments can therefore be used to elucidate the knowledge and rule bases underlying experts' judgements, to check the consistency of their claims and justifications, and finally to derive analytically evaluations of the claims in terms of their probabilities or credibilities.

If such an approach proves to be viable it provides a novel instrument for the accumulation of judgements by different experts. Novel insofar as the consensus opionion is derived from the matching of the experts' knowledge bases underlying their claims and not, as in traditional approaches (e.g. Kim & Roush (1980)), from the experts' global subjective probabilities. However, the knowledge bases of different experts can only be matched if they have been made explicit, checked for their internal consistency, and fed back to the individual experts. This is the only way to make sure that the experts accept the knowledge base analytically derived from their arguments as representative for their initial or revised belief system. In the following, an approach is sketched in which the arguments of experts are interactively analyzed for their structure by a decision-aiding system. The output of this process is an evaluation of the original or the revised claim together with the supporting evidence and the rebuttal of alternative arguments.

THE STRUCTURE OF ARGUMENTS

According to Toulmin, Rieke, and Janik (1979) an argument consists of the following parts:

(i) <u>claims</u> (in the model suggested here, claims are usually predictions)

(ii) <u>grounds</u> or reasons for believing the claims to be valid (the usual form is that of explicitly or implicitly quantified statements)

(iii) <u>warrants</u>, that is, statements about the relations between grounds and claims (e.g. causality, necessity, sufficiency, contingency)

(iv) <u>backing</u>, that is, commonly shared knowledge (Smith, 1982) which provides the rules for a combination of grounds and warrants in order to justify the claims (e.g. rules of syllogistic reasoning, or statistical inference)

(v) <u>modal qualifications</u> (e.g. possibly, usually, necessarily) which apply to to the propositions and to the inferential process (e.g. rules of fuzzy reasoning). The modal quantification of the ground together with the fuzzy truth values resulting from the inferential process determine the modal quantification of the claim.

(vi) <u>rebuttals</u> that is, alternative claims which can be inferred from the grounds, warrants, and the backing, too, because of the modal quantification (fuzziness) of propositions and inferential rules. Rebuttals can be overcome by either showing that they imply a smaller set of consistent propositions than the claim or by comparing the overall modal qualification of the rebuttals with the evaluation of the claims.

If these parts have been identified in an expert's argument and if the credibilities of subjective probabilities of the parts have been determined, it is possible to derive analytically the credibility of the totel claim. This identification process is done interactively with the decision-aiding system.

THE INTERACTIVE ANALYSIS OF ARGUMENTS

In situations for which the suggested model has been investigated, experts are asked for statements concerning a certain state of affairs (e.g. "What kind of critical incidents do you expect in the recently introduced USN steam engines?" or "How can accidents like the one at TMI be prevented?") However, since the described model had been initially developed in the framework of economic forecasting the introductory demonstration is taken from this domain, that is, the question "What will be the further development in the money market?" *) The answers usually consist in a claim (e.g. the prediction "The US Dollar will become even stronger until the election in the US. Afterwards it will remain on a high level if Reagan is reelected, but it will considerably drop if the Mondale ticket wins.") quite often accompanied by a list of grounds (e.g. "Until election day the prime rate will tend to increase." "In the case of Reagan's reelection the budget will be balanced" etc.). Using the technique described in Zimmer (1983) the intended meaning of

*) for details see Zimmer (1984 a). The analysis of expert argumentation about forecasting have been made in summer 1984.

terms like 'become stronger' or 'balanced' are analyzed resulting in elastic constraints (e.g. "becomes even stronger" means "increases in value at least as fast as during the last quarter and at most twice as fast"). These elastic constraints are fed back to the subjects either numerically or verbally. After the subjects have accepted the analytically derived fuzzy constraints as fitting their intuitive evaluations, the grounds for the given claim are checked by questions like "How does Reagan's reelection influence the exchange rate for the dollar?"

These questions are usually answered by fairly general statements (e.g. "He causes optimism in the economy") which in turn are scrutinized by 'how' questions until the expert is as specific as he or she can be. The final grounds (e.g. "He will reduce taxation according to the Shaffer model") together with the warrants (e.g. "Reduced taxation makes higher investments possible. Higher investments lead to a better productivity. High productivity combined with low taxation influences positively the GNP. If the GNP in the US is higher than in other countries, there will be an influx into the dollar market. That keeps the dollar strong"). In the given train of arguments the words indicating causative relations (e.g. 'influences') are underlined with broken lines. solid lines mark symmetric relations. Furthermore references like 'that' in the last line of the argument indicate equating relations which can be paraphrased by 'the essence of the given arguments is equivalent to the prediction "the dollar remains strong"'.

It is possible to discriminate three major types of valid causal argumentation: (i) a simple chain, (ii) a chain with alternative parallel paths, and (ii) a tree-like structure with multiple necessary but not sufficent causes. In normal discourse combinations of all three types occur. The most frequent invalid type of causal argumentation is characterized by loops, that is, antecedents are influenced by their consequences.

In describing and analyzing argumentative chains we use the term 'causative influence' for all relations indicating that a certain event (state or action) contributes to the final or an intermediate event. In accordance with the above mentioned valid types of causal argumentation there are two forms of causative influences: (i) causal influences, that is, the occurrence of event U is necessary for the occurrence of event V and (ii) moderating influences, that is, if a certain event W occurs or a disposition G is given, the relation between X and Y is modified. For instance, only if in a landing plane (X), the position of the wings is low (disposition G) the so-called surface effect (Y) occurs immediately before touch-down. This example stems from an investigation into the correspondence between the model underlying the knowledge of flight instructors and the models students have about the behavior of the plane used for training in the German Air Force (a Piaggio Trainer). One important difference is that the instructors know about the moderating influence of the wing position (G), whereas students regard the surface effect (Y) as a necessary consequence of slowing down the speed and approaching the ground (X). This misrepresantation leads to accidents or near-accidents if the students change to a high-wing plane (e.g. a DO 27 or a Cessna 152).

In the framework of fuzzy reasoning interpretations for the relations have been developed which allow to analyze the strength of the argument. The fuzzy constraints on the relations are evaluated and these local evaluations are integrated into a global judgment (see Figure 1 for some examples).

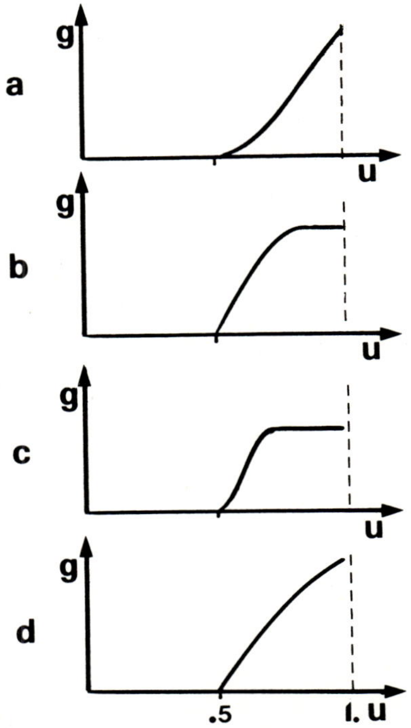

Figure 1 Influence functions; "g" is the weight for a quantified statement combined with "f" by means of a fuzzy multiplication or a similar operation, see Zimmer, 1984a. a) positive influence of X on Y, b) strong positive influence of X on Y, c) causal influence of X on Y, d) is a sufficient condition for Y. "U" is the 'Universe of Discourse'.

A special kind of relations are those with explicit (e.g. "Often X is influenced by Y") or implicit quantification (e.g. "Xs prevail if Y"). If the quantification and the meaning of the quantifier are unambiguous, the quantifier is represented by a fuzzy number in the interval [0, 100 %] giving the expected proportion of positive incidences of X (see Zadeh, 1984; Zimmer, 1984a,b). In the case of ambiguous quantifiers (e.g. usually) the meaning is interactively rendered so precise that the quantifier can be characterized by a unique fuzzy number (see Figure 2).

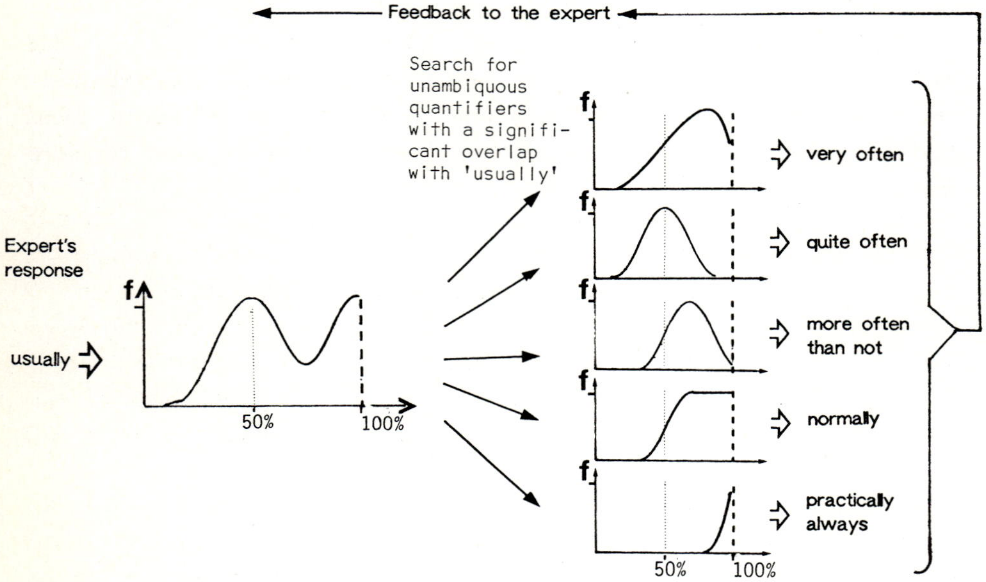

Figure 2 The disambiguation process for the quantifier "usually". The alternatives on the right hand are chosen because their possibility functions are single peaked and they are either entailed in the possibility function of "usually" or their intersections with it are large.

These local evaluations and the propagation of these interpretations to the final evaluation is comparable to the analysis of the backing in Toulmin's schema for arguments. In addition to the structural analysis of grounds and their interconnections it is checked how and if the reasons given are backed by factual knowledge (e.g. "On which observable instances do you base the statement 'X'?") and how reliable and generalizable these supporting facts are. For the final assessment of the argument's credibility the given ratings for the reliability of factual knowledge are integrated as fuzzy truth values modified by the relations between the grounds (see Zadeh, 1984; Zimmer 1984 b).

ELICITATION OF COUNTER ARGUMENTS

In normal discourse the strength of an argument does not only depend on the credibility of the facts underlying the given grounds and on the strength of the connection between the grounds and the claim but also on possibility of a rebuttal of plausible counterarguments.

In many claims counterarguments are explicitly referred to (e.g. "I predict X. One could also argue that Y. However, A cancels out Y as a possible course of development"). In other cases the absence of a certain ground is stated (e.g. "I predict that X, because A, B, C are given and D is not the case"). From a language pragmatic point of view giving negative grounds is not very informative and therefore this way of bringing forward arguments is unlikely to occur. However, if this is interpreted as an indirect refutation of a counterargument (e.g. "D is necessary for Y. D is not the case, therefore prediction Y can be cancelled out"), then stating the absence of facts or grounds makes sense. In the model the implied counterarguments are elicited by conversing the negative statement into the respective positive form and then asking the subject: "If D were given, what prediction would you make?"

The experimental evidence from the applications of the model indicates that in argumentative discourse the normal form of counterarguments is the contrary of the arguments and not its contradiction. From a logical point of view this seems curious. However, there are good reasons for this strategy from a language pragmatic point of view (Grice, 1975):

(i) Contraries preserve the context (universe of discourse). For instance, the counterargument for "A is small" is "A is large". Both arguments refer explicitly to size as the universe of discourse.

(ii) Contraries are usually more informative than contradictions. The negation of "A is small" can have the following meanings:
(a) size is not an applicable predicate for A, (b) A is less than small (minute) or more than small (moderately sized, large, huge, etc.).

(iii) If the linguistic labels in the universe of discourse are assumed to be those of natural language, then their meaning is vague or fuzzy. One property of fuzzy contradiction, however, is that the law of the excluded middle does not apply (Osherson & Smith, 1981). Therefore, the original argument and its contradictory counterargument have an overlap in meaning (see Figure 3), that is, there are instances which are in favor of the argument as well as its contradiction.

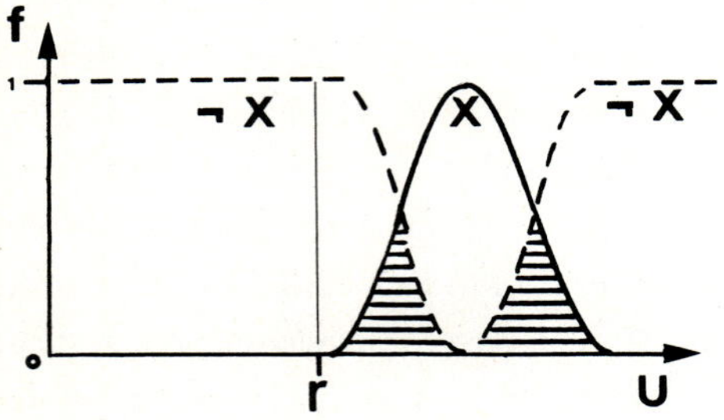

Figure 3 The membership function for X and for not X. The shaded area indicates the overlap in meaning of X and not X. Note that the reference point does not influence the characteristic of the counterargument.

This ambiguity of counterarguments vanishes if the contrary is chosen instead of the contradiction (see Figure 4).

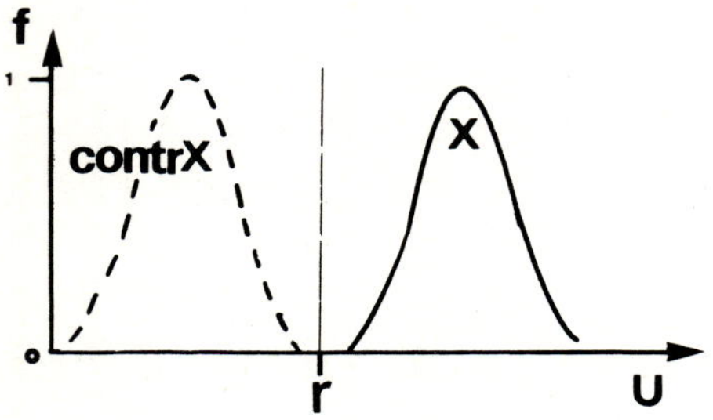

Figure 4 The membershipfunctions for X and the contrary of X. The position of contr(X) is determined by reflection of the membership function of X at the point of reference.

If counterarguments are neither stated explicitly nor implicitly, the experts are confronted with the contraries of circumstantial reasons they have given for their arguments (e.g. "If the Mondale/Ferraro tickets wins (the contrary of 'the Reagan/Bush ticket wins') what kind of development for the dollar would you predict?")

GLOBAL EVALUATION OF ARGUMENTS

After all the components of the argument have been elicited and after the meaning of the used predicates and the credibility of the ground, the warrant, the backing, and of rebuttals have been determined, the credibility of the claim is analytically derived from these values by means of fuzzy syllogistic reasoning (Zadeh, 1984; Zimmer, 1984 b). The credibility values are either transformed into labels (e.g. 'likely', 'unprobable' etc., see Zimmer, 1983) or are given in their original numerical form. If the subjective evaluation of the claim by the expert and the analytical evaluation are about the same, the interactive process ends. If, however, the expert disagrees, he or she is asked to give further grounds or to revise the credibility ratings for the facts given.

CONCLUSION

The interactive model for the elucidation of arguments underlying the claims (e.g. predictions, diagnoses) of experts on the one hand serves as a means for an unbiased probability assessment for claims. Insofar it resembles the procedure proposed by Henrion and Morgan (see Morgan, in press) and the investigation of Shafer & Tversky (1985) into subjective probabilities as formal languages for the expression of evidence and degrees of belief. On the other hand, however, it makes explicit the knowledge base on which the expert grounds his/her claim. The comparison of the knowledge bases underlying the predictions of different experts for the same event shows if these predictions are based on more or less same reasons or not. In the first case an accumulation of the evaluations made by different experts is admissible. In the other case, however, only those judgments can be pooled which are based on comparable knowledge bases.

REFERENCES

1 J.C. Flanagan, Psychological Bulletin, Bd. 51, 1954, pp. 327-355.
2 K. Forbus & A. Stevens, Project Steamer: III. Using qualitative simulation to generate explanations of how to operate complex physical devices, NPRDC TN 81-25, 1981.
3 D. Gentner & A. Stevens (Eds.) Mental models, L. Erlbaum, 1983, Hillsdale.
4 H.P. Grice, in: P. Cole and J.L. Morgan J.L. (Eds.), Syntax and Semantics, Academic Press, New York, 1975, pp.
5 P.N. Johnson-Laird, Cognitive Science,4 (1980) pp. 71-115.
6 K.H. Kim and F.W. Roush, Introduction to Mathematical Consensus Theory, Marcel Dekker Inc., New York, 1980, pp. 179.
7 G. Morgan, in: R. Uppuluri (Ed.) (ed.) Technology Assessment, North Holland, Amsterdam, in press.
8 D.N. Osherson and E.E. Smith, Cognition, 4, (1980) pp. 35-58.
9 J. Reason & K. Mycielska, Absent Minded? Prentice Hall, Englewood Cliffs, pp. 263.
10 G. Shafer & A. Tversky, Cognitive Science, 9, pp. 309-340.
11 S. Toulmin, R. Rieke and A. Janik (Eds.), An introduction to reasoning, Macmillan Publishing Co., Inc., New York, 1979, pp.
12 L.A. Zadeh, Syllogistic Reasoning in Fuzzy Logic and Its Application to Reasoning with Dispositions, UCB/EECS, Berkeley, Memorandum.
13 A. Zimmer, in: R.W. Scholz (Ed.), Decision Making Under Undercertainty, North-Holland, Amsterdam, 1983, pp. 159-182.
14 A. Zimmer, Int. Journal Man-Machine Studies ,20, 1984, pp. 121-134.
15 A. Zimmer, Annual meeting of the Psychometric Society, 1984, Santa Barbara.

Applications of Fuzzy Set Theory in Human Factors, edited by W. Karwowski and A. Mital
Elsevier Science Publishers B.V., Amsterdam, 1986 — Printed in The Netherlands

THE HUMAN OPERATOR: RELIABILITY AND LANGUAGE OF ACTIONS ANALYSIS

M. NOWAKOWSKI

Institute of Production Engineering and Management, Technical University of Wrocław, Wyb. Wyspiańskiego 27, 50-370 Wrocław (Poland)

ABSTRACT

The action of a human operator has been discussed using the linguistic theory of actions. The approach presented here is based on an assumption that human performance may be subjected to different constraints. Possibility measure is applied to numerical representation of human possibilities of actions. In some cases the evaluation of the possibilities of actions may be performed according to the theory of belief functions.

INTRODUCTION

Formal theories of actions present an opportunity to create a system in which it will be possible to formally present and describe a set of notions valid for description and analysis of a human operator's actions. This will be useful in design and organization. It also concerns the actions of a human operator, but more generally a man-machine system or a man-machine-environment system.

The action of a human operator will be discussed from the the linguistic theory of actions approach which was introduced by Maria Nowakowska (1973). This approach allows the grasping of the following aspects of actions:

- a structure of an action situation and its attitude toward limitations imposed on actions admissibility;
- limitations imposed on sets of possible choices by the features of the decision maker;
- relations between a goal structure and its achieveability through different choices;
- choice criteria from a set of possible alternatives;
- sequential character of actions.

The approach to a human operator action analysis presented here is based on an assumption that human performance may be subjected to different constraints (restrictions). Some limits are imposed by the situations, others are imposed by the psychophysical state of an operator, his capabilities, or grade of fatique. Additional constraints may result from his abilities, skills, motivations and so on. All these factors determine the level of a human operator's possibilities of actions at a given time.

When assuming that a man acts according to possibilities, understood in the above way, the present possibilities of actions and the level of human reliability may be related to one another. Numerical characterization of a possibility level is difficult to obtain. Hence we deal at the same time with uncertainty and imprecision. Therefore the application of the theory of fuzzy sets, theory of possibility and fuzzy measures becomes necessary.

SYNTAX AND SEMANTICS OF ACTIONS - PRIMARY CONCEPTS

A formal system of the linguistic theory of actions in question has been based on six primary concepts (Nowakowska, 1973):

$$\langle D, \#, \varphi, L, S, R\rangle . \tag{1}$$

An object described in terms of the system (1) is an arbitrary action situation, i.e. situation where a given person may perform one action from a number of actions.

D - is a set of elementary actions, i.e. out of which all considered actions can be built. The elementary actions may be decomposed further on. The degree of "atomization" of an action is a consequence of comfort of description and the aim of the analysis of an action situation which is carried out.

- is a distinguished element of the set D, interpreted as a "pause". Symbol # denotes any action which is not interesting from the view point of the situation being described, and thus it doesn't necessitate refraining from any other activity,

φ - helps to determine the time aspect: $\varphi(d)$ means quantity of time (expressed in appropriate units) necessary for performing action d. The same action may be performed at a different speed and it may be treated as different elements of the set D; $\varphi(\#) = 1$ by convention.

The interpretation of primary concept L may be fixed in a number of ways. The class D^{*} will be determined as a class of finite sequence built out of the elements of D (i.e. sequences in the form of $d_1, d_2, \ldots, d_n$, where $d_i \in D$, $i = 1,n$). An empty sequence e will be additionaly included in D^*. The elements of D^* may be connected by the concatenation operation: if $u = d_1 d_2 \dots d_n$ and $v = d'_1 d'_2 \dots d'_m$, then uv means a sequence $d_1 d_2 \dots d_n d'_1 d'_2 \dots d'_m$. The function φ on D^* will be defined as follows:

$$\varphi(e) = 0,$$
$$\varphi(d_1 d_2 \dots d_n) = \varphi(d_1) + \varphi(d_2) + \dots + \varphi(d_n). \qquad (2)$$

Formally, $L \subset D^*$ is a class of sequences of actions (i.e. a set of some complex actions). The formal structure is connected with the following linguistic analogy. D may be interpreted as "an alphabet", or "a dictionary" and L as "a language" (i.e. set of "grammatically correct" sequences). This analogy embraces all possible interpretations of the set L labelled later as the "language of actions". The description of L is representative of "action grammar" (i.e. generally speaking, a collection of rules characterizing sequences belonging to L). Thus we may assume, that L consists of all sequences of actions which are feasible or compatibile with psychophysical state of an individual, his personality, ethics and so on.

The first four primary notions $D, \#, \varphi$, L describe structural aspects of an action situation. Therefore by using the linguistic analogy we might talk about "action syntax". R and S are used in description of the results of actions, thus we might refer to them as "action semantics"

S - a set of possible results of actions: the elements of S may have different interpretations with respect to definite context of application.

R - a relation between sequences from L and results from S and moments of their succession.

If the set $N = \{0,1,\dots\}$ is a set of moments of time, then the $S \times N$ is the set of pairs (s, n) called time-results, or for the sake of simplicity, results. The pair (s, n) is interpreted as the sucession of the result s at the moment n.

The relation R binds the sequence of action with the results. Thus $u\,R(s,n)$ means that performing sequence u causes the result s at the moment n, starting from the initiating moment of sequence u. The sequence of actions may lead to many results and the same results also be caused by different sequences. Therefore the following may be stated:

$$R^{-1}(s,n) = \{u \in L : u R(s,n)\}, \tag{3}$$

that is, a set of all sequences of actions leading to result (s,n). The result (s,n) is achievable when $R^{-1}(s,n)$ is a non-empty set (i.e. $R^{-1}(s,n) \neq \phi$).

Usually, one concerns attainment, of a goal expressed as a configuration of results of the (s,n) type. For example, an alternative $(s,n) \vee (s,'n')$ and conjuction $(s,n) \wedge (s,'n')$ can be interpreted as the occurrence of s at a moment n and/or s′ at the moment n′ and a negation $\sim(s,n)$ can be interpreted as. A set of all sequences of actions leading to an alternative $(s,n) \vee (s,'n')$ is given by the $R^{-1}(s,n) \cup R^{-1}(s,'n')$ and the set of sequences leading to conjuction $(s,n) \wedge (s,'n')$ is the intersection $R^{-1}(s,n) \cap R^{-1}(s,'n')$. Actions which do not attain (s,n) are determined by $L \setminus R^{-1}(s,n)$.

Complex goals G may be configurations of the single results of (s,n) type and appropriate attainability sets are determined by operations of union, intersection and complement on sets of attainability for the single results (s,n).

For a given set $Q \subset S \times N$ we may define

$$R_w^{-1}(Q) = \bigcup_{(s,n) \in Q} R^{-1}(s,n), \tag{4}$$

and

$$R_s^{-1}(Q) = \bigcap_{(s,n) \in Q} R^{-1}(s,n). \tag{5}$$

The $R_w^{-1}(Q)$ is a set of all those sequences of actions, which lead to at least one result $(s,n) \in Q$. The $R_s^{-1}(Q)$ contains those sequences of actions which lead to all $(s,n) \in Q$. The goal Q is achievable in the weak sense if $R_w^{-1}(Q)$ is non-empty, and in a strong sense if $R_s^{-1}(Q)$ is non-empty.

At this point only the fundamental notions of the linguistic theory of actions have been presented. They are of importance in presenting the propositions connected with the discussion. The theory itself is more comprehensive (see, Nowakowska, 1973, 1980).

In Nowakowski (1980) some concepts were modified by introducing notions from the theory of fuzzy sets and possibility theory. The notion of a fuzzy language of actions and non-deterministic relation R were introduced there. A non-sharp formulating of goals was also discussed and goals achievability was invesigated. In Nowakowski (1981) the analysis was broadered by application of methods for solving fuzzy relations equations.

In the present work the main task is the analysis of a language of action or human possibilities of actions resulting from the extisting limitations of a human operator. When assuming that the relation which binds actions with results is deterministic, such analysis concerns achievability of given goals. Thus the discussion is connected with human reliability. One approach which may be taken is a numerical representation of human possibilities.

FEASIBILITY OF ACTIONS AND FUZZY LANGUAGE OF ACTIONS

For the sake of this discussion assume that for the analysed situation it is possible to determine a language of actions (L) which contains all correct actions (i.e. fuzzy admisible sequences of actions). Let L be an ideal language of actions such that it contains actions free from faults and errors. Therefore if the human operator always acted accordance with this language he would always act reliably.

The analysed situation may be determined in various ways and the desired, ideal language of action may be a broader or a narrower class of actions. One may also be concerned with the language of stimulus, signal, information state and so on.

In a real situation a human operator acts according to a language of actions, which is more or less distant from the ideal language of actions L. We assume that the real language of actions is a result of all factors limiting the human operator, which have been mentioned before. These factors form the so-called "fuzzy grammar" of a real language of action. This language is subject to individual differences as well as differences in time intervals.

Let's assume that the real language of actions L_t (index t may describe a moment or a person) is a "fuzzy version" or a result of "fuzzification" of an ideal language L i.e.

$$L \underset{t}{\rightsquigarrow} L_t, \qquad (6)$$

where $\underset{t}{\rightsquigarrow}$ is a "fuzzy grammar" determined by a set of limitations. Formally the fuzzy language of actions L_t is defined by the membership function:

$$f_t : L \rightarrow [0, 1] \; , \qquad (7)$$

where $f_t(v)$ is a grade up to which the action v is admissible, grammatical and feasible. The fuzzy language of actions L_t may be investigated at different levels of feasibility, admissibility α. The language of actions at the level of feasibility α denoted as $L_t(\alpha)$ will be defined by:

$$L_t(\alpha) = \{ u : f_t(u) \geq \alpha \}. \tag{8}$$

$L_t(\alpha)$ is a class of those sequences of actions which are feasible in at least the α grade.

In this class a special language may be designated. It may be called the language with positive feasibility denoted as L_t^+ and defined by:

$$L_t^+ = \{ u : f_t(u) > 0 \}. \tag{9}$$

L_t^+ consists of actions where the feasibility is greater than 0.

When analysing the function of limitations which form the real language of actions, the class of actions belonging to the ideal language of actions, which is unfeasible at any given moment, becomes very interesting.

The class denoted as L_t is determined by:

$$L_t^- = L \setminus L_t^+ . \tag{10}$$

POSSIBILITIES OF ACTIONS AND PHYSICAL VIEW OF POSSIBILITY MEASURE

The expression "possibilities of actions" has been used previously in the present work. The idea of a fuzzy language of actions is based on reflecting "possibilities of actions" understood as an ability to perform or the case of attaining an action with respect to a number of restrictions. It is quite natural to associate "possibility of action" with the theory of possibility (Zadeh, 1978) in order to determine possibilities of action numerically. There arises a question as to whether the measures of possibility (i.e. its formal properties), capture intuitions connected with evaluation of action possibilities.

The possibility measure Π determined on a set of elementary actions $D = \{d_1, d_2, \ldots, d_n, \#\}$ satisfies the following axioms

$$\Pi(D) = 1; \quad \Pi(\phi) = 0, \tag{11}$$

where D is a sure event and ϕ an impossible event. Fulfillment of the axiom $\Pi(\phi) = 0$ does not require further explanation. Also $\Pi(D) = 1$ is acceptable, because it means that there is an elementary action with possibility of performance equally 1. In a particular case we may assume that the possibility of performing any of the "relevant" actions (i.e. $d_1, d_2, \ldots, d_n$) or obstaining from action (performing $\#$) is also 1. In other words,

$$\max\left[\Pi(\#),\ \Pi(\{d_1, d_2, \ldots, d_n\})\right] = 1. \tag{12}$$

The next axiom is as follows

$$\forall A, B \subset \check{D} \quad \Pi(A \cup B) = \max(\Pi(A), \Pi(B)), \tag{13}$$

or in terms of elementary actions

$$\forall d_i, d_j \in \check{D} \quad \Pi(d_i \cup d_j) = \max(\Pi(d_i), \Pi(d_j)). \tag{14}$$

In the formula $\Pi(d_i)$ is short for $\Pi(\{d_i\})$.

The property (14) agrees with the intuition concerning possibilities of action in the above presented interpretation because the realization $d_i \cup d_j$ becomes easier to attain (perform). It may be stated, that the possibility measure in its physical interpretation is consistent with properties of possibilities of actions and may be used for numerical representation of physical possibilities. Thus the possibility meassure determined on the set of elementary actions D, will determine "elementary possibilities of actions", where $\Pi(d_i), \forall d_i \in D$ represent possibility of performing an action d_i understood as the ease of performing (attainment) d_i with respect to the existing restrictions.

If we return to the concept of the fuzzy language, and assume that the present elementary possibilities of an action given by the possibility distribution

$$\Pi_t : D \rightarrow [0,1] \tag{15}$$

are known, then membership (admissibility) of an action $u = d_1 \ldots d_r$ in fuzzy language of actions L_t is determined according to

$$f_t(u) = \min(\Pi_t(\check{d}_1), \ldots, \Pi_t(d_r)). \tag{16}$$

The formula (16) reflects the principle of the "weakest link" as postulated in Nowakowska (1980). As she explained, the value $\Pi_t(u)$ is equal to the least possibility of performing elementary actions which form the sequence u.

We now return to the evaluation of achievability of goals of actions. Let G mean the goal which may be simple or complex near or far in time, requiring simple actions or a complex sequence of actions. As it has been shown for each goal G we may determine a set $R^{-1}(G)$ consisting of those actions which lead to goal achievement. If we know the fuzzy language of actions L_t we may ask a question: what is the achievability of the goal G under L_t?

Firstly, the goal G is achievable if $L_t^+ \cap R^{-1}(G) \neq \phi$. Thus for different α which suit different levels of action possibilities we can determine classes of actions leading to achieving the goal G according to the formula:

$$R_\alpha^{-1}(G) = L_t(\alpha) \cap R^{-1}(G). \tag{17}$$

Thus R_α^{-1} consists of those actions which achieve G and have at least α possibilities of performance. It is interesting at which level α of action possibilities the given goal G is achievable. That is for which α, $R_\alpha^{-1}(G)$ is non-empty. If for high α (i.e. near to 1) the classes of actions $R_\alpha^{-1}(G)$ are empty it will be reflected in low reliability of a human operator. Thus the level of reliability of a human operator may be connected with the level of action possibilities. A numerical characterization of this relation may be the so-called maximum achievability of the goal G denoted as MA(G) and understood trough the following

$$MA(G) = \max(\alpha : R^{-1}(G) \neq \phi). \tag{18}$$

IDENTIFICATION OF ACTIONS POSSIBILITIES

The numerical evaluation of action possibilities becomes a significant problem. The task can be formulated in the following way: if we have some knowledge concerning human features and environmental factors (i.e. evidence), what can be said about the possibility of man's performing a given action in a given situation? More precisely, which actions belonging to a set of elementary actions are feasible (may be performed by the man)? The analysed situation may be more or less complex. The scope of atomization or the necessity of a macroscopic definition of the situation is dependent upon the goal of analysis and the method of conceptualization.

Let's discuss a very simple situation to focus our attention on and recall intuitions. Assume that when given information appears a human operator is to perform an action d. Assume also that the time of performance is important. We can then determine a set of elementary actions $D = \{d_1, d_2, \ldots, d_n, \#\}$, and each action $d_i \in \{d_1, \ldots, d_n\}$ is identical but performed different times (i.e. lasting 1,2,...,n units of time), and # means the lack of action. In this case an evaluation of action possibilities is reduced to the evaluation of possibility of an action d at any given time i.e. 1,2,...,n units.

Now let's try to use the apparatus given by Shafer (1976) in his theory of belief functions. This evaluation of some propositions from different points of view, on the basis of different evidence is difficult to perform in a holistic way. That is why Shafer suggests decomposing evidence and then formulating probabilistic judgments (i.e. determining numerical grades of support of some proposition given by each item of evidence). The elementary evaluations are there aggregated according to Dempster's rule of combination.

The so-called basic probability assignment is one of the fundamental tools. It is the mass of evidence allocated to an event $A \subset D$, is denoted $m(A)$ and

$$\sum_{A \subset D} m(A) = 1, \tag{19}$$

$$m(\phi) = 0.$$

Events $A \subset D$ for which $m(A) > 0$ are focal elements.

It is comfortably, when each item of evidence points out to a subset of a set D. In this case Shafer (1976) suggests applying simple support function focused on A_o and defined by

$$\begin{aligned} m(A_o) &= \check{s}, \\ m(D) &= 1 - s, \end{aligned} \tag{20}$$

for $A \neq A_o$ and $A \neq D$, $m(A) = 0$.

This means that an item of evidence supports a proposition concerning feasibility of actions from a class A_o in the grade s. For instance $A_o = \{d_1, \ldots, d_r\}$ means performing action d in the time interval $[1, r]$.

A particular case of a simple support function is a vacuous belief function. There is only one focal element (i.e. D). This concept has been used to represent total ignorance. With the help of the vacuous belief function we can model a situation, where on the basis of evidence (or lack of it) we can only say that a human operator will perform an action from D and notking more precisely can be added.

We can aggregate evidence according to Dempsters rule of combination, and we have m_1 and m_2 (related to an item of evidence) we can determine m, by formula:

$$m(c) = \frac{\sum m_1(A) \cdot m_2(B),\ A,B \subset D,\ A \cap B = C}{\sum m_1(A) \cdot m_2(B),\ A \cap B \neq \phi} . \qquad (21)$$

Evidence may be combined by (21) iteratively.
For a given basic assignment m the probability P(A) of an event $A \subset D$ belongs to a sub-interval $[P_I(A),\ p^u(A)]$ of the unit interval. $P_I(A)$ (resp: $p^u(A)$) is called the lower (resp: upper) probability. The boundaries are defined from m as follows

$$P_I(A) = \sum_{\phi \neq B \subset A} m(B) , \qquad (22)$$

(i.e. it is the total mass of evidence which lies on focal elements implying A)

$$p^u(A) = \sum_{B \cap A \neq \phi} m(B) , \qquad (23)$$

(i.e. it is the total mass of evidence which lies on focal elements making the occurence of A possible).

Shafer (1976) interpreted $P_I(A)$ as a degree of belief or a degree of support for A. Dubois and Prade (1982) referred to this as credibility. The degree of plausibility ($p^u(A)$) is related to a degree of support of the opposite event since $\forall A,\ p^u(A) = 1 - P_I(\bar{A})$.

Based upon the aggregation of evidence obtained by Dempster's rule of combination, the feasibility of actions from D can be evaluated. For instance, the probability of an event $A = \{d_1, ..., d_r\}$ belongs to an interval $[P_l(A),\ p^u(A)]$. If a reliable, correct action requires performing action d at time $[1, r]$ then $[P_l(A),\ p^u(A)]$ is an evaluation of a reliable action.

When focal elements $A_1, ..., A_p$ are nested (i.e. $A_1 \subset A_2 ... \subset A_p$) then the possibility measure is a particular case of the upper probability (plausibility), and the dual measure designated N (necessity measure) by Dubois and Prade (1982) is a particular case of lower probability.

Using the above example it can be said that if evaluations concerning the time of action d create a nested structure, then

$$\Gamma(d_i) = \sum_{\{d_i\} \cap A \neq \phi} m(A). \qquad (24)$$

In Dubois and Prade (1982) a one-to-one correspondence between probability distribution and possibility distribution is introduced on finite domains. Denoting $p_i = p(d_i)$ and $\Pi_i = p^u(d_i)$ $i = 1, n$ yields, using (24)

$$\forall_i, \ p_i = \sum_{k=1}^{n} \frac{1}{k} (\Pi_k - \Pi_{k+1}), \qquad (25)$$

where the n+1 element of D are supposed to be ordered according to the decreasing values of Π_k (i.e., $\Pi_1 \geqslant \Pi_2 \geqslant \dots \Pi_{n+1}$) and $\Pi_{n+2} = 0$ by convention. As shown in Dubois and Prade (1982) this transformation preserves the following inequalities

$$\forall A \subseteq D \quad N(A) \leqslant P(A) \leqslant \Pi(A). \qquad (26)$$

Particular evidence may be quite explicitly represented in the form of possibility distributions or fuzzy sets. In the discussed example the time of action may be defined by a fuzzy set corresponding to an expression "about a". In this case the aggregated possibility as explained by Dubois and Prade (1983) is given by

$$\forall d_i \in D \ \Pi(d_i) = \frac{\min(\Pi_1(d_i), \Pi_2(d_i))}{\max \ \min(\Pi_1(d_i), \Pi_2(d_i))} \qquad (27)$$

Obviously in real situations we must also consider complexity, and it will be necessary to apply a number of tools for numerical representation of actions possibilities.

POTENTIALS AND OPEN QUESTIONS

Thus for the discussion has been focused on constraints of actions imposed by features of the human operator and the external factors. A formal theory of a human operator would however require some broadening of the discussion with the remaining aspects which were mentioned in the introduction.

The basic propositions have been sketched in the interpretative rather than technical sense. Possibilities of actions may be understood differently in various situation and there will be different problems concerning their determination. In some cases, it will be possible to determine them by function through the results of laboratory investigations. In other cases, however their determination may be connected with results of random experiments. Additionally, some will require an evaluation given by human experts.

One may also doubt whether the numerical evaluation of action possibility should be expressed by a scalar. It is important to present fluctuation of action possibilities properly. They may be modelled with the help of a probabilistic set of extended fuzzy expressions (Hirota, 1979), stochastic fuzziness (Norwich and Turksen, 1982) or random sets (Goodman, 1982). Besides, the physical view of possibility measures requires additional notions connected with independence, noninteractivity, conditioning.

Whether the discussion will go beyond formal speculations is dependent upon practical applications of the proposed approach. It may appear in the shape of a simulation of a human operator. Some useful tools are given by Chanas and Nowakowski (1985).

Of additional worth is the mention that the discussion and the proposals included in this text deal with a broader class of situations rather than just the situation of a human operator's actions.

REFERENCES

1 S. Chanas and M. Nowakowski, Single value simulation of fuzzy variable, Tech. Univ. of Wrocław, Rep. PRE 25 (1985) (submitted to Fuzzy Sets and Systems).
2 D. Dubois and H. Prade, On several representations of uncertain body of evidence, in: Fuzzy Information and Decision Processes, (M.M. Gupta et.al.eds.), North-Holland, 1982.
3 D. Dubois and H. Prade, Upper and lower possibilities induced by a multivalued mapping, Proc. IFAC Symp. on Fuzzy Information, Knowledge. Representation, Marseille, 1983.
4 I.R. Goodman, Fuzzy sets as equivalence class of random sets, in: Fuzzy sets and Possibility Theory, R.R. Yager (Eds.), Pergamon Press, 1982.
5 K. Hirota, Extended fuzzy expressions of probabilistic sets, in: Advances in Fuzzy Set Theory and Applications, M.M. Gupta et. al. (Eds.), North-Holland, 1979.
6 M. Nowakowska, Language of Motivation and Language of Actions, The Hague, Mouton, 1973.
7 Nowakowska M., New Ideas in Social Sciences, Ossolineum - PAN, 1980 (in Polish).
8 M. Nowakowski, Possibility distributions in the linguistic theory of actions, Int. J. of Man-Machine Studies, 12, (1980), 229-239.

9 M. Nowakowski, Possibilities of actions and their linguistic analysis. Prakseologia 4, (1981), (in Polish).
10 A.M. Norwich, I.B. Turksen, Stochastic Fuzzines, in Approximate Reasoning in Decision Analysis, M.M. Gupta et.al. (Eds.), North-Holland, 1982.
11 G. Shafer, A Mathematical Theory of Evidence, Princeton University Press, 1976.
12 L.A. Zadeh, Fuzzy sets as a basis for the theory of possibility, Fuzzy Sets and Systems, 1, (1978), 3-28.

Applications of Fuzzy Set Theory in Human Factors, edited by W. Karwowski and A. Mital
Elsevier Science Publishers B.V., Amsterdam, 1986 — Printed in The Netherlands

THE FRAME MODEL OF RELIABILITY

EDWARD A. FRANUS

Department of Psychology, Jagellonian University, Cracow (Poland)

ABSTRACT

This paper presents a graphical model of relations between conditions of instability and the reliability of man's activity in steering a system. The author analyses the two ways that reliability can be conceived: 1. as an errorless performance, described in the scheme: "done - not done", 2. as an fuzzier, more graded way, described in the scheme: "done errorlessly - almost errorlessly - satysfying - not done". Both schemes could be useful in the research. For the estimation of man's reliability it profits greatly to use the notion "task". Each task can be described exactly, and clear requirements for its performance recognised.

Comparing the model of man's reliability with the models of a machines reliability it is possible to ascertain the considerable differences, reflected between a machine and a man (as enumerated in part 2). The number of errors increase as optimal conditions are exchanged for extreme conditions. In part 8 the author proposes a net of tasks for research into the two groups of determinants of reliability.

INTRODUCTION

The reliability of any system in which man appears as a subsystem in the role of an operator is admittedly dependent upon the reliable activity of man (Lomow 1967, Franus 1977). That is the underlying cause of why in ergonomics the problem of man's reliability (not of a machine) has continued. The research resulting from this problem has proven more complicated than originally invisioned. The dependence on many factors, the lack of an univocal definition (Kilinski 1971) and the lack of a clear plan for research have brought about few achievements.

The model of reliability and the net of tasks presented in the paper may be used for planning and organizing the research. It should be remembered, however, that man's final reactions to all information attained at entry level become at last a resultant response on outlet. Therefore we must make allowance for the category of fuzzy influences of all factors that determine the reliability of man-machine system, or broadly speaking of a man-technics system.

1. DEFINITION OF RELIABILITY

In human engineering there are two possible ways of defining reliability: quantitative and qualitative. The former treats reliability as a probability and the latter treats it as a capability to perform the intrusted functions under stated conditions and time (Lomow 1963, Niebylicyn 1966, Milerian 1971, Meister 1973, Ratajczak 1974, Franus 1975, Embrey 1978).

There is a difference of opinions on the performances of functions. Some authors see reliability as a capability for "errorless" performance (Bobniewa 1966). Others describe it in a fuzzy way as a "satisfactory" performance of action in definite conditions and time (Lomow 1963), or as the "most acceptable" maintenance of the stable work parameters on various complicated work places (Niebylicyn 1966).

Reliability as an errorless performance may be described in the scheme: "done - not done" and "so - no" respectively.

Alternatively the fuzzy way of reliability comprehension may be displayed in the scheme more graded: "done errorlessly - almost errorlessly - satisfying - not done".

Both methods of reliability comprehension and both schemes could be useful in the research.

However, as they now appear they are not completely in agreement. Therefore we can do nothing but assume the principle that the former (scheme) is adequate to foolproof work situations, exacting man's simple reactions. The second way (scheme) is adequate to complex work situations, exacting more complicated courses of action. The bound between these two different approaches, using scopes of reliability conception is also fuzzy. Consequently, we can assume that man's reliability becomes a fuzzy notion according to the complication of man's behavior (when it becomes more and more unstable in the more and more complex work situations). Of course the complication of the system depends on the changes of his elements and the relations among them. These changes and relations are a complex of reliability factors. For this reason using a model of reliability it is necessary to emphasize the changes of the reliability factors.

Reliability as a parameter of action (of man or of man-machine system) reveals its three characteristics: 1. it is a probable feature (not necessary), 2. it is gradual and unstable (we can describe it by a coefficient), 3. it is relative (dependent on conditions and time). Consideration of these is useful in the research of reliability.

2. DIFFERENCES BETWEEN MAN AND A MACHINE

Although cybernetics created a very useful language and, consequently, the convenient plane for communication between technicians and humanists, both partners should remember that differences between man and machine exist. The differences are:

1. The details of a machine are mutually independent; between them are not such compensation mechanisms that mean a good deal in man's actions (Niebylicyn 1964).

2. The function of a machine has two phases: travel and immobility. Man's activity is characterized by very fuzzy stages (called transitional stages), such as drowsiness, distempers of awareness, nervous instability, fainting, and tiredness. These stages can be the causes of errors (Nowakowska 1977, 1979).

3. The function of a machine has a linear appearance, that is to say, the end result is a sum of answers of all signals obtained on entrance. Man's reactions, instead, are in the end the resultant of all information obtained on entrance.

4. A machine makes its operations quickly and evenly, the operator works slowly and unevenly.

5. A machine can function very long and correctly, the operator gets tired and makes errors (Jethon 1977).

6. An automatic machine functions through the given program, but a man-operator in a damage situation can depart from the program and adapt his own intervention to the changed conditions. In unusual situation a man's reactions can be intelligent (dynamical and elastic).

7. Human errors are fundamentally different from machine errors and so they cannot necessarily be treated mathematically in the same fashion (Adams 1982, Wickens 1984).

For this reason the resulting reliability of man-machine system is dependent upon the reliable activity of man, not of machine.

3. THE ROLE OF TASKS IN THE ESTIMATION OF RELIABILITY

For the estimation of man's reliability it is necessary to understand the notion of "performance". There are often doubts exempli gratia: Is a man reliable if he performes the action less fine, or if he performes it fine but not at exactly the right time? The number of doubts increases when we assume more criteria for the estimation, such as usefulness, aesthetics, originality, etc.

We can avoid these doubts if we assume two limitations:

1. under the notion "performance" we put the notion "task",
2. the notion "realiability" we shall treat like a parameter of action by the

designation: "done - not done".

Both these limitations should be considered together. The result from the application of the notion "task" is considerable, because we can describe each task exactly and establish clear requirements for its performance. In order to estimate reliability we examine each requirement by the designation "done - not done".

However for the rational control of reliability and for its anticipation we must know physical and psychical conditions that are important determinants.

4. RELIABILITY AND A WORKLOAD

All work is connected with load. Therefore it can be said that the load is an attribute of the work (action) and leads to tiredness, which can be the cause of errors. The tired worker is deceptive.

Tiredness is an unpleasant feeling and workers try to avoid it. But the work is not avoidable. For many workers it is an autotelic value or a means to an end (Vroom 1964).

If the work is necessary for people, it is difficult if not impossible to free them from it. However we may try to describe the degree of load, which is best tolerated. Consider a zone between minimal and maximal load an "optimal condition zone", we can then define the minimum and the maximum of load as two "extreme condition zones" that are usually undesirable for people. These categories should be used in the construction of reliability model.

What is an optimal load? We can define it as a relation between the degree of load and the degree of man's capacity to perform a task, best tolerated by him (Goldstein 1978). Because individual differences of load tolerance are considerable and variable, we must treat the "optimal conditions" of work as a fuzzy notion (Nowakowski 1979).

5. MODELS OF MACHINES RELIABILITY

The reliability of any machine is a relative notion. It is dependent on a function of a machine in the performance of a task. First one must understand it as an errorless function (e.g. at computer) and at some other time as a graduate feature (e.g. at internal combustion engine). For these two ways of comprehending reliability a different model is necessary.

The model of computer reliability has a sharp transition between the reliability zone and damage zone, according to transition between optimal conditions and disadvantageous (nontolerated conditions, shown in Figure 1). The reliability zone is narrow because the function of the computer must be errorless (Card 1983).

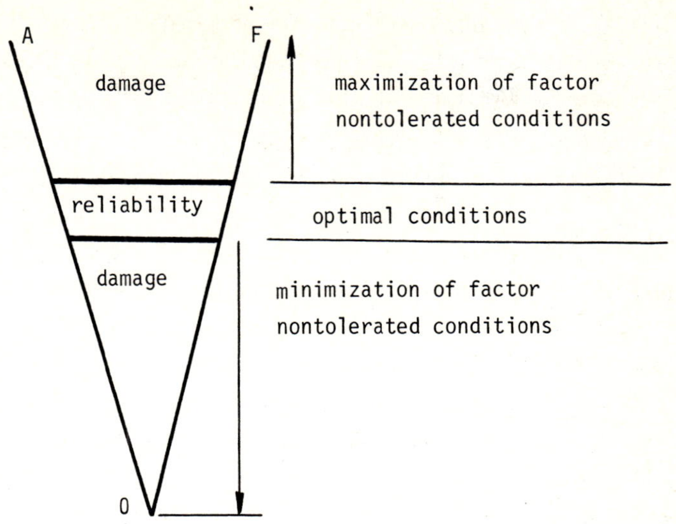

Fig.1. Model of computer reliability

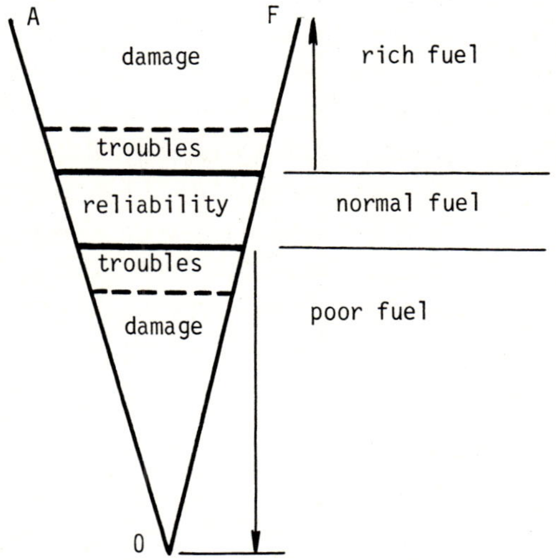

Fig.2. Model of engine reliability

Legend: AOF - an area of a machine potential function

The reliability of an engine is treated more mildly (tolerently). Even if the carburetor is not carefully regulated the engine is able to continue its function. Therefore in the model of its reliability we find between reliability and damage (zones) two zones of trouble (Figure 2). The bounds between these zones are fuzzy (Zadeh 1965, 1977).

6. MODEL OF MAN'S RELIABILITY

There are numerous differences between a machine and man. But for the purposes of modelling man's reliability the most important of them are human abilities (adaptation, or intelligent intervention in changed conditions) and human weakness. These human qualities cause much broader zones of conditions, errors and reliability rules in the model of man's reliability. Moreover they must be treated more dynamically in comparison with a machine (see Figure 3).

As shown area of man's activity between two straight lines AO and OA is divided into three zones: one zone of reliability and two zones of errors. They are relatively broader than analogous zones shown in Figure 2. The reliability zone is adequate for the interval of optimal conditions. The excess of their limits (ULR or ILR) lead to the intervals of extreme conditions and, at the same time, to the zones of errors. In the process of maximizing or minimizing loads the number of errors increases to the limits of action possibility (ULA, ILA). The refusal of activity follows at two intervals: super extreme conditions and sensorial deprivation (ZAR). All limits of the zones are labile. Under the influence of different factors (motivation, emotion, tiredness, adaptation, compensation, learning) they can be changed a little up and down.

Comparing the model of man's reliability with the reliability models of machines considerable differences between intervals and zones is evidenced. Man's wide abilities for adaptation to difficult conditions are reflected in the wider zones.

7. ROLE OF ERRORS

Applying the gauge of man's reliability we can use the scheme "done - not done" and the errors indicator. Many authors have addressed the topic of errors (Keele 1973, Swain 1977, 1980, Danaher 1980, Norman 1981), but in this writing we are interested only in two of their features: greatness and quantity. A fuzzy division between microerrors and macroerrors, or between a low number and a high number of errors can be defined, according to the requirements of the task (Goldstein 1978).

Microerrors do not have any direct influence on a system function in the main. But they can appear even in the reliability zone and then we can treat them as a prognostic of near inefficiency, especially when their number increases.

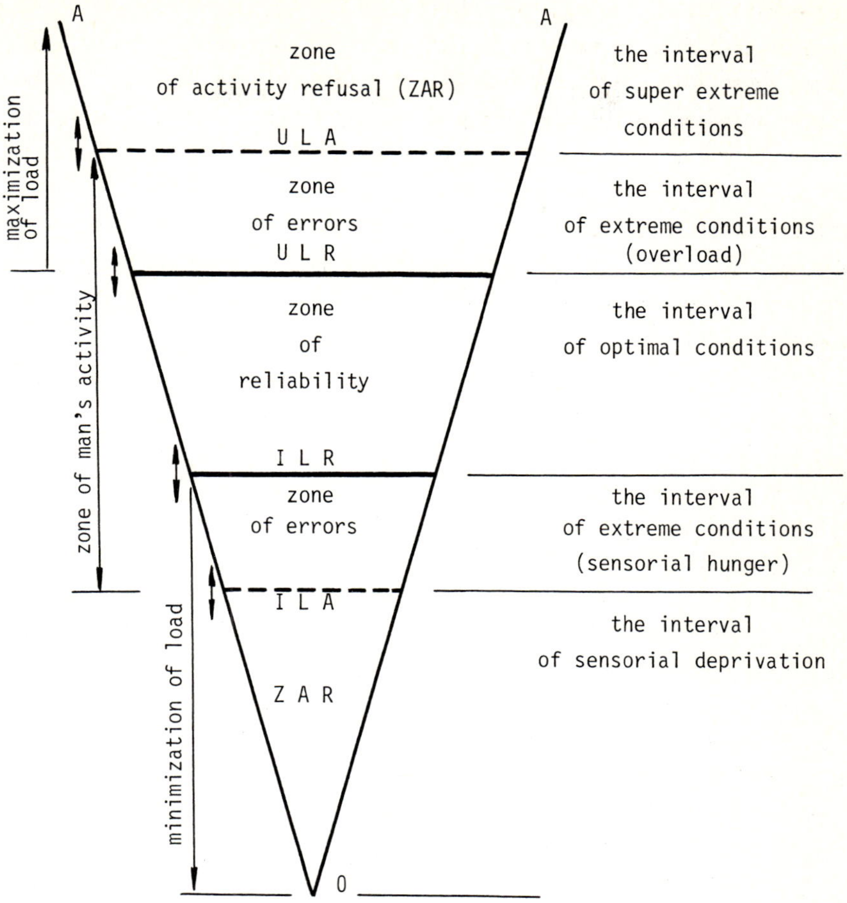

Fig.3. Model of man's reliability

Legend: AOA - an area of man's potential activity, ULR - upper limit of reliability, ILR - inferior limit of reliability, ULA - upper limit of activity, ILA - unferior limit of activity, ZAR - zone of activity refusal.

Macroerrors usually bring about changes in the system function and therefore they are dangerous. Even one macroerror can cause some damage. Because the quantity passes into quality so also do we allowances for the possibility that the increased number of microerrors will at any moment convert into macro-errors.

What to do towards the reliability of system? The model shown in Figure 3 indicates that the main problem depends on the assurance of the optimal conditions for man's activity (Grobelny 1981). We have in mind the external conditions (workingstand and environment) and internal conditions (physiological and psychological).

As a result of the interesting theory of W.D. Niebylicyn (1966), we can examine man's abilities and experiences in optimal conditions. In these conditions man functions best, but there are no possibilities to check reliability. In the opinion of the same author, reliability can be examined only in the extreme conditions, where the vocational capacities lose their meaning while the nervous and organic factors begin to prevail. Man's reliability is dependent on such factors as: nervous endurance of overload, psychological and physical resistance to lasting effort, resistance to monotony, resistance to disturbances, and resistance to pernicious influences of environmental. Individual differences however are quite large.

In order to determine the actual and potential possibilities for man's activity, investigations of his behavior in optimal and extreme conditions is necessary (Kocowski 1972, Franus 1977). With regard to responsible workstanding, it is very important to know not only man's abilities and capacities, but also his over all possibility limits of action in extreme conditions (when he begins to make errors, lose pace with reaction, breaks off the work etc. - Goldstein 1978).

8. NET OF TASKS FOR RESEARCH THE RELIABILITY

All determinants of man's reliability can be divided into two groups: 1. physical and technical work conditions, 2. physiological and psychological features of operator's personality. However, both groups include more specific determinants. Presented below is a proposed net.

A. <u>Tasks for adaptation of work conditions to operator</u>

I. General relations include the following:

1. optimize the function division between a machine and man,
2. assure the correspondence between signal and steering devices on their function fields,

3. assure the correspondence between steering system and the object on their function fields.

II. Signal devices include the following:
1. optimize the signal legibility,
2. localization of the signal devices,
3. code system and signal serving tempo.

III. Steering devices include the following:
1. optimize the catches form,
2. devices location,
3. compositeness of steering system,
4. resistance of steering devices.

IV. Safety, comfort and organization include the following:
1. assure the safety conditions of work,
2. optimize the physical environment conditions (climate, colour and so on),
3. perfect the organization of work process,
4. improve human relations.

B. <u>Tasks for research of personality features</u>

I. Competences include the following:
1. intelligence and special abilities,
2. capacities and skills,
3. ability to get training,
4. professional experience.

II. General physical features include the following:
1. anthropometrical marks,
2. sense of perception,
3. health.

III. Motor efficiency include the following:
1. speed of reaction,
2. kinesthetic sensitivity,
3. precision movements,
4. coordinated movements,
5. individual tempo.

IV. Intellectual processes include the following:
1. perception of colours, shapes and situations,
2. features of attention, memory and imagination,
3. features of problem solving,

4. efficiency of coding and decoding.

V. Special features of personality include the following:
1. features of temperament,
2. features of character,
3. dominants of personality,
4. motivation system (ideas, value system, aspirations, morale, aptitudes).

VI. Functional features include the following:
1. nervous endurance on overload,
2. prolonged endurance,
3. resistance to disturbances,
4. resistance to pernicious influences of environment factors (chemical, temperature, dust, humidity),
5. resistance to monotony.

The presented net may be treated only as a general frame for detailed themes, resulting from the research development. The research can lead towards two purposes: 1. elaboration of norms and a directive lines system for ergonomic engineering and organization, 2. elaboration of norms for the selection of operators for different jobs and functions.

The adequate selection of the features which are prominent (dominant) for man's real action (job, activity) is an important aspect of using of the net in practice. Dominants can be revealed by analyzing methods in the work processes. It is normal that in different jobs, different dominants which determine reliability will appear.

The presented model of reliability and the net of tasks seem to be useful as the means to indicate the direction and the scope of ergonomic (multidisciplinary) research.

REFERENCES

Adams, J.A., 1982, Issues in human reliability, Human Factors, 24, 1-10.
Bobniewa, M.I., 1966, Tiechniczeskaja psichologija, Moskwa.
Card, S., Moran, T.P., Newel, A., 1983, The psychology of humancomputer interactions, Hillsdale, NJ: Erlbaum Associates.
Danaher, J.W., 1980, Human error in ATC system operations, Human Factors, 22.
Embrey, D.E., 1976, Human reliability in industrial systems: an overview, VI Congress IEA.
Embrey, D.E., 1978, The human operator and system reliability, College House, Gosta Green, Birmingham JET.
Franus, E., 1975, Niezawodnosc czlowieka jako kryterium dostosowania maszyny i warunkow pracy, IOS, Krakow.
Franus, E., 1977, Model niezawodnosci czlowieka i jego znaczenie dla ergonomii,

Przeglad Psychologiczny, 1, t.XX.
Goldstein, I.L., Dorfman, P.W., 1978, Speed stress and load stress as determinants of performance in a time-sharing task, Human Factors, 20.
Grobelny, J., Nowakowski, M., 1981, Mozliwosci wykorzystania teorii zbiorow rozmytych w opisie i modelowaniu ukladu czlowiek-technika, Ergonomia, t.4, nr. 1-2.
Jethon, Z., (red.), 1977, Zmeczenie jako problem wspolczesnej cywilizacji, PZWL, Warszawa.
Kilinski, A., 1971, Definicje opisowo-analityczne i wartosciujaco-normatywne podstawa pojec teorii niezawodnosci, Prakseologia, 38.
Kocowski, T., 1972, Czynniki okreslajace niezawodnosc czlowieka jako operatora na stanowiskach krytycznych, Prace Naukowe Instytutu Organizacji i Ekonomiki, Seria: Studia i Materialy, 4, Politechnika Wroclawska.
Lomow, B.F., 1963, Czielowiek i tiechnika, Leningradskij Gosudarstwiennyj Uniwersytet, Leningrad.
Lomow, B.F., 1967, Czielowiek w sistiemach uprawlenija, Moskwa.
Meister, D., 1973, A critical review of human performance reliability predictive methods, IEE, Transactions on Reliability, 22.
Milerian, A.F., 1971, O nadioznosti operatora w razlicznych rezimach raboty, Woprosy Psichologii, 4.
Niebylicyn, W.D., 1964, Nadieznost raboty operatora w sloznoj sistiemie uprawlenija, Sb., Inzeniernaja psichologija, Moskowskoj Gosudarstwiennyj Uniwersytet, Moskwa.
Niebylicyn, W.D., 1966, Osnownyje swoistwa nierwnoj sistiemy czielowieka, Proswieszczenije, Moskwa.
Norman, D.A., 1981, Categorization of action slips, Psychological Review, 88.
Nowakowska, M., 1977, Methodological problems of measurement of fuzzy concepts in the social sciences, Behavioral Science, 22.
Nowakowska, M., 1979, Teoria dzialania, PWN, Warszawa.
Nowakowski, M., 1979, Rownania rozmytych relacji w ocenie niezawodnosci czlowieka i doborze zawodowym, Raporty Instytutu Organizacji i Zarzadzania Politechniki Wroclawskiej, Seria PRE, 81.
Ratajczak, Z., 1974, Psychologia inzynieryjna, Uniwersytet Slaski, Katowice.
Sheridan, T., 1981, Understanding human error and aiding human diagnostic behavior in nuclear power plants. In J. Rasmussen and W.B. Rouse (Eds): Human detection and diagnosis of system failures, New York, Plenum Press.
Swain, A.D., 1977, Error and reliability in human engineering. In B. Wolman (Ed.): International encyclopedia of psychiatry, psychology, psychoanalysis and neurology, Vol.4, New York, Von Nostrand Reinhold.
Swain, A.D., Guttman, H.E., 1980, Handbook of human reliability analysis with emphasis on nuclear power plant application, Washington.
Vroom, V.H., 1964, Work and motivation, New York, Wiley.
Wickens, C.D., 1984, Engineering psychology and human performance, Columbus, Ohio, C.E. Merill Publishing Co.
Zadeh, L.A., 1965, Fuzzy sets, Information and Control, 8.
Zadeh, L.A., 1977, Fuzzy sets as a basis for theory of possibility, Fuzzy Sets and Systems, 1.

Applications of Fuzzy Set Theory in Human Factors, edited by W. Karwowski and A. Mital
Elsevier Science Publishers B.V., Amsterdam, 1986 — Printed in The Netherlands

IDENTIFICATION OF LEARNING CURVE BASED ON POSSIBILISTIC CONCEPTS

J. WATADA[1], H. TANAKA[2] and T. SHIMOMURA[3]

[1] Faculty of Business Administration, Ryukoku University, Fukakusa, Fushimi, Kyoto 612 (Japan)

[2] Department of Industrial Engineering, College of Engineering, University of Osaka Prefecture, 591 Osaka, Sakai, Mozu-Umemachi 4-804 (Japan)

[3] Department of Industrial Engineering, Osaka Electro-Communication University, 572 Osaka, Neyagawa, Hatsumachi 18-8 (Japan)

ABSTRACT

In this paper, the learning phenomenon is discussed within the framework of possibility. The learning phenomenon is generally found in repetitive human tasks or the production process in an organization. A curve depicting the learning phenomenon is called a learning curve. The learning curve expresses the decrease of time per cycle or of the cost per cycle for a large number of repetitive cycles caused by learning effects. This decrease is caused by the learning ability of a person or the adjustability of an organization. It is natural to interpret the learning effects as the possibility of human ability.

The possibilistic regression model is employed to analyze the learning phenomenon. This approach enables us to understand two characteristics of a learning effect. One is the trend of the learning effect, and the other is the width of possibility at each time sampled.

The possibilistic model has been applied to analyze the learning phenomenon in data entry jobs. This analysis indicates that learning effects have been found not only on its trend, but also on the possibility width. The learning effect on the trend is a decrease in time of the job cycle caused by repetitive performance. This repetitive performance also causes a stable narrowing of the possibility width of each cycle. That is, when the subject is experienced in the performance of some task, he can do the task at stable speed.

INTRODUCTION

The learning phenomenon is generally found in repetitive human tasks or production processes in an organization. A trend curve which depicts the learning phenomena is called a learning curve. The learning curve expresses the decrease of time per cycle or of the cost per cycle for a large number of repetitive cycles caused by a learning effect. This decrease is explained in terms of the learning ability of a person or the adjustability of an organization. Especially in industries, the identification of learning is important to set production standards or to determine the potential for cost reduction of high-volume production(ref.1).

Various models have been proposed to represent and analyze the learning phenomena. Factors such as a learning coefficient and a learning rate are calculated from the learning curve. These coefficients are useful in setting production standards. In a realistic setting, fluctuation of learning effects is ordinarily found in a person or an organization when each value for a learning cycle is plotted on a graph. Therefore, the mean method of accumulated values is employed in smoothing the variation of the learning values. It should be noted that the learning phenomenon is related to human nature and fluctuation of the values in cycles exists in the learning phenomenon in nature. Therefore, we have to properly account for learning characteristics in terms of both the learning curve and deviation of learning values from the trend of the curve. From this view point, Nakayama and Sakai have statistically discussed the human learning phenomenon(ref.2).

In this paper, we view the fluctuation of learning as possibilities of human ability. Therefore, we employ the possibilistic concept to express human learning characteristics, where various characteristics relating to human nature are considered to have possibilistic intervals. This concepts of possibility, proposed by L.A. Zadeh(ref.3), has widened applicable fields in which Fuzzy Set Theory can be applied. The possibility is closely related to a fuzzy number. While most statistics deal with probability measure of events, possibility is treated through a possibility measure. This paper discusses the human learning ability within the framework of the possibilistic concepts. We identify the learning curve through a possibilistic regression model(ref.4) whose parameters are fuzzy numbers and which expresses possibility of learning curve.

BASIC CONCEPTS

The basic concepts are explained here in preparation for dealing with a learning curve in the framework of possibility. The possibilistic concept in a fuzzy set theory has been discussed in many articles since its proposal of L.A. Zadeh (ref.3) in 1978. In the possibilistic regression model(ref.4) parameters agreed upon as fuzzy numbers have their possibility distributions. Furthermore, we can evaluate the possibility of the value in a fuzzy function from the view point of a possibility measure(ref.7). The possibilistic regression analysis is formulated by a fuzzy function to model the possibilistic structure of a system. The fuzzy function can be calculated from possibility distributions of parameters through the extension principle(refs.7,8), which can be also explained by terms of possibility measure.

Possibility distribution and possibility measure

L.A. Zadeh has proposed a possibilistic interpretation of a fuzzy set(ref.3), where the possibility distribution is defined by a membership function of a

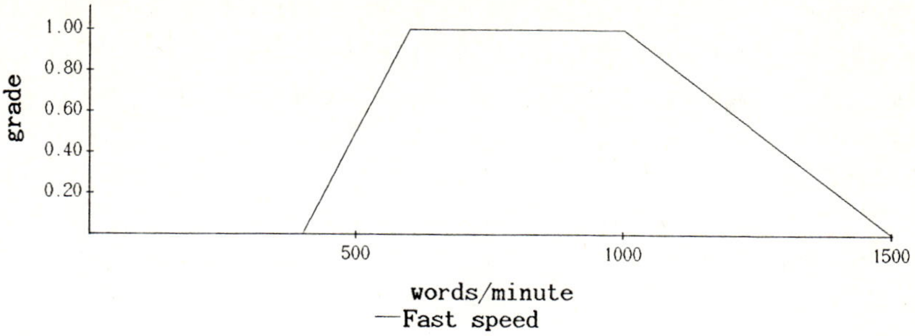

Fig. 1. Fuzzy set: Fast speed in typing.

fuzzy set.

Let F be a fuzzy subset of a universe X whose membership function is denoted by $f_F(x)$. Given the proposition "u is F", the degree of possibility that u=x is denoted by $\Pi(x)$ and is defined to be numerically equal to the membership of x to F, i.e.,

$$\Pi(x) = f_F(x) \tag{1}$$

where x is a variable on a universal set X.

Example 1. Let us consider the proposition "John types fast", where the concept "fast" is defined as a fuzzy set in R^1 characterized by the trapezoidal membership function as shown in Fig. 1. Utilizing Fig. 1, let us consider a typing speed of 550 words per minute, whose grade of membership in the fuzzy set "fast speed" is 0.75. This grade 0.75 can be interpreted as the degree of possibility that John's typing may be 550 words per minute, given that proposition

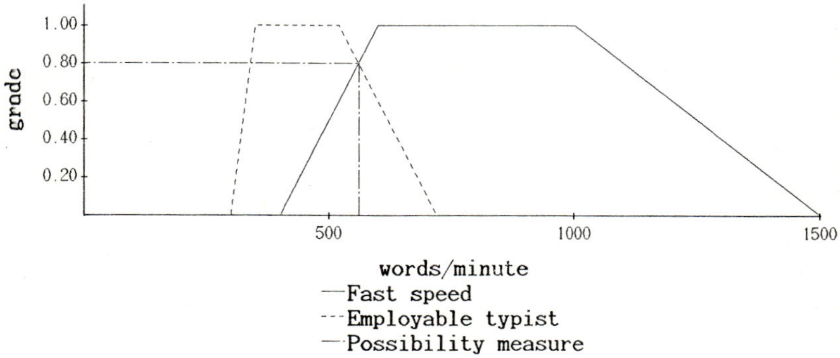

Fig. 2. Fuzzy set: Employable typists.

"John's typing speed is fast".

Given a possibility distribution $\Pi(x)$, a possibility measure of a crisp set E is defined as

$$\Pi(E) = \sup_{x \varepsilon E} \Pi(x) \tag{2}$$

In a similar way, the possibility measure of a fuzzy set A is defined as

$$\Pi(A) = \sup_{x \varepsilon X} \{f_A(x) \wedge \Pi(x)\} \tag{3}$$

Example 2. A certain company intends to employ typists. Employable typists may be expressed by a possibility distribution illustrated in Fig. 2. If a typist has very good ability to type, he/she will require a sufficient wage. On the other hand, a typist who has very poor ability will not be appropriate for the job. What is the possibility that an applicant with a fast typing speed will be employed? Using Equation 3 to calculate the possibility measure we find this measure to be 0.8.

A possibility measure has the following properties:

$$(i) \quad \Pi_X(\phi) = 0, \quad \Pi_X(X) = 1 \tag{4}$$

$$(ii) \quad \Pi_X(A_1 \cap A_2) = \Pi_X(A_1) \wedge \Pi_X(A_2) \tag{5}$$

$$(iii) \quad A_1 \subset A_2 \longrightarrow \Pi(A_1) \leqq \Pi(A_2) \tag{6}$$

(iii) derived from *(ii)* shows that Π_X is monotonic. Let us denote a possibility space(refs.9,10) as $(X, \mathcal{F}(X), \Pi_X(\cdot))$, where $\mathcal{F}(X)$ is a set of all fuzzy subsets of X. Given two universal sets X and Y, and a function g: $X \rightarrow Y$, a possibility space $(Y, \mathcal{F}(Y), \Pi_Y(\cdot))$ can be induced from the given possibility space through the function g as follows:

Denoting a set E as $E = \{x|y=g(x)\}$, a possibility distribution of y is induced from $\Pi_X(\cdot)$ as

$$\Pi_Y(y) = \Pi_X(E) \tag{7}$$

which can be rewritten by the definition of a possibility measure as

$$\Pi_Y(y) = \sup_{\{x|y=g(x)\}} \Pi_X(x) \tag{8}$$

Hence, we have for $B \in \mathfrak{F}(y)$

$$\Pi_Y(B) = \sup_y \{f_B(y) \wedge \Pi_Y(y)\} \tag{9}$$

which leads to a possibility space $\{Y, \ (Y), \Pi_Y(\cdot)\}$. Eqn. 8 leads to the following relations:

$$g(A) = B : \quad f_B(y) = \begin{cases} \sup\limits_{\{x|y=g(x)\}} f_A(x) & ; \{x|y=g(x)\} \neq \phi \\ 0 & ; \text{otherwise} \end{cases} \tag{10}$$

$$g^{-1}(B) = A : \quad f_A(x) = f_B(g(x)) \tag{11}$$

Fig. 3 explains briefly the above-mentioned relations. Relating to the function $g: \mathfrak{F}(X) \to \mathfrak{F}(Y)$, we have the properties:

(*i*) Given a fuzzy set A in X, for the fuzzy set B in Y such that $g(A)=B$, we have $\Pi_X(A) \leq \Pi_Y(B)$. If A is a crisp set, $\Pi_X(A) = \Pi_Y(B)$.

(*ii*) Given a fuzzy set B in Y, for the fuzzy set A in X such that $g^{-1}(B)=A$, we have $\Pi_X(A) = \Pi_Y(B)$.

(*i*) and (*ii*) are proved as follows:

(*i*)
$$\begin{aligned} \Pi_Y(B) &= \sup_y \{f_B(y) \wedge \Pi_Y(y)\} \\ &= \sup_y \{f_{g(A)}(y) \wedge \Pi_Y(y)\} \\ &= \sup_y \{ \sup_{y=g(x)} f_B(y) \wedge \sup_{y=g(x)} \Pi_X(y)\} \\ &\geq \sup_y \sup_{y=g(x)} \{f_B(y) \wedge \Pi_Y(y)\} \\ &= \sup_x \{f_A(y) \wedge \Pi_X(y)\} \\ &= f_X(A) \end{aligned}$$

(*ii*)
$$\begin{aligned} \Pi_Y(B) &= \sup_y \{f_B(y) \wedge \Pi_Y(y)\} \\ &= \sup_y \{f_B(y) \wedge \sup_{y=g(x)} \Pi_X(x)\} \\ &= \sup_x \{f_{g(A)}(x) \wedge \Pi_X(x)\} \\ &= f_X(A) \end{aligned}$$

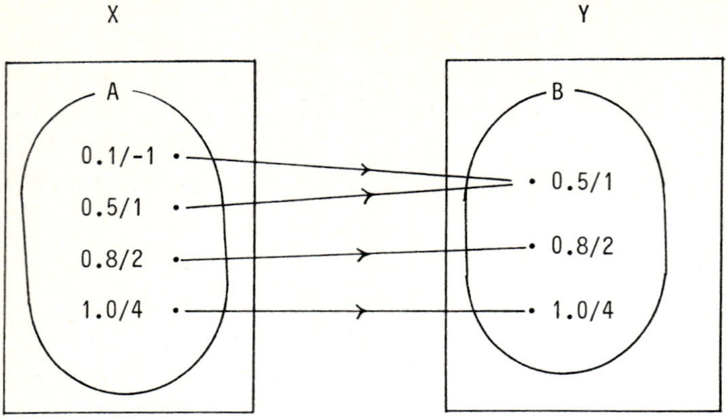

$$y = g(x) = |x|$$

$$A=0.1/-1+0.5/1+0.8/2+1.0/4 \longrightarrow g(A)=0.5/1+0.8/2+1.0/4$$

$$\cap \qquad\qquad\qquad\qquad \parallel$$

$$g^{-1}(B)=0.5/\pm1+0.8/2+1.0/4 \xleftarrow[g^{-1}]{} B$$

Fig. 3. Mapping of a fuzzy set.

A fuzzy function whose parameters are fuzzy numbers is denoted by g(X,A). By the above-mentioned relation 10, the fuzzy number of Y=g(X,A) can be calculated as

$$f_Y(y) = \sup_{y=g(X,A)} f_A(A) \tag{12}$$

where $f_Y(y)=0$ for the case of $\{A|y=g(X,A)\}=\phi$.

Operation on fuzzy numbers and extension principle

The extension principle introduced by L.A. Zadeh(ref.7) plays a central role in dealing with a possibilistic model. This method extends nonfuzzy mathematical concepts to fuzzy quantities. An explanation of the extension principle for handling a fuzzy function is given below.

Let g be a mapping from a cartesian product of universes X_1 and X_2, $X_1 \times X_2$ to a universe Y such that $y=g(x_1,x_2)$. The fuzzy function of fuzzy numbers $A_1 \varepsilon X_1$ and $A_2 \varepsilon X_2$ is defined as

$$Y = g(A_1,A_2) \;;\quad f_Y(y) = \sup_{g(a_1,a_2)=y} \{f_{A_1}(a_1) \wedge f_{A_2}(a_2)\} \tag{13}$$

In the case that g is a binary operation #

$$Y = A_1 \# A_2 \; ; \quad f_Y(y) = \sup_{a_1 \# a_2 = y} \{f_{A_1}(a_1) \wedge f_{A_2}(a_2)\} \tag{14}$$

Now, let us define the form of a triangular fuzzy number considered in this paper.

A triangular fuzzy number A_j is denoted by (α_j, c_j) and defined as

$$A_j = (\alpha_j, c_j) \tag{15}$$

where α_j and c_j stand for the center and width of fuzzy set A_j, respectively and the membership function of fuzzy number A_j is written as

$$f_{A_j}(a_j) = 1 - \frac{|a_j - \alpha_j|}{c_j} \tag{16}$$

Fig. 4 illustrates a triangular fuzzy number A_j.

Example 3. Assume that "about 5" and "about 3" may be defined by triangular fuzzy numbers (5,2) and (3,1), respectively. Then let us calculate the summation of "about 5" and "about 3".

$$Y = \text{"about 5"} + \text{"about 3"}$$

$$= (5,2) + (3,1)$$

$$= (8,3)$$

$$= \text{"about 8"}$$

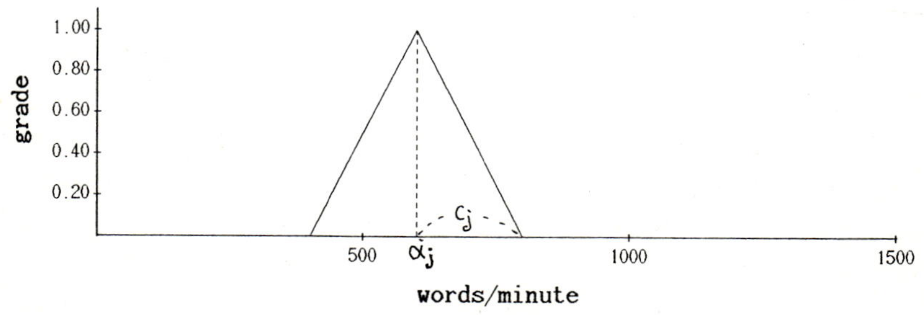

Fig. 4. Fuzzy number: $A_j = (\alpha_j, c_j)$

A function whose parameters are fuzzy number A_j $(j=1,...,n)$ defined as

$$Y = g(X,A_1,...,A_n) \; ; \quad f_Y(y) = \sup_{y=g(\mathbb{X},\mathbb{A})} \{ f_{A_1}(a_1) \wedge \cdots \wedge f_{A_n}(a_n)\} \tag{17}$$

Fuzzy parameters are defined by such fuzzy sets as illustrated in Fig. 4.

Given a fuzzy parameter vector $A = (\alpha,c)$, $j=1,..,n$ the fuzzy linear function

$$Y = A_1x_1 + \cdots + A_nx_n = \mathbb{A}\,\mathbb{x} \tag{18}$$

is obtained as the following membership function(ref.4.):

$$f(y) = \begin{cases} (\{y - \mathbb{X}'\alpha\}/\mathbb{c}'|\mathbb{x}|) & ; \mathbb{x}\neq 0 \\ 1 & ; \mathbb{x}=0, y=0 \\ 0 & ; \mathbb{x}=0, y\neq 0 \end{cases} \tag{19}$$

where $\mathbb{x} = (|x_1|,...,|x_n|)^t$ and $f_Y(y)=0$ when $\mathbb{c}'|\mathbb{x}| < |y-\mathbb{x}'\alpha|$.

This type of fuzzy linear functions can be symbolically rewritten as

$$(\alpha_1,c_1)\, x_1 + \cdots + (\alpha_n,c_n)\, x_n = (\alpha'\mathbb{x},\mathbb{c}'|\mathbb{x}|) \tag{20}$$

In the contrast to the above, the linear function whose parameters are gaussian random variables can be symbolically written as

$$N(\mu_1,\sigma_1^2)\, x_1 + \cdots + N(\mu_n,\sigma_n^2)\, x_n = N(\mu'\mathbb{x},(\sigma^2)'\mathbb{x})$$

where the vectors of variances σ_j^2 and mean μ_j are denoted as $\sigma^2 = (\sigma_1^2,...,\sigma_n^2)$ and $\mu = (\mu_1,...,\mu_n)$, respectively.

The above calculations are similar in form, but the meanings are different. Hence, the fuzzy output Y is calculated in the view of possibility of the parameter vector $\mathbb{A}$.

IDENTIFICATION OF LEARNING CURVE

The learning phenomenon is discussed in psychology, ergonomics and industrial engineering. Although learning is inherent in human nature, the learning phenomenon is not limited to the working individual. It is also evident in each company, industry and economy which consists of many human workers. When considering a company or industry, the learning curve is referred to as a manufacturing progress function, an improvement curve or a production progress function.

Several models of a learning curve have been proposed in analyzing the learning phenomenon. Typical learning models include the log linear model, the log model, the exponential model and the power series model.

(i) Log linear model

$$y = a\ x^{b} \tag{21}$$

(ii) Exponential model

$$y = a\ e^{bx} \tag{22}$$

(iii) Power series model

$$y = a_0 + a_1 x^{-1} + a_2 x^{-2} + \ldots + a_k x^{-k} \tag{23}$$

The log linear model is usually employed to analyze the human learning phenomenon.

In this section, we employ the possibilistic concepts mentioned above to cope with the human learning phenomenon, whose curve is modeled by the several types of function. Since the learning data are possibly realized, the possibilistic linear model will be fitting to the given data. Thus possibilistic linear models which have fuzzy parameters in learning models 21 to 23 are considered as regression models.

Possibilistic regression model

Fuzzy regression analysis(ref.4) has been proposed on the basis of the Fuzzy Set Theory. In this fuzzy regression model, the difference between observed and estimated values is not considered as statistical error, but is assumed to result from the fuzziness of a system structure itself. Regarding the deviation between the data as the reflection of fuzziness on system parameters. In other words, in the possibilistic linear regression model, spread of data is supposed to be due to the possibility of the parameters of the linear function.

The following possibilistic linear function is considered as a general input-output data shown in Table 1:

$$Y_i = A_1 X_1 + \ldots + A_K X_K \qquad ; \ i=1,\ldots,n \tag{24}$$

The problem here is to determine the fuzzy parameter vector A so that the output y_i may be contained in the estimated fuzzy set $Y_i = A\ x_i$ at more than a certain degree.

Table 1. Input-output data.

Sample number	Output	Input			
	y_j	x_{j1}	x_{j2}	...	x_{jK}
1	y_1	x_{11}	x_{12}	...	x_{1K}
2	y_2	x_{21}	x_{22}	...	x_{2K}
.	.	.	.		.
.	.	.	.		.
.	.	.	.		.
N	y_N	x_{N1}	x_{N2}	...	x_{NK}

The following are supposed to obtain a possibilistic linear regression model:

(i) A possibilistic linear model is written in Eqn. 24.

(ii) Fuzzy parameter A_j is a triangular fuzzy number denoted as (α_j, c_j).

(iii) Given input-output relations (x_i, y_i) and a threshold h where i=1,...,N, it must hold that

$$f^*_{Y_i}(y_i) \geqq h \qquad ; \; i=1,\ldots,N \tag{25}$$

This means that the estimated fuzzy set Y^*_i containes the observed values y_i at least at the degree of more than h.

(iv) The criterion for evaluating the possibilistic linear model is

$$S = \sum_{i=1}^{N} \sum_{j=1}^{K} c_j \, |x_{ij}| \tag{26}$$

where $\sum_{j=1}^{K} c_j \, |x_{ij}|$ is width of the Y_i and therefore S is the sum of width of $\{Y_i\}$.

The possibilistic linear regression model is obtained by fuzzy parameters $\mathbb{A} = (\alpha, \mathbf{c})$ that minimizes S in (iv) subject to the constraints (iii). Given a threshold h, this results in the following linear programming problem.

$$\min_{\alpha,c} S = \min_{\alpha,c} \sum_{i=1}^{N} \sum_{j=1}^{K} c_j |x_{ij}| \qquad (27)$$

subject to

$$(1-h) \sum_{j=1}^{K} c_j |x_{ij}| + x_i \alpha \geqq y_i \quad ; i=1,\ldots,N \qquad (28)$$

$$(1-h) \sum_{j=1}^{K} c_j |x_{ij}| - x_i \alpha \geqq -y_i \quad ; i=1,\ldots,N \qquad (29)$$

$$c_1, c_2, \ldots, c_K \geqq 0 \qquad (30)$$

Possibilistic parameters $A_j = (\alpha_j, c_j)$, $j=1,\ldots,K$, can be easily obtained by solving the above linear programming problem. As for a threshold h, it should be noted that if a sufficient number of samples is given, we can set h=0 according to their possibilities. In this case, only the possibility included in the data is considered. If the given sample number is half the ideal, it is suggested that h=0.5. This means that the Y value has double the possibility width than one obtained at h=0.

Identification of learning curve

Parameters of a possibilistic learning function are fuzzy numbers. For the sake of simplicity, our discussion is restricted to triangular fuzzy numbers.

Let us discuss with learning models (i) to (iii) described before. These models can be regraded as fuzzy linear functions with respect to parameters. For instance, log linear models can be changed into the linear form (24) by taking lograithm in both sides of Eqn. 21. i.e.,

$$\log y = \log a + b \log x \qquad (31)$$

Setting Y = log y, A = log a and X = log x, we have

$$Y = A + B X \qquad (32)$$

According to the discussion of possibilistic regression model, the possibilistic learning curve can be obtained by solving the following LP problems corresponding to both a log linear model and a power series model.

(1) Log linear model.

$$\min_{\alpha,c} S = \min_{\alpha,c} \sum_{i=1}^{N} \{c_0 + c_1 |x_i'|\} \tag{33}$$

subject to

$$(1-h)\{c_0 + c_1 |x_i'|\} + \alpha_0 + x_i' \alpha_1 \geqq y_i \quad ; i=1,\ldots,N \tag{34}$$

$$(1-h)\{c_0 + c_1 |x_i'|\} - \alpha_0 - x_i' \alpha_1 \geqq - y_i \quad ; i=1,\ldots,N \tag{35}$$

$$c_0, c_1 \geqq 0 \tag{36}$$

where $y' = \log y$ and $x_i' = \log x_i$ $(i=1,\ldots,N)$.

(2) Power series model.

$$\min_{\alpha,\mathbb{C}} S = \min_{\alpha,\mathbb{C}} \sum_{i=1}^{N} \sum_{j=0}^{K} c_j |x_{ij}'| \tag{37}$$

subject to

$$(1-h) \sum_{j=0}^{K} c_j |x_{ij}'| + \boldsymbol{x}_i' \alpha \geqq y_i \quad ; i=1,\ldots,N \tag{38}$$

$$(1-h) \sum_{j=0}^{K} c_j |x_{ij}'| - \boldsymbol{x}_i' \alpha \geqq - y_i \quad ; i=1,\ldots,N \tag{39}$$

$$c_0, c_1, \ldots, c_K \geqq 0 \tag{40}$$

where $x_{ij}' = x_i^{-j}$ $(i=1,\ldots,N;\ j=0,1,\ldots,K)$.

Fig. 5 illustrates a possibilistic learning curve obtained by the power series model in the case K=1.

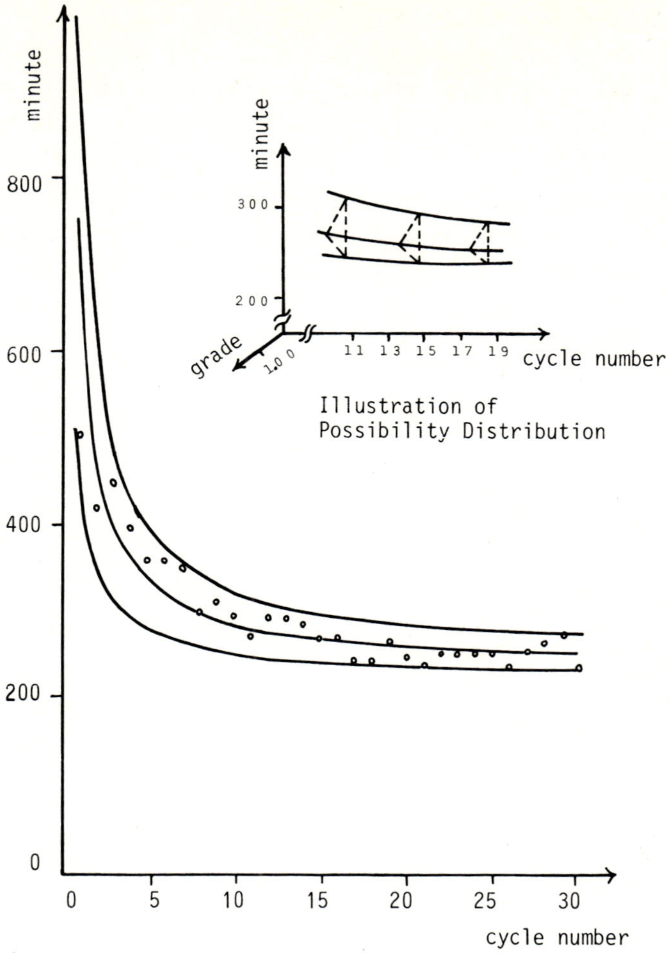

Fig. 5. Illustrative Possibilistic Learning Curve.

LEARNING CURVE IN DATA ENTRY JOB

Data entry is an important job concerning information processing. The success of information processing is dependent upon the accuracy of the data entry. Here we analyze the learning phenomenon in data entry operations for word processing.

The word processor employed was Fujitsu OASIS-100 and the subjects were three students who had no experience on any type of word processor including a typewriter. One cycle performance of the job consisted of the input of 20 letters of Japanese alphabet through an English-Japanese alphabet conversion. That is, every Japanese phonetic alphabet letter can be expressed by using two or one English letters. When English letters are input in the word processor, every two/one English alphabet letters are automatically converted into the corresponding Japanese alphabet letter through the English-Japanese alphabet conversion function.

The time of each operation cycle was measured for 30 complete cycles. The subjects had ten minutes rest interval between cycles. In the experiment mistyped characters were not counted as input characters.

Results

As illustrated in Fig. 6, the cumulative mean of the data shows the decreasing hyperbolic relation of learning curves. Therefore, we assumed a possibilistic regression function as

$$Y = A_0 + A_1 X^{-1} \tag{41}$$

Because of the sufficient number of the observed data, the possibilistic regression model was analyzed at h=0. For subjects A, B & C, the obtained possibilistic learning curves are as follows:

$$Y_A^* = (241.5,\ 13.1) + (511.8, 226.2)\ X^{-1} \tag{42}$$

$$Y_B^* = (283.3,\ 38.7) + (291.4,\ 32.7)\ X^{-1} \tag{43}$$

$$Y_C^* = (251.9,\ 35.1) + (264.2,\ \ 0.0)\ x^{-1} \tag{44}$$

Fig. 7 illustrates a possibilistic learning curve for subject A. As shown in Fig. 7, the estimation of each cycle time is calculated in a fuzzy number which expresses a possibilistic interval. The learning effect could be recognized in the data entry job through the decreasing possibilistic function obtained from the data each of three objects. Especially, subjects A and B have a large learning effect. On the other hand, the result of subject C in the data entry

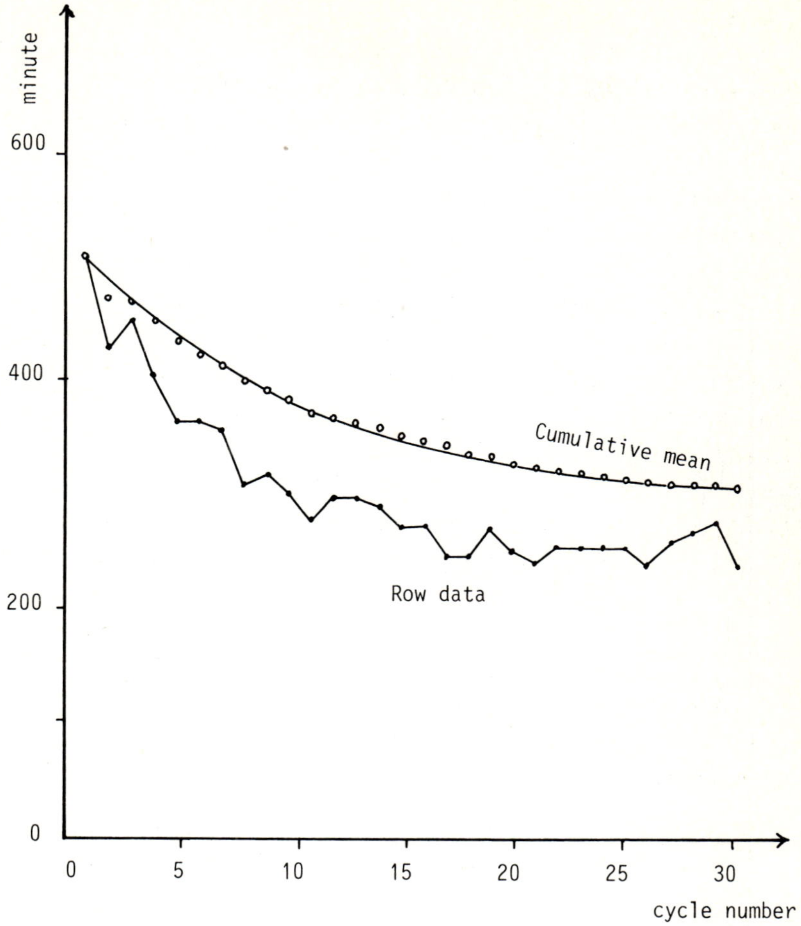

Fig. 6. Learning Curve of Cumulative Mean.

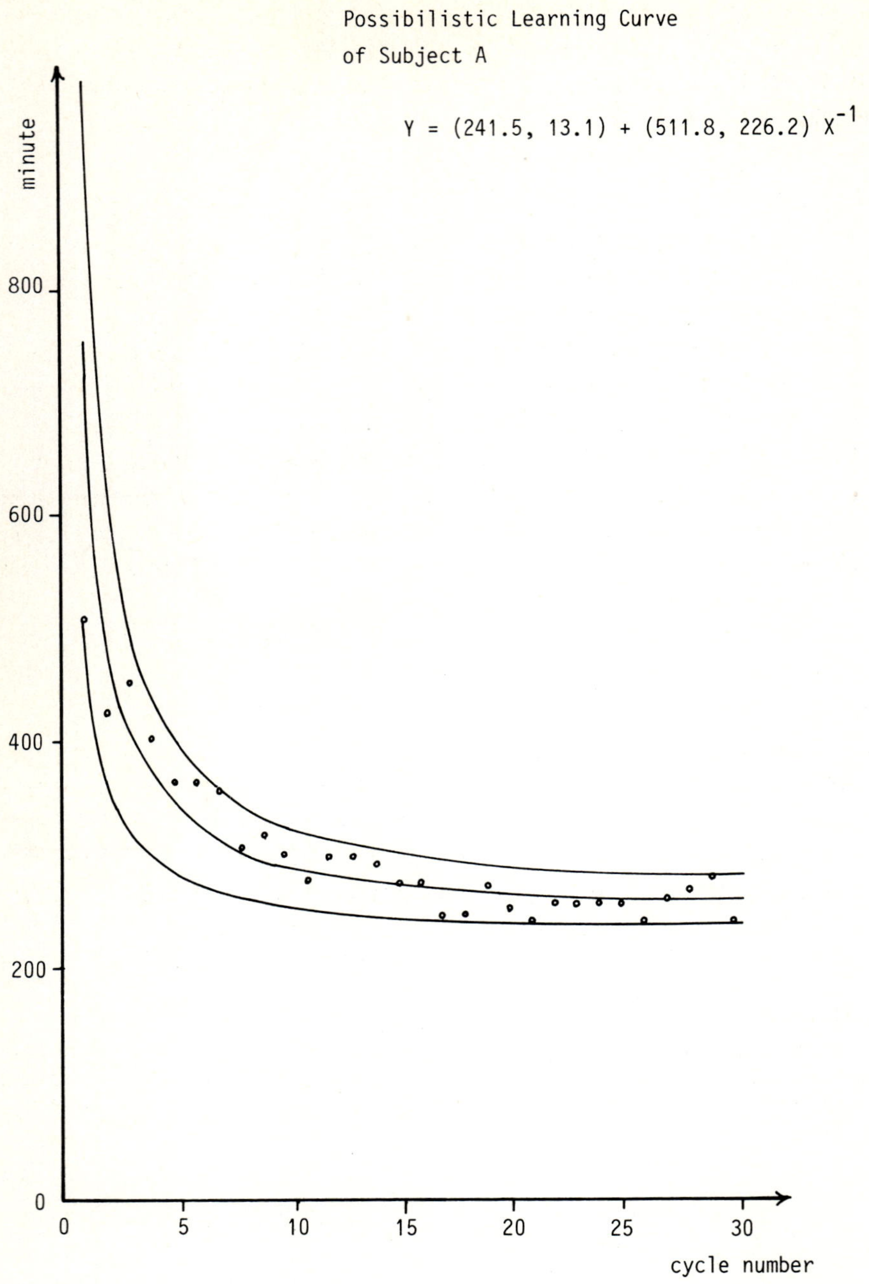

Fig. 7. Possibilistic Learning Curve of Subject A.

job has fluctuated widely over the trials but it does not show a sufficient learning effect. Furthermore, Fig. 7 explains that subject A has a learning effect even on the possibility interval. That is, as he became experienced in the job, he could perform the task in a stabler speed.

CONCLUDING REMARKS

In this paper, the learning phenomenon has been discussed using a possibilistic model. The decrease of time or cost through repetitive performances of a human task or of a production process is caused by a learning effect. The learning effect is found not only in an individual's performance on a task, but also in the performacne of each organization such as a company, industry or an economy in which human performance plays a central role.

Emphasis should be placed on the learning phenomenon which is brought out through the possibility of human ability and also that it is natural to interpret the learning phenomenon within the framework of the possibility concept. Therefore, the possibilistic regression model can be employed to analyze the learning phenomenon.

This possibilistic approach enables us to grasp the two characteristics of learning effects. One is related to the trend of learning effect and the other is related to the possibility interval at each time.

Through analyzing the learning phenomenon in data entry job, we are able to understand that the learning effect even on the possibility width is apparent in the repetition of a job. That is, when persons become experienced in a task, they can perform the task with a stable speed. Therefore, it can be concluded that our possibilistic learning model can deal properly with the learning phenomena in terms of its trend and also its possibilistic interval.

REFERENCES

1 W.M. Hancock and F.H. Bayha, The learning Curve, in Handbook of Industrial Engineering ed. by G. Salvendy, John Wiley, New York, 1982, pp 4.3.1-4.3.13.

2 S. Nakayama and R. Sakai, The Evaluation on the Achievement of Workers, J. of Japan Industrial Management Association, Vol. 32, NO. 5, (1981), pp 372-377 (in Japanese).

3 L.A. Zadeh, Fuzzy Sets as a Basis for a Theory of Possibility, Int. J. Fuzzy Sets and Systems, Vol. 1, (1978), pp 3-28.

4 H. Tanaka, S. Uejima and K. Asai, Linear Regression Analysis with Fuzzy Model, IEEE Trans. on SMC, 12 (1984), pp 903-907.

5 J. Watada, H. Tanaka and K. Asai, Fuzzy Quantification Theory Type I, The Japanese Journal of Behaviormetrics, 11 (1984), pp 66-72 (in Japanese).

6 H. Tanaka, T. Shimomura, J. Watada and K. Asai, Fuzzy linear regression analysis of the number of staff in local government, FIP-84 at Hawaii on July 22-26, 1984.

7 L.A. Zadeh, The Concept of a Linguistic Variable and its Application to Approximate Reasoning-I, Inform. Sci., 8 (1975), pp 199-249.

8 D. Dubois and H. Prade, Fuzzy Sets and Systems: Theory and Application, Academic Press, 1980, pp 36-67.

9 S. Nahmias, Fuzzy Variables, Fuzzy Sets and Systems, 1, 2 (1978) pp 97-111.
10 M. Sugeno, Fuzzy theory IV (Lecture note), J. of SICE, 22 (1983) pp 554-449 (in Japanese).
11 D. Duboir and H. Prade, Additions of Interactive Fuzzy Numbers, IEEE Trans. on Automatic Control, AC-26, 4 (1981) pp 928-936.
12 M. Mizumoto and K. Tanaka, Some Properties of Fuzzy Numbers, Advances in Fuzzy Set Theory and Applications ed. by M. M. Gupta, R. R. Ragade and R. R. Yager, North-Holland (1979) pp 153-164.

Part IV.

DESIGN AND ANALYSIS OF WORK SYSTEMS

Applications of Fuzzy Set Theory in Human Factors, edited by W. Karwowski and A. Mital
Elsevier Science Publishers B.V., Amsterdam, 1986 — Printed in The Netherlands

AN APPLICATION OF FUZZY METHODS IN THE COMPLEX ERGONOMICS DIAGNOSTICS OF INDUSTRIAL PRODUCTION SYSTEMS

L.M. PACHOLSKI

Technical University of Poznan, Strzelecka 11, 60-965 Poznan, (Poland)

ABSTRACT

At the Technical University of Poznan has been worked out and put into practice at several productive plants, the basic methodical conception of ergonomics diagnostics combine the ergonomic diagnosing and implementation procedure in one methodical system. The diagnostic variable have been interpreted as the membership measures according to fuzzy set theory. Owing to the multiaspectual character of ergonomics diagnostics symptoms formulated in multiobjectual production system scale, the basic ergonomics diagnosis assumes the matrix form in accord with criterion of fuzzy relation between symptoms and man-machine units. The above mentioned fuzzy set has been interpreted as the base of ergonomicity degree valuation of different variants of the production system potential modernization. The valuation has been realised in convention of fuzzy logic with implementation of the meta-implicative base of infernence. The model of the infernence procedure has been based on the composition equation of fuzzy relation.

INTRODUCTION

When discussing the problem of conditioning application of ergonomics in the production practice, the formal-procedural and organizing aspect of the problem is most frequently emphasized.

Whereas particular premises of applying ergonomics, resulting from the essence itself and specifity of this field of knowledge, treated in categories of applied science, are rarely exposed. This group of conditionings, of endogenous character, is connected with a scientific thorough study of the methodological problem and objective criterion of the present ergonomics. The above mentioned fact constitutes the basic premise which determines the way and point of reference to realize ergonomics projects in practice.

The treatment of the subject of investigations and ergonomics methodology in scientific categories requires the creation of a theory formed in a formalized way, including descriptions and functional dependences as well as logical rules etc.

The application of the formal modelling, as the instrument of scientific study, creates such possibilities. In the case of ergonomics, subjective treatment of man, both within the methodological problem and the objective criterion (man-machine unit), favours the great indeterminacy and complexity of the description, relations and rules.

This inaccuracy and fuzziness, and therefore "softness" from the point of view of formal modelling, characterizing phenomena, processes and methodological as well as objective systems of modern ergonomics, can be set to some extent against "hardness" of methodological and objective classical approaches, applied in engineering. Besides, it should be emphasized that that the classical methods of formal modelling occured in principle for the very need to form descriptions, dependences and logical rules concerning the "hard" technical systems.

The usefulness of classical methods of formal modelling of technical systems became rooted in the mentality of scientists to much a degree that these methods also began to be transformed into the field of knowledge, comprising within range the human factor apart from the technical one. Such transformation of "hard" methods to investigate phenomena, processes and systems determined by the cooperation of the human factor with the technical one, are criticized more and more frequently as well by theoreticians as by practicians engaged in fields of knowledge whose subject is man.

Here, inadequacy of the character of "technical" modelling classical methods, for the "humanizing" character of investigated problems, is the basic critical argument. The variables, entering the "humanizing" model, can be not always expressed in the form of numerical parameters. Therefore, in the stage of the model conceptualization the variables, the values of which cannot be expressed in the form of numbers, are rejected or exchanged artificially for the measurable one.

The consequences of such formal manipulations, performed on input variables, can be often very essential for the final shape of investigated phenomena, processes and systems determined by the cooperation of the human factor with the technical one.

The above mentioned difficulties, connected with the satisfaction of demands for measurability of variables, entering, the model which characterizes the methodological and objective systems of modern ergonomics, created the necessity to search for suitable formal solutions on the ground of the theories of fuzzy sets.

THE METHODOLOGICAL PROBLEM AND THE OBJECTIVE ASPECT OF ERGONOMICS

In the practice of methods of complex realizations of ergonomics undertakings, the problem of interrelations and connections, between the technical knowledge, the economic-organizational one and the knowledge of man, is many a time interpreted differently.

Here, three basic methodological formulae can be separated:

- The first one depends on strict compliance with the specific division of competence among specialists, representing different fields of knowledge co-operating with ergonomics. Particular specialists, working separately, present their own one-sided investigative elaborations, concerning the analysed problem, in order to determine its interdisciplinary solution, if necessary within the limits of joint consultations.

- The second eventuality is based on the postulate of such an enlargement of one's own speciality that it can comprise the remaining "ergonomics fields". This form of realization of ergonomics undertakings seems to be more effective than the previous one. Both these solutions are likewise often met with.
- The third formula, predominating unquestionably over the preceding ones, consist in current interdisciplinary collaboration of specialists, representing separate fields of knowledge, which co-operate on the ground of ergonomics. The abolition of barriers, mainly artificial, between people instructed primarily in particular basic disciplines, seems to be the most proper and modern approach to the methodological problem of the present-day ergonomics.

The problem of methodological specifity, beside the objective aspect, is the basic correlative with it criterion of the distinct feature of ergonomics as applied science.

In ergonomics, the determination of the object of investigations does not present more serious difficulties, since it is the man-machine unit. When treated elementarily, this unit is the constituent part of a larger organizational whole of the production system.

The occurence of groupings, the elementary man-machine units is typical for definite practical conditions of an industrial enterprise. The science of organization and management, conceived classically, numbers these groupings among the categories of probabilistic systems and understands them as a notion of the production system: static as well as dynamic connections of human, physical and financial resources, processing the input sopplies (work, the media and objects of work as well as information) into the output states, expressed in material form (industrial goods, material services) or information.

Uniformity and compexity are the basic features of the production systems. Uniformity means subordination of all the constituent elements of the system to one purpose - achieving a definite result by means of a set of optimized outputs, when the input actions are given.

Complexity of production systems is connected with the fact of the dependent connexion of all the parameters of the system.

The humanizing approach to the production system as the object of ergonomics investigations, in the light of the conception of the possibility theory (treating the possibility as a mathematical construct - analogous measure to the probability measure), indicates the usefulness of reconsidering the formerly quoted classical qualification for groupings of the elementary man-machine units.

The fuzzy character of many variables, characterizing the functioning of production systems, suggests the usefulness of applying the notional category of possibility systems in this case.

Further considerations will deal with the methodological problem, connected with the presentation of the basic methodic conception, the so-called ergonomics diagnostics, because the increased knowledge of the objective aspect of ergonomics, to which above attention has been drawn, can make an extensive subject by itself for a publication.

This conception has been implemented in a dozen or so Polish production enterprises; this signifies the practical verification of a definite methodological tool of present-day ergonomics.

BASIC CONCEPTION OF ERGONOMICS DIAGNOSTICS

The conception of basic ergonomics diagnostics assumes as starting point the existence of a natural mechanism of ergonomics stability in the production systems. Equilibrium, existing in the processes of system functioning, is the result of permanent, dynamic oscillations of the ergonomicity, of its particular man-machine units.

These oscillations, just as in natural systems, can be considered as attribute of the production system functionig, provided that the amplitude of "deviations" does not exceed the limits of system control. Going beyond the above mentioned limits is revealed by disturbances in the system.

By way of example, the following belong to external factors which can cause states of instability: too excessive tasks set to the system or its objects, simultaneus summing up of a series of tasks, constant overload of the system or its objects.

The following can be numbered among exemplifying internal factors: disturbances in the co-ordination of the system performance, wearing out of objects and connections in the system, "inborn" or "acquired" defects and irregularities of the system.

We most frequently make use of diagnostic technique to know the state of stability of a system, connected with its "treatment", prophylaxis and prognosis of further efficiency.

The diagnostic methods have been developed particulary with reference to formally determined technical systems, in which the variables included in the diagnostic model can be expressed in the form of the actual numerical parameter.

As it has already been mentioned in the introduction, the specifity of the man-machine units is their inaccuracy and fuzziness, and therefore "softness" of the system, resulting from the features of man as the object receiving and processing informations as well as making decisions and doing definite activities.

That is why, in diagnostics of production systems, determined by man's participation, there exists the necessity of elaborating a complex methodological tool, functioning under conditions of the fuzzy description of these processes.

Consequently, on approaching the problem of ergonomic diagnosing of the production system in categories of the fuzzy sets theory, we can formulate the following definition of the "soft" diagnostics procedure:

If a possible set of activities "D" is given and if one or more activities "d_i" included in this set can lead to the definition of the actual state of the "s_i" system, then the selected activity, arranged in a logical sequent, is called diagnostic procedure and the defined in this way possible state of the "S" system is called diagnostics.

The essence of the mental diagram and logic sequent of actions, composing the construction of the ergonomics diagnosing model, can be formulated as follows:

1. The assumption of the production system as the object as well as of the diagnostics procedures as the method accounting for the derivation of the conception of the model from the diagrams of the organizational procedure and of the diagnostics of natural systems. Obvious convergences can be noticed in both diagrams shown in figure 1.
2. Procedure, consisting in the determination, verification and checking of hypotheses, ought to be derived from the general diagram of the investigative activity. Therefore, detailed elements, resulting from the comparisons of organizational activity and diagnostics of natural systems, should be "built in" into the analytical procedure and the synthetic one, characteristic for this diagram.
3. The principal ergonomics aim of diagnostics procedure, in reference to the production system is to state the character of inconformity of the system objects to the interdisciplinary requirements of ergonomics. Consequently, it ought to be elucidated what does not correspond to the requirements of ergonomics as well as in what conditionings these inconformities occurred.
4. The answer to the question thus formulated is connected in the analytical part with the regard for two fundamental stages of diagnostics procedure. The first one, connected with the problem of indexes of ergonomicity and requirements, on the basis of which inconformity to the premises of ergonomics can be stated, and the other one which aims at answering the question on what this inconformity depends and therefore at isolating the diagnostics symptoms.
5. Acceptance of the procedure sequent, consisting in successive answers to the

questions: under what conditions? and what? is the principle of the correct diagnostics procedure. The attempt to answer the question, concerning the consequences of the state of incompatibility of the system objects with the requirements of ergonomics as well as the prospects of the diagnosed system, is also an essential stage of the production system diagnosis.

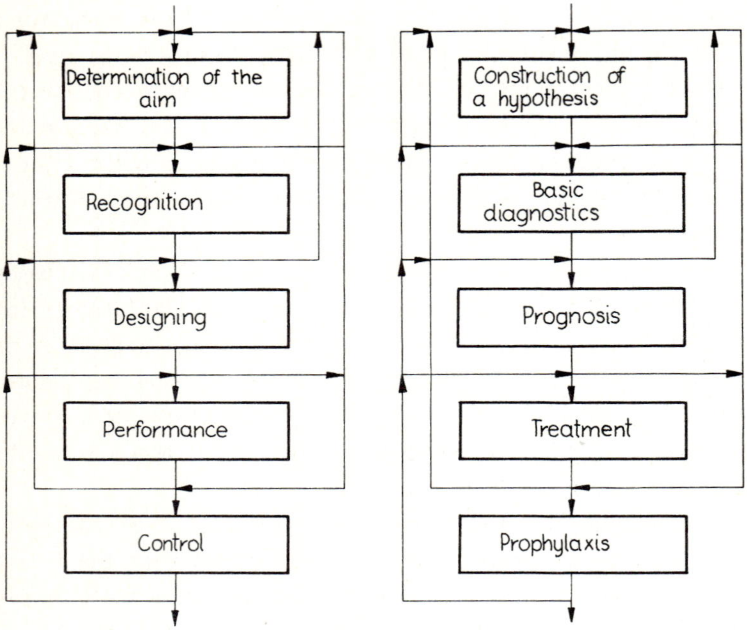

Fig. 1. Convergences in the diagrams of organizational procedure and diagnostics of natural systems.

6. Two stages create the synthetic part of diagnostics. The first one of them includes the elaboration of the program and implementation of the solution reforming the state of incompatibility with the requirements of ergonomics. Reformation can concern the stable systems - then it has the form: of liquidation of causes, liquidation of symptoms or aided proof. In the case of technological and organizational modernization, corrective ergonomics has to satisfy a much more important and more dynamic range of duties.
   The other stage of diagnostic synthesis is connected with counteraction to the process of occurrence of system incompability with the requirements of ergonomics.
   In practice, this kind of preventive activity is realized by current

ergonomics control of the production system as well as by the elaboration and implementation of such new (prospective) constructional, technological and organizational solutions which meet the requirements of present-day ergonomics.

Thus, on taking into account the above mentioned remarks, the output model of ergonomics diagnostics will assume the form shown in figure 2.

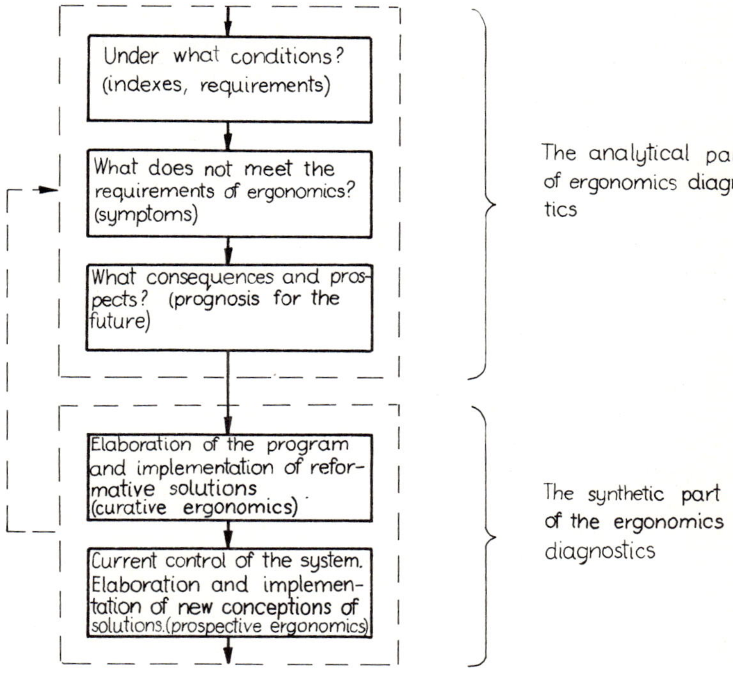

Fig. 2. The output model of ergonomics diagnostics.

Further development and formalization of the model of ergonomics diagnostics consisted in taking into account the specific of analytical and implemental undertakings, typical for modern ergonomics. These problems are described in: "Methodology of ergonomics diagnosing" (Pacholski L., 1977, Poznan) and shortly in: "Proceedings of the 1984 International Conference on Occupational Ergonomics" (p. 167-170).

The model shown in figure 3 seems to be the most useful of the variants, issued in English, in publications for practical purposes. This model can be acknowledged as illustration of the basic conception of methodical ergonomics diagnostics, sufficiently useful for further considerations presented in this paper.

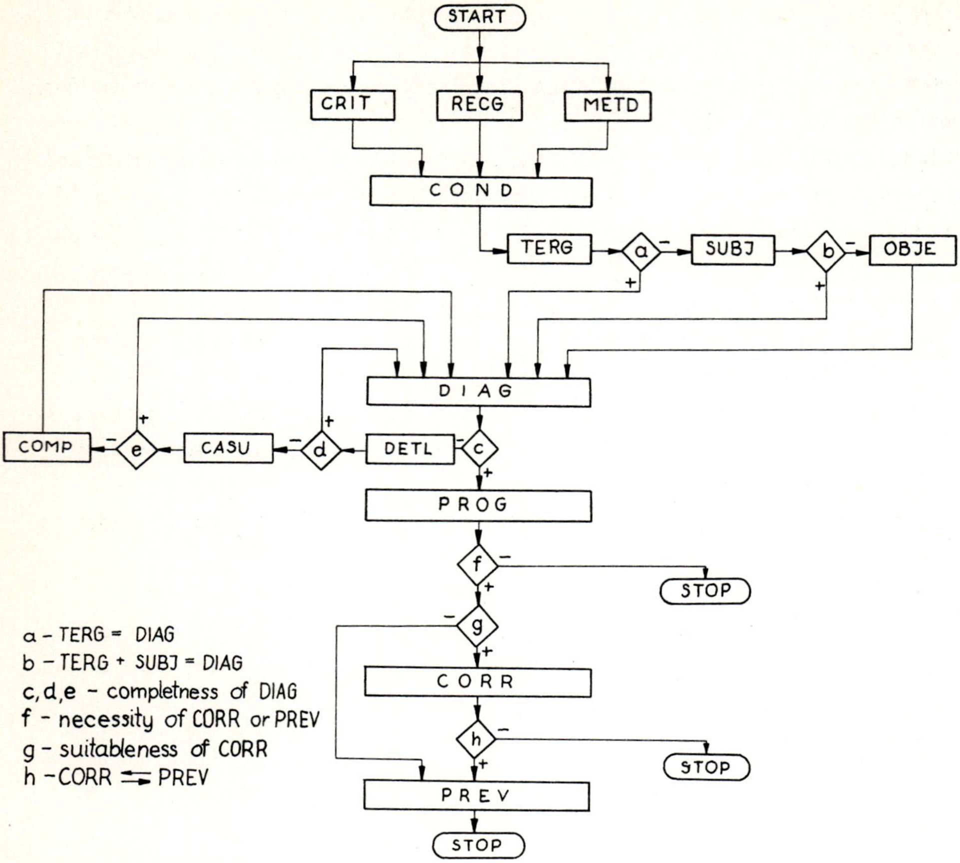

Fig. 3. The ergonomics diagnostics model.

THE FUZZY CHARACTER OF ERGONOMICS DIAGNOSTICS VARIABLES

The formulation of the basic ergonomics diagnosis of the complex production system is based on the so-called diagnostic symptoms, determined by the method of experts and connected with the set of detailed ergonomics requirements.

A list of 150 indexes of ergonomicity, elaborated by the author of this paper for the demands of the presented method, is the basis of the construction of the above mentioned set of symptoms.

The following four principial thematic groups create the structure of the set:

1. Space and equipment,
2. Man-machine interaction,

3. Work environment,
4. Work design and organization.

Each of the mentioned main thematic groups contains from four to seven problems. Whereas, detailed indexes, which in sum create the set of 150 components, are included within each of these problems. The indexes have been connected with detailed ergonomics requirements.

The ergonomics variable, describing the so-called ergonomicity of the diagnosing man-machine unit, can be the evaluation of the degree of conformity of this unit to the above mentioned requirements.

The ergonomics diagnostics parameters have been interpreted as measures of membership, in accordance with the fuzzy set theory.

The fuzzy character of the standard, reflecting a definite ergonomics requirement, is the fundamental methodical premise of such treatment of ergonomics diagnostics parameters.

On the analogy of the technical field, ergonomics made attempt to standarize the above mentioned standards.

However, these standards, constituting a system of references for variables describing the ergonomicity of the object, have a fuzzy character, because they can be located as well in the zone of ergonomics optimum as in the zone approximated to this optimum.

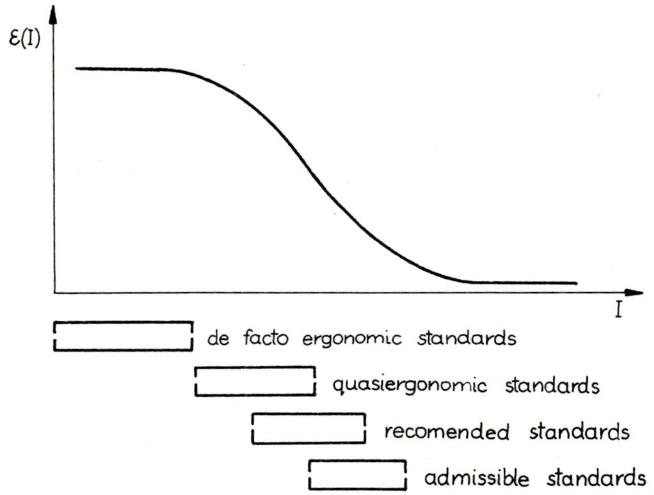

Fig. 4. The fuzzy character of the edge boundaries of ergonomics standards. (I - degree of intensity of negative action of the factor included in the ergonomics index, ε(I) - degree of ergonomicity of the evaluated object in the range of the above mentioned index).

The problem of the fuzzy character of ergonomics standards, constituting the reference system to evaluate the variables describing the so-called ergonomicity of the diagnosed object, is illustrated visually in figure 4.

In addition, it should be noticed that on account of its humanized character all sorts of ergonomics standards show natural dynamics, consisting in gravitating towards the zone of standards, de facto ergonomics ones.

The difficulties, connected with the fulfilment of the condition of precision as well as that of measurability of all the ergonomics requirements, brought about the necessity of treating the above mentioned 150 component set of indexes as a class of elements with fuzzy limits, i.e. in which there is no sharp boundary between the elements belonging and those not belonging to this class.

The above formulation is an informal definition of the fuzzy set.

Therefore, if $E = \{e\}$ will be a set of system indexes, then the fuzzy set W in E is characterized by the membership function f defined on E and assuming the values in the $[0, 1]$ interval, i. e.:

$$f : E \rightarrow [0, 1] \qquad (1)$$

The W fuzzy set of diagnostics variables is represented as a set of ordered pairs:

$$W = \{ (f(e), e) \; ; \; e \in E \} \qquad (2)$$

The value of functions from the $[0, 1]$ interval, determining the value of membership of element e to the W set, is established on the basis of subjective evaluations of the ergonomics team, according to the premised form of membership function.

DIAGNOSTICS MATRICES OF FUZZY RELATIONS

On account of the character, having many aspects, of the ergonomics diagnostics estimate, formulated in the scale of the multiobjectual production system, the input form of the basic ergonomics diagnosis has the form of a matrix recording of 150 rows and number of columns corresponding to the number of elementary man-machine units.

The above matrix constitutes a formal reflection of the R relation of the form:

$$R \subset W \times U \qquad (3)$$

The R fuzzy relation between W and U corresponds to the basic diagnostics information on the influence of a definite form of index W on the ergonomicity of performance of man-machine units in the production system.
Therefore, the membership function:

$$f_R(w,u) \in [0\ ,\ 1] \qquad (4)$$

defines the rank of diagnostics usefulness of a definite parameter of ergonomicity, whereas the function of possibilities distribution:

$$\pi_R(w,u) \in [0\ ,\ 1] \qquad (5)$$

defines the possibility with which parameter w influences the performance of a definite man-machine unit, in accordance with the requirements of ergonomics.

The matrix of relation between w and u can constitute both the output point of ergonomics prognosing and the basis of concrete undertakings, reforming the production system in ergonomics respect.
However, from the formal point of view, it is imperative to process earlier this matrix into the matrix of the so-called ergonomics corrective undertakings. In order to attain this, algorythmical matchings of set of 150 indexes of the w ergonomicity with the set of 100 respective corrective undertakings k were performed.

Processing of the matrix of the ergonomicity indexes ito the matrix of ergonomics corrective undertakings is performed with the help of the computer program, elaborated at the Technical University of Poznan, to process diagnostic data.
The printout has the form of a matrix of 100 rows and number of columns corresponding to the number of diagnosed man-machine units.
The above matrix constitutes a formal reflection of the fuzzy relation Q of the form:

$$Q \subset K \times U \qquad (6)$$

The form complementing this relation, defined as follows:

$$f_{\bar{Q}}(k,u) = 1 - f_Q(k,u) \quad , \quad k \in K\ , \quad u \in U \qquad (7)$$

can constitute the basis to define the priorities of ergonomic correction of particular objects of the diagnosed production system.
The numerical parameters, stating precisely the rank of a definite corrective

undertaking, are obtained as a result of diagnosing concrete values of the membership function.
As a result of processing the above mentioned diagnostic data, it is therefore possible to determine the formalized sequent of undertakings, limiting casually the concrete symptoms of the production system ergonomics instability (block 4 in fig. 2 as well as block CORR in fig. 3).

APPLICATION OF THE META-IMPLICATION PRINCIPLE IN ERGONOMICS DIAGNOSTICS

The fuzzy set, reflecting the relation of diagnostics parameters and objects of the production system, characterized by the membership function f_R can constitute the basis of ergonomics evaluation of usefulness of different variant of potential modernization of the production system.

In the production practice of industrial enterprises, these variants most frequently occur as a result of the action of natural mechanisms of the technological and organizational progress.
Most of these solutions has a reformative character in relation to the current state. Their realization is undertaken as well on the basis of corrective formulae as of prospective undertakings.

Multi-variantivity of technological and organizational solutions is one the one hand a characteristic feature of corrective and prospective technological modernization in the industrial practice, whereas on the other hand the treatment of the human factor requirements as of secondary importance.
The above fact involves the necessity of testing ergonomics usefulness of these variants in the context of typical ergonomics diagnostics procedure, connected with the stage of synthesis (fig. 2 and fig. 3). This evaluation concerns the humanizing aspect.
Consequently, it has not an absolute character but constitutes some kind of prognosis, based on the attestation of different variants of the potential modernization of the production system.
Here, infernence can be realized in the convention of fuzzy logic, on applying the meta-implicative principle of infernence (the so-called soft infernence).

The model of selective procedure was based on the following equation of three fuzzy relations composition:

$$M = R \circ T \tag{8}$$

where:

R - the relation between the index of ergonomicity and the elementary man-machine units, creating the diagnosed production system

$$R \subset W \times U \tag{9}$$

T - the relation between the variant of the technological-organizational modernization of the production system and the indexes of ergonomicity

$$T \subset V \times W \tag{10}$$

M - the relation between the variants of the technological-organizational modernization of the production system and the elementary man-machine units, creating the diagnosed production system

$$M \subset V \times U \tag{11}$$

The above mentioned equation (8), recorded in categories of membership function

$$f_M(v,u) = \max_{w \in W} \left[f_R(w,u) \wedge f_T(v,w) \right] \tag{12}$$

can constitute a rational premise of ergonomics optimization of the implementation decisions, connected with the technological-organizational modernization of production systems in practice.
In the practical meaning, this premise is a concrete recording of M the fuzzy relation in the matrix form.

The designing ot the M matrix as well as the search for the maximum values of connections, within the relation between the variants of the technological-organizational modernization of the production system and its already functioning work-stands, can be realized as an element of computer data processing, connected with the previously described model of diagnostics.
The above solution completes essentially the complementarity and practical usefulness of the complex ergonomics diagnostics.

RECAPITULATION

The presented paper aims principally at showing the basic conception of practical applications of fuzzy methods and techniques in complex ergonomics diagnostics of production systems.
The model of ergonomics diagnostics, elaborated at the Technical University of Poznar, is an attempt to define a method of co-operation, on the ground of the interdisciplinary ergonomics, of experts realizing the complex ergonomics undertakings under concrete conditions of practice.

The elaboration of exactly such methods, specific for this field of knowledge, of complex realization of ergonomics undertakings, can constitute one of the fundamental conditions to make ergonomics as an interdisciplinary applied science.

Its methodological and subjective specific is mainly defined by man's subjectivity.

The occurrence of the difficult to measure and elastic humanizing aspect of ergonomics, apart from the determined and measurable technical aspect, causes that the category of the so-called ergonomicity evades any attempts of "hard" parametrization.

The application of the calculus of probability rules to the methodological and subjective interpretation of phenomena and ergonomic processes is often not very reliable, in spite of using time-consuming and expensive statistic experiments.

The opinion that the category of ergonomicity is of the not statistic or not quite statistic nature, has many advocates in the circle of scientists and practicians, engaged in the field of ergonomics.

As things are, the use of the fuzzy sets formula, to describe the methodological forming of phenomena and processes, treated complexly and connected with the category of ergonomicity; is a desirable alternative and a rational one under these conditions.

The theoretical usability and practical usefulness of applying the fuzzy methods for the present-day ergonomics methodology requires, however, both experimental verification and numerous implementation undertakings, realized under concrete conditions of practice.

REFERENCES

1 M. Gupta, A Survey of Process Control Applications of Fuzzy Set Theory, in Proceedings of the 1979 IEEE Conference of Decision and Control, 1979.

2 W. Karwowski and M.M. Ayoub, Fuzzy Modelling of Stresses in Manual Lifting Tasks, Ergonomics, vol. 27, no. 6, 1984, pp. 641-649.

3 M. Nowakowska, Methodological Problems of Measurement of Fuzzy Concepts in the Social Sciences, Behavioral Science, 22 (1977), pp. 107-115.

4 L. Pacholski, An Ergonomic Method Diagnosing Complex Production Systems, Ergonomics, vol. 25, no. 6, 1982, pp. 568-569.

5 L. Pacholski, An Methodical Problems of Ergonomic Diagnosing, in Proceedings of the 1984 International Conference on Occupational Ergonomics , Attwood & Mc Cann, Toronto, 1984, pp. 167-170.

6 E. Sanchez, Solutions in Composite Fuzzy Relations Equations. Application to Medical Diagnosis in Brouwerian Logic, in Fuzzy Automata and Decision Processes, M. Gupta, G. Saridis, B. Gaines, Elsevier North Holland, New York, 1977.

7 M.A. Sinclair,and C.G. Drury, On Mathematical Modelling in ergonomics, Applied Ergonomics, 10 (1979), pp. 225-234.

8 R.M. Tong, The Construction and Evaluation of Fuzzy Models, in Advances in Fuzzy Set Theory and Applications, North Holland, Amsterdam, 1979.

9 L. Zadeh, Fuzzy Sets, Information and Control, 8 (1965), pp. 338-353.
10 L. Zadeh, Outline of a New Approach to the Analysis of Complex Systems and Decision Processes, IEEE Transactions on Systems, Man and Cybernetics, 3 (1973), pp. 28-44.
11 L. Zadeh, Fuzzy Sets as a Basis for a Theory of Possibility, Fuzzy Sets and Systems: International Journal, 1 (1978), pp. 3-28.
12 H. Zimmermann, Fuzzy Programming and Linear Programming with Several Objective Functions, Fuzzy Sets and Systems, 1 (1978), pp. 45-55.

Applications of Fuzzy Set Theory in Human Factors, edited by W. Karwowski and A. Mital
Elsevier Science Publishers B.V., Amsterdam, 1986 — Printed in The Netherlands

APPLICATIONS OF APPROXIMATE REASONING IN RISK ANALYSIS

W. KARWOWSKI[1] and A. MITAL[2]

[1]Department of Industrial Engineering, University of Louisville, Louisville, Kentucky 40292 (U.S.A)

[2]Department of Mechanical and Industrial Engineering, University of Cincinnati, Cincinnati, Ohio 45221 (U.S.A)

ABSTRACT

This paper discusses applications of approximate reasoning techniques in risk analysis. Vagueness and imprecision in mathematical quantification of risk are equated with fuzziness rather than randomness. The concept of fuzzy risk evaluation, using linguistic representation of the likelihood of the occurrence of a hazardous event, exposure, and possible consequences of that event, is proposed.

INTRODUCTION

The problems of risk, broadly defined as an undesirable implication of uncertainty (ref.1, 2 and 3), and its analysis can be classified into two distinct categories. The first category deals with uncertainties which are essentially random, while the second one involves problems which are not really probabilistic, but cause uncertainty due to complexity of the systems under investigation and vagueness of the natural language.

Unfortunately, at present the techniques used for calculating risk in the area of systems safety equate such uncertainties with randomness only, in order to "compensate for the fact that the given nominal value of any parameter, even if explicitly defined, is rarely known with absolute precision" (ref.4). Therefore, the quantification is usually obtained by assigning to each set of events a probability measure. However, in many cases it is virtually impossible to determine precisely the value of the probability of a given event. Such a situation may be due either to lack of evidence or to the inability of the safety engineer to make a significant assessment of the frequency of occurrence of an event (ref.5). In other words, the probability of an event may be ill-defined, and one would simply say that a given event is "more or less likely," "remotely possible," etc., where terms such as "likely" and "possible" are vague and imprecise descriptors which are often used by risk analysts (ref.6). One must remember, however, that use of such value judgments introduce uncertainty due to fuzziness, not randomness.

The theory of fuzzy sets make it possible to study the complex and ill-defined systems where uncertainty is caused by fuzziness, or degree vagueness (ref.7). In this paper, a fuzzy logic-based approach to risk analysis is proposed as an alternative to the techniques currently used in the area of systems safety, and linguistic variables (ref.8) are applied to analyze hazardous situations using approximate reasoning methods.

THE USE OF NATURAL CATEGORIES IN RISK ASSESSMENT

According to Fine (ref.6), the risk (or uncertainty of loss) imposed by a particular hazard increases with the likelihood of occurrence of the event (L), exposure (E), and the possible consequences (C) of that event. It is also assumed that although all hazards can never be completely eliminated, the associated risks from the hazards can be reduced (ref.2). In a traditional approach, the calculation of the value of risk (S) is based on the product of numerical values of likelihood, exposure and consequences. This product is called ´risk score.´ Using experts´ judgments, several risk scores can be obtained and ordered with respect to seriousness of their risks. A risk score summary is then presented to safety managers in order to undertake specific action and reduce the existing hazards.

Development of a practical risk analysis system (ref.2) indicates that engineers have long recognized the imprecise nature of risk evaluations and the importance of judgments based on sound experience (ref. 9, 10 and 11). As stated by Feagans (ref.12), risk is a fuzzy concept in the sense that there does not exist a unique risk that a hazardous event will occur in a given period of time.

Since risk analysis deals then with quantities whose future values are uncertain, such values must be based on subjective judgments of human experts. It should be noted here that although humans are quite unsuccessful in quantitative predictions, they are comparatively efficient in qualitative forecasting (ref.13). In fact, the knowledge of experts usually consists of qualitative verbal expressions as evidenced by recent developments in knowledge-based expert systems (ref.14).

Although the meaning of verbal descriptors is vague and it may be difficult to find their numerical representations (ref. 15, 16, 17, 18, and 19), system safety analysts have worked out a method for risk assessment which is primarily based on human judgment and experience. Several verbal descriptors of the risk score were derived, and their approximate numerical representations were found. The descriptors relate to all three factors, i.e.: (1) the likelihood that some hazardous event will occur, (2) the exposure to that particular hazardous situation, and (3) the possible consequences should the hazardous event actually occur (see Table 1). The objective of this paper was to develop the fuzzy linguistic model of the above practical risk analysis system.

LINGUISTIC VARIABLES REPRESENTATION OF RISK FACTORS

An assignment of numerical values to the components of the ´risk score´ is subjective by the nature of human judgment. For example, although the concept of probability (likelihood) is well defined, it does not provide for the sharp probability estimates needed to generate adequate risk estimations. Instead, the quantification of ´likelihood´ uses verbal descriptors such as quite possible, unusual but possible, very unlikely, etc. An event may be clearly stated, but its probability is vaguely defined.

Therefore, in this paper, the probability P will be treated as a linguistic variable (ref.8) with the typical values (P_i) such as likely, very likely, more or less likely, very unlikely, etc., and with the understanding that likely is synonymous with probable. Since the likelihood of occurrence of the hazardous event is related to the probability that it might actually occur (ref.2), the numerical variable ´probable,´ with values $0 \leq P_i \leq 1$, is the base variable for the ´likelihood.´

A linguistic value such as likely is interpreted as a label for the fuzzy restriction (characterized by its compatibility function) on the values of the base variable. Typical values of the linguistic variable contain primary terms, such as likely and unlikely; hedges such as very, more or less, quite, extremely, and somewhat; and fuzzy connectives such as and, or, and either; as well as negation, not. The connectives, negation, and hedges are treated as modifiers of the operands (primary terms) in a context-dependent situation (ref.8).

Interpretation of the linguistic values

Although the meaning of the linguistic values proposed in table 1 are open to individual interpretation, the differences in subjective assessments can be resolved by extending the precision of associated verbal definitions through discussion among the experts. It is very important that the structure of verbal descriptors does not cause misunderstanding (ref.3 and 12), and this can be prevented if the agreed upon definitions are provided. As indicated by many researchers (ref.20, 21 and 22), primary linguistic values should have an intuitive appeal and be easily differentiated. For that reason, the values "likely' and "unlikely" were proposed to represent verbal descriptors which are most frequently used by risk analysts, i.e., "quite possible" and "only remotely possible," respectively.

TABLE 1. Linguistic representation of risk factors.

Descriptors used in risk analysis after [2]	Corresponding fuzzy linguistic values
1. LIKELIHOOD (of the accident-sequence)	- [very likely]
- might well be expected	- [likely]
- quite possible	- [more-or-less likely]
- unusual but possible	- [unlikely]
- only remotely possible	- [very unlikely]
- conceivable, but highly unlikely	- [very very unlikely]
- practically impossible	- [impossible]
2. EXPOSURE (to the hazardous event)	
- continuous (many times daily)	- [very high]
- frequently (once a day)	- [high]
- occasionally (one per week or month)	- [moderate]
- monthly (one per month/year)	- [more-or-less low]
- rarely	- [low]
- very rare	- [very low]
3. CONSEQUENCES (of the accident)	
- catastrophe (extensive damage, over $\$10^7$; many fatalities)	- [extremely high]
- disaster (10^6-10^7, fatalities)	- [very high]
- very serious ($\$10^5$-$10^6$, a fatality)	- [high]
- serious ($\$10^4$-$10^5$, serious injury	- [medium]
- important ($\$10^3$-$10^4$, disability)	- [more-or-less medium]
- noticeable ($\$10^2$-$10^3$, first aid)	- [low]

One must still keep in mind that the assessor himself is an essential source of fuzziness, since the same hazardous event may be perceived differently depending upon the experience and individual preferences in risk acceptability (ref.11).

The compatibility functions for the chosen linguistic values are represented here by a string of numbers (see table 2) rather than a continuous function (ref.23 and 24). However, one can also derive the representations of the primary linguistic terms by using the canonical forms of the S and π functions (ref.25), and adjusting the appropriate parameters.

TABLE 2. Compatibility functions for the linguistic representations of risk factors.

Individual term	Compatibility function					
high	0	0	0.1	0.3	0.7	0.9	1
medium	0	0.2	0.7	1.0	0.7	0.2	0
low	1	0.9	0.7	0.3	0.1	0	0
unknown	1	1	1	1	1	1	1
undefined	0	0	0	0	0	0	0
more or less high	0	0	0.3	0.5	0.85	0.95	1
very high	0	0	0	0.1	0.5	0.8	1
likely	0	0.1	0.5	0.7	0.9	1	1
unlikely	1	1	0.9	0.8	0.5	0	0
not likely	1	1	0.5	0.3	0.1	0.1	0

The degree or severity of consequences (C) of the particular event due to some hazard conditions, and the exposure (E) to such hazard were defined in a way similar to the likelihood (L) of the event. The base variable for the degree of consequences was represented by the extent of property damage, and/or by the seriousness of the injuries, expressed by the amount of loss, in a range from \$10 to $\$10^7$ (see Table 1). The primary terms of the variable consequences (C) are high, low, and medium, with the graphical interpretations depicted in Figure 1. Similarly, the primary terms for the variable exposure (E) were defined. The base variable was defined numerically by the relative frequency of occurrence of the hazard events, in days of operation (see Figure 1).

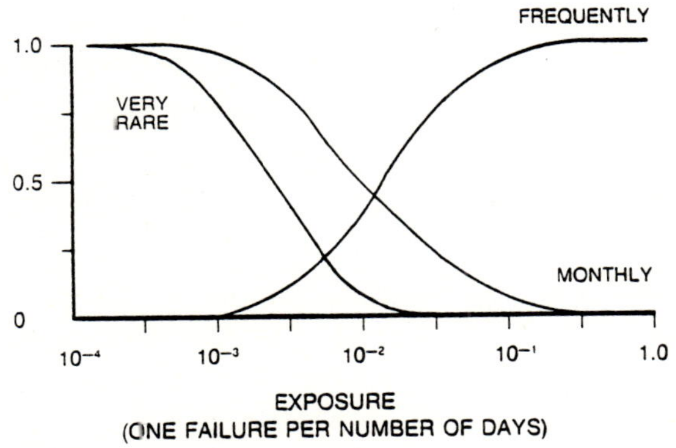

Fig. 1. Linguistic values of exposure.

As indicated above, most of the primary terms for the linguistic variables can be derived based on empirical data and the experience of safety experts, as evidenced by the numerical reference points associated with each of the descriptive (verbal) estimates of the magnitude of likelihood, exposure, and consequences.

APPROXIMATE REASONING METHODS AND RISK SCORE CALCULATIONS

In a traditional approach to risk analysis (ref.2 and 6), the seriousness of the risk due to a recognized hazard (risk score) is calculated as the product of numerical ratings assigned to each of the three factors. The risk score is then compared with the Risk Score Summary. This is done to make a recommendation with respect to an appropriate action to be undertaken in order to reduce or eliminate existing hazards. Although in the first step of the above procedure the analyst is required to select some verbal descriptors of each of the three risk factors, in the second step these are translated into single numbers, and their product (risk score) is translated back into the action column, with verbal description of the seriousness of risk. Also, the risk scores do not have unique recommended actions, but a broad range of risks; for example, between 270 and 1500 is considered to constitute the highest hazards for which an immediate corrective action is required.

In our view, the risk analysis problem can be represented by the following premise: IF exposure = (e) and likelihood = (l) and consequences = (c), THEN risk = (s); where s stands for the linguistic variable with such values as very high, high, substantial, or slight risk. Although the above linguistic values are based on the numerical magnitude of the risk score, the meaning of the above propositions are vague, and therefore the rules of logic are not applicable since classical logic implies that the meaning of propositions is unambiguous (ref.13).

People usually organize their world knowledge by causal relationships, and in reasoning, people apply what they are most familiar with, i.e. the rules underlying conversation and language (ref.26). Because all the factors in risk analysis are vague and imprecise, an alternative method must be used, allowing for approximate reasoning from the vague inputs. The method used here is based on fuzzy reasoning (ref. 25 and 27).

FUZZY REASONING MODELS IN RISK ANALYSIS

As stated by Zadeh (ref.28), approximate (fuzzy) reasoning refers to the process by which an imprecise conclusion is deduced from a collection of imprecise premises, and such reasoning is qualitative rather than quantitative in nature. In this paper, we propose to base the risk analysis on the method of approximate reasoning (ref. 24 and 27) which utilizes fuzzy conditional statements and compositional rules of inference.

The fuzzy conditional statement: IF A THEN B, or $A \rightarrow B$, where the antecedent (A) and consequent (B) are fuzzy sets, describes a fuzzy relation R between two fuzzy variables A and B. If A is a fuzzy subset in a universe of discourse U, and B is a fuzzy subset of a universe of discourse V, then the Cartesian product of A and B is defined as a fuzzy relation R:

$$A \times B = \sum_{R=U \times V} (f_A(u) \wedge f_B(v))/(u,v)), \qquad (1)$$

where R is usually given in the form of a matrix, and Σ stands for the union.

According to Mamdani's conjunctive logic (ref.24), if the fuzzy relation R, from U to V, is known, and A is a fuzzy subset of U, then the fuzzy subset B of V, which is induced by A, is given by the composition of R and A (A is a unary relation) as follows:

$$B = A \circ R, \qquad (2)$$

where B is given by the max-min product of A and R:

$$B = A \circ R \rightarrow f_B(v) = \sup_u (f_A(u) \wedge f_R(u,v))/(v). \qquad (3)$$

For example, if it is known that EXPOSURE = "high," and the relation between EXPOSURE and RISK (where risk is also defined as a linguistic variable) is R, then the value of RISK can be found using the compositional rule of inference.

Although there are many different ways of interpreting conditional propositions for the purposes of fuzzy reasoning (for a review see ref.29, 30 and 31), in this paper only Zadeh's maximum rule (ref.28) was used. The elementary models of approximate reasoning from conditional propositions were taken from Baldwin and Pilsworth (ref.30).

Verbal rules in risk analysis

A number of simple production rules, which most likely would be perceived in a similar way by a majority of risk analysts, can be relatively easily identified. For example, if exposure is known to be "very high," likelihood is "very likely," and consequences are "very high," then risk could be defined as "very very high" or "extremely high." However, in many other cases derivation of the risk value is not that obvious and hence derivation may be very difficult.

For the purpose of this study, two different examples of risk estimation in hypothetical situations are considered. The first example refers to the situation in which both exposure and likelihood are more or less constant and can be easily estimated, but the potential consequences may vary considerably, and therefore the value of risk will also change. The second example deals with the estimation of risk in a situation where originally the relationships between the given factors and risk, and risk value are known, and where the risk factors change inducing a change in the original value of risk.

Example 1.

Suppose the exposure or frequency of occurrence of the hazard event that could start an accident sequence is "high," meaning that the hazard event occurs daily. The likelihood that once the hazard event occurs, the complete accident sequence of events will follow, is perceived as "likely" or quite possible. Although these two factors are considered to be relatively constant, the consequences of the hazard event may change, depending upon the time of the day, and therefore may be perceived differently at different times. The question to be answered is how would the value of risk (S) be affected by changes in potential consequences from "more-or-less medium" to "high"?

Suppose the following universes of discourse and relevant propositions (see Table 2 for the definitions of linguistic values) are defined:

Exposure $X_E = \{e_1, e_2, e_3, e_4, e_5, e_6, e_7\}$,

Likelihood $X_L = \{l_1, l_2, l_3, l_4, l_5, l_6, l_7\}$,

Consequences $X_C = \{c_1, c_2, c_3, c_4, c_5, c_6, c_7\}$, and

Risk $Z = \{z_1, z_2, z_3, z_4, z_5, z_6, z_7\}$,

and:

P_E: E = "high" = (0, 0, 0.1, 0.3, 0.7, 0.9, 1.0)

P_L: L = "likely" = (0, 0.1, 0.5, 0.7, 0.9, 1.0, 1.0)

P_{c1}: C_1 = "more-or-less medium" = (0, 0.45, 0.84, 1.0, 0.84, 0.45, 0)

P_{c2}: C_2 = "very high" = (0, 0, 0, 0.1, 0.5, 0.8, 1.0)

P_s: S = "?"

where: $P_E \subset X_E$; $P_C \subset X_C$; $P_L \subset X_L$; and $S \subset Z$. P_i, for i = (E, C, L and S) are fuzzy propositions, and X_E, X_C, X_L and Z are universes of discourse.

The fuzzy relation R between $E = P_E$ and $L = P_L$ is the product of P_E and P_L in the following form:

$$R_{E \times L} = \begin{bmatrix} 0 & 0 & 0 & 0 & 0 & 0 & 0 \\ 0 & 0 & 0 & 0 & 0 & 0 & 0 \\ 0 & 0.1 & 0.1 & 0.1 & 0.1 & 0.1 & 0.1 \\ 0 & 0.1 & 0.3 & 0.3 & 0.3 & 0.3 & 0.3 \\ 0 & 0.1 & 0.5 & 0.7 & 0.7 & 0.7 & 0.7 \\ 0 & 0.1 & 0.5 & 0.7 & 0.9 & 0.9 & 0.9 \\ 0 & 0.1 & 0.5 & 0.7 & 0.7 & 1.0 & 1.0 \end{bmatrix}$$

Since the risk $S = P_c \circ R_{ExL}$, then according to equation (3):

$S_1 = (0, 0.1, 0.5, 0.7, 0.7, 0.7, 0.7)$, and

$S_2 = (0, 0.1, 0.5, 0.7, 0.9, 1.0, 1.0)$ or "high."

In other words, if the consequences are "very high" then the risk is "high." An interpretation of S_1 when consequences are "more-or-less medium" will be discussed below.

Example 2.

Consider the hazardous situation in which the following fuzzy conditional statement, given by a human expert, applies: IF exposure = "very high," consequences = "more or less high," likelihood = "likely" THEN risk = "high." We will define the following propositions:

P_E: E = "very high" = (0, 0, 0, 0.1, 0.5, 0.8, 1.0)

P_C: C = "more or less high" = (0, 0, 0.3, 0.5, 0.85, 0.95, 1.0)

P_L: L = "likely" = (0, 0.1, 0.5, 0.7, 0.9, 1.0, 1.0)

P_S: S = "high" = (0, 0, 0.1, 0.3, 0.7, 0.9, 1.0)

The above statement, concerning the value of risk under the described hazardous condition, can be written using fuzzy logic (ref.30) as follows:

$$\text{IF } (E = P_E) \cap (C = P_C) \cap (L = P_L) \text{ THEN } S = P_S \tag{4}$$

subject to a propositional assertion $P = P_E \cap P_C \cap P_L$. Such a conditional proposition defines a relation D, where:

$D \subset X_E \times X_C \times X_L \times Z.$

Relation D can be expressed using the max-min definition as follows:

$$D = D_E \cap D_C \cap D_L \tag{5}$$

and primitive conditional propositions are in the form of:

$$\left.\begin{array}{l} \text{IF E THEN S } (D_E) \\ \text{IF C THEN S } (D_C) \\ \text{IF L THEN S } (D_L) \end{array}\right\} \quad \begin{array}{l} D_i \subset X_i \times Z \\ \text{for } i = (E,C,L) \end{array} \tag{6}$$

or equivalently:

IF $E = P_E$ THEN $S = P_S \rightarrow D_E = P_E \times P_S$

IF $C = P_C$ THEN $S = P_S \rightarrow D_C = P_C \times P_S$

IF $L = P_L$ THEN $S = P_S \rightarrow D_L = P_L \times P_S$.

The relationship between likelihood and risk ($X_L \times Z$), exposure and risk ($X_E \times Z$), and consequences and risk ($X_C \times Z$) can be established by human experts, as is done in the traditional approach. Considering the primitive conditional propositions given by (6), we derive the following fuzzy relations D_E, D_C, and D_L using formula 3.

$$D_E \subset X_E x Z = \begin{bmatrix} 0 & 0 & 0 & 0 & 0 & 0 & 0 \\ 0 & 0 & 0 & 0 & 0 & 0 & 0 \\ 0 & 0 & 0 & 0 & 0 & 0 & 0 \\ 0 & 0 & 0 & 0.1 & 0.1 & 0.1 & 0.1 \\ 0 & 0 & 0.1 & 0.3 & 0.5 & 0.5 & 0.5 \\ 0 & 0 & 0.1 & 0.3 & 0.7 & 0.8 & 0.8 \\ 0 & 0 & 0.1 & 0.3 & 0.7 & 0.9 & 1.0 \end{bmatrix}$$

$$D_C \subset X_C x Z = \begin{bmatrix} 0 & 0 & 0 & 0 & 0 & 0 & 0 \\ 0 & 0 & 0 & 0 & 0 & 0 & 0 \\ 0 & 0 & 0.1 & 0.3 & 0.3 & 0.3 & 0.3 \\ 0 & 0 & 0.1 & 0.3 & 0.5 & 0.5 & 0.5 \\ 0 & 0 & 0.1 & 0.3 & 0.7 & 0.85 & 0.85 \\ 0 & 0 & 0.1 & 0.3 & 0.7 & 0.9 & 0.95 \\ 0 & 0 & 0.1 & 0.3 & 0.7 & 0.9 & 1.0 \end{bmatrix}$$

$$D_L \subset X_L x Z = \begin{bmatrix} 0 & 0 & 0 & 0 & 0 & 0 & 0 \\ 0 & 0 & 0 & 0 & 0 & 0 & 0 \\ 0 & 0 & 0.1 & 0.3 & 0.5 & 0.5 & 0.5 \\ 0 & 0 & 0.1 & 0.3 & 0.7 & 0.7 & 0.7 \\ 0 & 0 & 0.1 & 0.3 & 0.7 & 0.9 & 0.9 \\ 0 & 0 & 0.1 & 0.3 & 0.7 & 0.9 & 1.0 \\ 0 & 0 & 0.1 & 0.3 & 0.7 & 0.9 & 1.0 \end{bmatrix}$$

Let us assume now that the conditions specified above have changed, and one has to deal with a new hazardous situation which induces a different value of risk. We will use the above model of approximate reasoning (fuzzy compositional rule of inference) to answer the following question: given (E) <u>exposure</u> = "medium," (C) <u>consequences</u> = "very high," and (L) <u>likelihood</u> = "unlikely," what is the value of risk (S)?

The compositional rule of inference (ref.30) (as the least restrictive inference S Z of P from D), given by: S = P o D, becomes now:

$$S = \bigcap_{i = (E,C,L)} (P_i \circ D_i) \qquad (7)$$

where P_E = "medium," P_C = "very high," and P_L = "unlikely" (see Table 2), and the above formula can be written as follows:

$$s(z) = \bigwedge_{i=\{E,C,L\}} \left[\bigvee_{x_i \in X_i} [p_i(x_i) \wedge d_i(x_i;z)] \right] \tag{8}$$

According to equation (8) the value of risk will then be equal to:

$$S = (P_E \circ D_E) \cap (P_C \circ D_C) \cap (P_L \circ D_L). \tag{9}$$

After performing relevant computations, one can obtain a numerical interpretation of the linguistic value of risk (S) in a new hazardous situation. From equation (9) we have:

$$S_3 = (0,0,0.1,0.3,0.5,0.5,0.5) \cap (0,0,0.1,0.3,0.7,0.9,1.0) \cap$$
$$\cap (0,0,0.1,0.3,0.5,0.7,0.7) = (0,0,0.1,0.3,0.5,0.5,0.5),$$

which could be approximated by "[(more or less high) and (more or less medium)]".

LINGUISTIC APPROXIMATIONS AND INTERPRETATIONS OF THE RISK SCORES

Verbal models (ref.16 and 32), or quantitative models with linguistic variables and verbally formulated relations between the variables, may be very useful in the area of systems safety in general (ref.9 and 10), and, as shown above, in risk analysis in particular. This is primarily due to the fact that human experts are capable of making knowledgable and reliable verbal statements about intuitively perceived phenomena of the real world (ref.34). Since the linguistic values are naturally vague, they allow greater flexibility than single numbers.

According to Wenstop (ref.16), by introducing linguistic variables as the inputs into models, one takes a step toward meeting the demand for validity. The problem remains, however, how to translate the fuzzy outputs into a meaningful set of linguistic values. Such a process, called linguistic approximation, is essential in risk analysis to decide on the corrective action that may need to be associated with a particular fuzzy value of risk. As seen in the example stated above, it is not a trivial task to find a label for such a fuzzy set (of risk) as $S_1 = (0, 0.1, 0.5, 0.7, 0.7, 0.7)$.

Fortunately, some methods are presently available to deal with this problem. The simplest method, called the "best fit" method (ref.33), is usually applied when the set of possible linguistic expressions is small, and it is computationally easy to calculate a distance from the fuzzy output to the fuzzy sets representing the available linguistic values. The natural term whose fuzzy set is the closest to the output is then selected as its meaning. A more advanced method of "successive approximations" is based on the evaluation of the endpoint "brackets" and hedges which are being replaced as the expression "in progress" gets closer to the fuzzy set being approximated (ref.22). Although a large number of natural expressions can be efficiently evaluated, the method requires that all fuzzy sets be convex. The LAMS system (ref.18) does not require normality of fuzzy sets, and allows an assignment of linguistic values to a fairly complex fuzzy spread by labelling its segments. In this "piecewise decomposition" technique, the linguistic expressions chosen for each interval are then combined using the fuzzy connectives "and" or "or."

Wenstop (ref.16 and 23) proposed a context independent, quantitative analysis with linguistic values utilizing an APL auxiliary language. The linguistic approximation method, implemented by an APL´s LABEL function, is based on two parameters of the fuzzy set to be labeled, i.e., its imprecision (the sum of its membership values) and its location (the center of gravity). One of the 56 linguistic labels (spread out in a location-imprecision system) with the shortest distance to the coordinates of the fuzzy set to be labeled is chosen as the representation of the fuzzy output from the model. The main concern of this method is to ensure that the input-output values are acceptable by the standards of natural language, provided that the linguistic expressions are used systematically and diligently.

The above discussion indicates that the meaningful interpretation of the fuzzy outputs of approximate reasoning models can be efficiently performed with assistance of computers.

FINAL REMARKS

Although the theoretical considerations presented here are of a preliminary nature, it is believed that the proposed approach will be very useful in the analysis of risks in many industrial environments. Further work is under way to experimentally verify the theses outlined in this paper. Examples of the questionnaire forms which are presently used for such verification are shown in the Appendix.

REFERENCES

1 W.H. Heinrich, D. Peterson and N. Ross, Industrial Accident Prevention, McGraw Hill, New York, 1980.
2 K.T. Graham and G.F. Kinney, A Practical Safety Analysis System for Hazards Control, Journal of Safety Research 12 (1980) 13-20.
3 C.B. Chapman and D.F. Cooper, Risk Analysis: Testing Some Prejudices, European Journal of Operations Research 14 (1983) 238-247.
4 S.W. Malasky, Systems Safety, Hyden, New Jersey, 1974.
5 C.B. Brown, A Fuzzy Safety Measure, Journal of the Engineering Mechanics Division 5 (1979) 855-872.
6 W.T. Fine, Mathematical Evaluations for Controlling Hazards, in: Selected Readings in Safety, Academic Press, Macon, 1973, pp. 68-85.
7 M. Smithson, Applications of Fuzzy Set Concepts to Behavioral Sciences, Mathematical Social Sciences 2 (1982) 257-274.
8 L.A. Zadeh, The Concept of Linguistic Variable and its Application to Approximate Reasoning, Information Sciences, Part I: 8 (1975) 199-249, Part II: 8 (1975) 301-357, Part III: 9 (1975) 43-80.

9 D.I. Blockley, Analysis of Subjective Assessments of Structural Failures, International Journal of Man-Machine Studies 19 (1978) 185-195.
10 T.T.P. Yao, Damage Assessment of Existing Structures, Journal of the Engineering Mechanics Division 4 (1980) 785-799.
11 Y. Shinohara, Fuzzy Set Concepts of Risk Assessment, International Institute for Applied Systems Analysis (Unpublished Report WP-76-2), Laxenburg, Austria, 1976.
12 T.B. Feagans and W.F. Biller, Fuzzy Concepts in the Analysis of Public Health Risks, in Wang and Chang (Eds.), Fuzzy Sets, Plenum Press, New York, 1980, pp. 391-404.
13 A.C. Zimmer, Verbal Versus Numerical Processing, in R. Scholz (Ed.), Individual Decision Making Under Uncertainty, North Holland, Amsterdam, 1983.
14 C.V. Negoita, Expert Systems and Fuzzy Systems, The Benjamin/Cummings Co., Menlo Park, California, 1985.
15 A.C. Zimmer, A Model for the Interpretation of Verbal Predictions, International Journal of Man-Machine Studies 20 (1984) 121-134.
16 F. Wenstop, Quantitative Analysis with Linguistic Values, Fuzzy Sets and Systems 4 (1980) 99-115.
17 A.K. Nath and T.T. Lee, On the Design of a Classifier with Linguistic Variables as Inputs, Fuzzy Sets and Systems 11 (1983) 265-286.
18 F. Eshragh and E.H. Mamdani, A General Approach to Linguistic Approximation, International Journal of Man-Machine Studies 11 (1979) 501-519.
19 F. Wenstop, Exploring Linguistic Consequences of Assertions in Social Sciences, in M.M. Gupta, R.K. Ragade, and R.R. Yager (Eds.), Advances in Fuzzy Set Theory and Applications, North Holland, Amsterdam, 1979, pp. 501-518.
20 T.W. Cooley and T.O. Hicks, Jr., A Fuzzy Set Approach to Aggregating Internal Control Judgment, Management Science 29 (1983) 317-334.
21 F. Narashimhan, Goal Programming in a Fuzzy Environment, Decision Science 11 (1980) 325-336.
22 D.P. Clements, Fuzzy Ratings for Computer Security Evaluation, Ph.D. Dissertation, University of California, Berkeley, 1977.
23 F. Wenstop, Quantitative Analysis with Linguistic Values, Fuzzy Sets and Systems 4 (1980) 99-115.
24 E.H. Mamdani, Advances in the Linguistic Synthesis of Fuzzy Controllers, International Journal of Man-Machine Studies 8 (1976) 669-678.
25 L.A. Zadeh, A Fuzzy-Algorithmic Approach to the Definition of Complex and Imprecise Concepts, International Journal of Man-Machine Studies (1976) 249-291.
26 I. Begg, On the Interpretation of Syllogisms, Journal of Verbal Learning and Verbal Behavior 21 (1982) 595-620.
27 L.A. Zadeh, Fuzzy Logic and Approximate Reasoning, Synthese 30 (1975) 407-428.
28 L.A. Zadeh, Outline of a New Approach to the Analysis of Complex Systems and Decision Processes, IEEE Transactions on Systems, Man and Cybernetics, SMC-3 (1973) 28-44.
29 Y. Tsukamato nd T. Terano, Failure Diagnosis by Fuzzy Logic, in: Proceedings of the IEEE Conference on Decision and Control, 1977, pp. 1390-1395.
30 T.F. Baldwin and B.W. Pilsworth, A Model of Fuzzy Reasoning Through Multi-Valued Logic and Set Theory, International Journal of Man-Machine Studies II (1979) 351-380.
31 T. Whalen and B. Schott, Alternative Logics for Approximate Reasoning in Expert Systems: A Comparative Study (Unpublished Report), Georgia State University, Atlanta, Georgia, 1984.
32 G.C. Oden, Integration of Fuzzy Linguistic Information in Language Comprehension, Fuzzy Sets and Systems 14 (1984) 29-41.
33 K.T. Schmucker, Fuzzy Sets, Natural Language Computations, and Risk Analysis, Computer Science Press, Rockville, 1984.

34 H.M. Hersh and A. Caramazza, A Fuzzy Set Approach to Modifiers and Vagueness in Natural Language, Journal of Experimental Psychology: General 105 (1976) 101-107.

35 M. Kochen, Applications of Fuzzy Sets in Psychology, in L.A. Zadeh (Ed.), Fuzzy Sets and Their Applications to Cognitive and Decision Processes, Academic Press, New York, 1975, pp. 395-408.

APPENDIX - An experimental verification of the fuzzy model of risk analysis.

Given below are examples of the questionnaire forms which were recently sent to 980 safety engineers with the purpose of practical verification of the proposed fuzzy models of risk analysis.

SITUATION #2

A maintenance employee has just finished servicing an air conditioner which is located on the roof of a building. The employee has used a 20 foot extension ladder for gaining access to the 14 foot high roof top. Before descending down the ladder, the employee extends the ladder while standing on the roof. Then, the employee descends.

The employee weighs 279 lbs. and the ladder is rated for 250 lbs. The ladders' locking mechanism or device does not always return to locked position. The ladder is not secured at ground level by stakes or by having another individual hold the ladder.

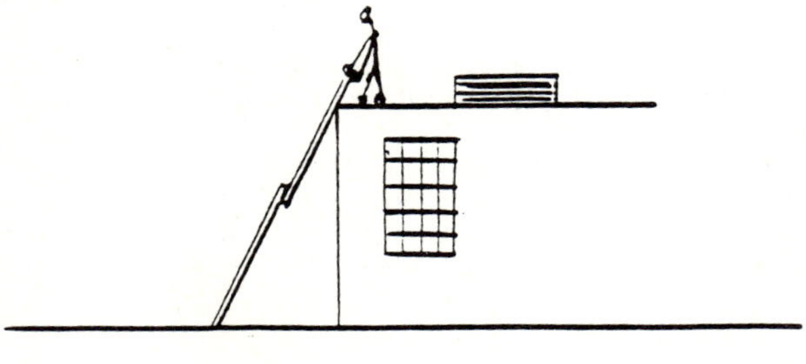

SITUATION #4

A utility company is completing a 1.1 million dollar service upgrade project. Nine (9) electrical linemen are preparing to place insulators on all six of the projects previously erected transmission towers prior to placement or hanging of electrical transmission wires. The towers are 116 feet in height and are placed 50 feet apart. Each tower is supported by guy wires. A boom truck (aerial platform) operated by a rigger operator is being used to lift the linemen to the top of the towers. After the first man is placed on Tower No. 4, the rigger retracts the boom and drives the truck down and around Tower No. 1 in order to lift additional linemen into placement on top of Tower No. 1.

The man placed on Tower No. 4 is using his safety belt. The towers are located in a grass field. No road surfaces are near the towers. No guards or warning signs mark the tower guy wires.

EXPOSURE

EXPOSURE is defined as the frequency of occurence of the hazardous event.

Instructions: Assess the situation for EXPOSURE as defined above, select one of the terms from the list below which you think best describes the level of EXPOSURE for the situation. Then, place an "X" on the scale below where you feel your selected term would be located on the scale. Place the number of the selected term directly above the "X" mark.

Never |————————————————————| Continuous

1. Very High
2. High
3. Moderately High
4. Moderate
5. Moderately Low
6. Low
7. Very Low

HUMAN CONSEQUENCE

HUMAN CONSEQUENCE is defined as the level of human injury from a hazardous event.

Instructions: Assess the situation for HUMAN CONSEQUENCE as defined above, and select one of the terms from the list below which you think best describes the level of HUMAN CONSEQUENCE for the situation. Then, place an "X" on the scale below where you feel your selected term would be located on the scale. Place the number of the selected term directly above the "X" mark.

None |————————————————————| Catastrophe

1. Multiple Fatalities
2. A Fatality
3. Disabling Injury (injury is permanent)
4. Serious Injury (injured will recover)
5. Noticeable Injury (first-aid required)

LIKELIHOOD

LIKELIHOOD is defined as the likelihood of occurrence of a hazardous event as related to the mathematical probability that it might actually occur.

Instructions: Assess the situation for LIKELIHOOD as defined above, and select one of the terms from the list below which you think best describes the level of LIKELIHOOD for the situation. Then, place an "X" on the scale below where you feel your selected term would be located on the scale. Place the number of the selected term directly above the "X" mark.

Never |————————————————————| Certainty

1. Very High
2. High
3. Moderately High
4. Moderate
5. Moderately Low
6. Low
7. Very Low
8. Highly Unlikely

RISK

RISK is defined as the chance that injury or damage will result from a particular hazard.

Instructions: Assess the situation for RISK as defined above, and select one of the terms from the list below which you think best describes the level of RISK for the situation. Then, place an "X" on the scale below where you feel your selected term would be located on the scale. Place the number of the selected term directly above the "X" mark.

None |————————————————————| Extreme

1. Very High
2. High
3. Moderately High
4. Moderate
5. Moderately Low
6. Low
7. Very Low
8. Insignificant

Applications of Fuzzy Set Theory in Human Factors, edited by W. Karwowski and A. Mital
Elsevier Science Publishers B.V., Amsterdam, 1986 — Printed in The Netherlands

ASSESSMENT OF RESISTIVE LOAD SENSATION FOR PROTECTIVE RESPIRATORY EQUIPMENT

MOJ-CHAO, MA[1], LI-FAN, ZHANG[2] and XING-YU, WU[3]

[1]Institute of Psychology, Academia Sinica, Beijing [2,3]Fourth Military Medical College, Xi´an, People´s Republic of China

ABSTRACT

The analysis of empirical data gathered in this study revealed that the certainty of response in each grade category under each loading phase had a possibilitistic distribution. It appears from this finding that when the total resistance is kept constant, the perceived magnitude is dependent upon the phase of the breathing cycle loaded by externally added resistance. The degree of sensory experience for expiratory loading was lightest. Moreover, it is possible to obtain the magnitude of physiological stimuli corresponding to each grade category on the scale. The above results are of practical importance for establishing, and revising, a standard of the permitted external respiratory work load on men wearing a respiratory apparatus.

INTRODUCTION

Studies of perceived magnitude of added loads to breathing were conducted as early as 1940. The classical psychophysical method has successfully been used to detect the absolute threshold and differential threshold of respiratory sensation produced by added resistive or elastic load for the purpose of protective equipment, and to elucidate the mechanism of dyspnea (ref. 3,4). It has also been suggested that the psychophysical function for suprathreshold respiratory mechanical load conforms to Steven´s power law (ref. 4,5). Recently it was proved that the perceived magnitude of added loads to breathing is a direct power function of both peak mouth cavity pressure and duration, and is only indirectly related to added resistive loads (ref. 6).

The results obtained by category scale, on the other hand, can be easily applied for establishing and revising standards of permitted external respiratory work load on men wearing respirators (ref. 7-9). The procedure, however, does not accurately reflect the growth of sensation (ref. 10). In this work, we assume that some kind of sensation, such as the sensation of effort or discomfort, might be fuzzy in nature. Therefore, their occurrence is a problem of possibility, and not of probability. This may very well be the case as indicated by a fuzzy set model for category judgement. When a category judgment is made, there are reactions, with different magnitude in each grade category, on an ordinal sensory category scale. The aim of this study was to apply a new kind of psychophysical category scale to quantifying sensations associated with resistive load to breathing.

METHOD

Subjects

Thirty-seven healthy young male students, ranging in age from 18 to 23 years, participated in the study. All subjects had normal spirometry and had no history of cardio-pulmonary disease.

Apparatus

The resistance manifold consisted of a series of resistors made of nylon net, arranged in a Lucite tube. Rubber stoppers were inserted in ports between the resistors to produce the desired loads. Subjects, wearing an oro-nasal mask, sat behind a screen and inhaled room air through a two-way respiratory valve which had its inspiratory and expiratory ports attached to the resistance manifolds. Measurements of respiratory flowrate and mouth cavity pressure (mask cavity pressure) were made with the assistance of a respiratory flow and a pressure measuring instrument (xy-1), whose Fleisch type pneumotachograph was connected between the two-way valve and the mask (Fig. 1). The analog output was recorded by a magnetic data recorder (A69, Sony). Off-line computation and analysis of experimental data were accomplished by a microcomputer (MDR Z-80).

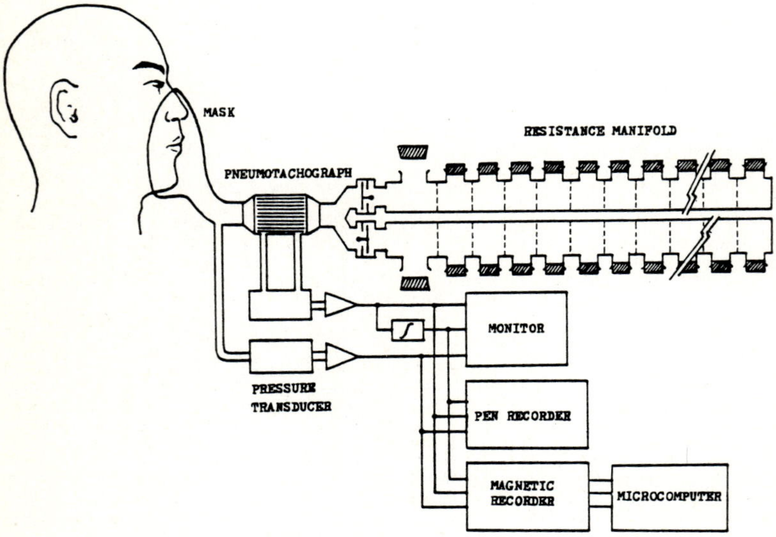

Figure 1. Experimental set-up and apparatus.

Psychophysical Scale

A new type of category scaling, the Multistage Evaluation Method (MEM), which was developed on the basis of a fuzzy set model for category judgement (ref. 11-13), was adopted to estimate effort sensations associated with added resistance to breathing.

TABLE 1

MEM answer sheet for effort sensation associated with breathing

attitude	categories / certainty	none	light	moderate	heavy	severe
	very				X	
positive	considerable					
	slightly			X		
	slightly					
negative	considerable					X
	very	X	X			

As shown in Table 1, the five grade categories of the respiratory resistive sensation were: (1) none (not noticeable): (2) light (noticeable, but with no discomfort): (3) moderate (breathing with effort and some discomfort, but tolerable): (4) heavy (difficulty in breathing and intolerable for a long time): (5) severe (the sense of total suffocation). The first and second columns of Table 1 show the subject's attitude (positive or negative) and the degree of certainty (confidence), respectively.

Experimental Design

A two-factors factorial design (6x3) was used. The first independent variable (factor A) corresponding to the total resistive load had six levels: A_1 (40), A_2 (80), A_3 (160), A_4 (240), A_5 (320), and A_6 (480) (mm H_2O [Litre* second]). The second independent variable (factor B), associated with load phases of the breathing cycle, had three levels: B1-inspiratory, B2-expiratory, and B3-combination. A total of 18 different respiratory resistive loads were, thus, obtained through various combinations of A and B. The order of presentation of the different combinations was random. The order of load phases within each load level was arranged in a 3x3 Latin Square design.

Protocol

Prior to the experiment, subjects were instructed in the procedure. Particular care was taken to ensure that each subject understood the basic concept of the MEM and the objective of the experiment. Some practical sessions were allowed. Once it was determined that the subject was ready, a respirator was donned. The subject seated himself in an adjustable chair. Each load condition was run for three minutes. The respirator flow rate and oral pressure changes were recorded at the end of the second minute. At the end of the third minute, the subject recorded his sensation due to the effect of the added respiratory load. This procedure was repeated for all remaining load conditions. The entire experiment for each subject lasted for one and a half hours. During this 90 minute period, subjects were allowed to rest twice.

Procedure

The subject was first asked to select the answer which was the closest discription of his sensation to the resistive load to his breathing. At the same time, he was asked to indicate the grade of certainty of the reaction by marking the appropriate cell in that category. Next, he was required to indicate grade of certainty in categories next to the category which was the closest description of his sensation to the added loading. This was accomplished by marking the appropriate cells.

Estimation of Sensory Magnitude

The following relationship used to estimate the sensory magnitude which is elicited from a given loading quantity u_i $(i=1,2,...,n)$:

$$\mathrm{MEM} = F(u_i) = \frac{1}{T}\left(\sum_{j=1}^{m} Y_{ij}\, W_j\, U_j \Big/ \sum_{j=1}^{m} Y_{ij}\, W_j\right) \tag{1}$$

where T is a constant representing the loading quantity which may cause the most intolerable respiratory sensation. According to Pope´s results (ref. 13), let T=1100 such that the psychological magnitude is normalized. The parameters Y_{ij}.W_j and U_j can be separately explained as follows: Y in equation (1) is the weighting coefficient related to the grade certainty C_{ij}. By this coefficient the subject responded to a given loading u in each sensory category r_j $(j=1,2,...,m)$. Value C_{ij} comes from Table 1 in which it is necessary to transform empirical evaluation into numerical data. The relation of the transformation is the numerical system {3,2,1,} which corresponds to a set of the grade certainties {very, considerably, slightly}. Symbol ´+´ was used to represent the positive certainty and symbol ´-´, the negative certainty. In order to take C_{ij} from a unit interval, let:

$$Y_{ij} = (3 + C_{ij}) / 6 \tag{2}$$

Another weighting coefficient W_j in equation (1) was determined as follows:

The weighting value in category or categories corresponding to negative certainty of the response was designated 1.0; within other category, or categories, related to positive certainty, weighting coefficient in 'peak' category was designated 3.5; weighting coefficient in category, or categories, in close vicinity to 'peak' category was designated 3.3; weighting coefficient in category farther out 'peak' was designated 3.0. W_j, thus, reflects the relative importance of each category r_j ($j=1,2,...m$) on the scale.

U_j in equation (1) is the linguistic response of the category r_j on the sensory scale. The numerical value of U_j corresponds to particular loading quantity I_j $(1,2,...n)$ which may elicit the maximum possibility of the reaction at r_j over u_i ($i=1,2,...n$), i.e.:

$$Y_{I_j j} = \max_{1 \leq i \leq n} Y_{ij} \qquad (3)$$

As shown in Fig. 3, the particular loading quantity I_j ($j=1,2,...,m$) related to the peak value on each curve is 80 (light), 240 (moderate) and 480 (heavy) respectively.

RESULTS AND ANALYSIS

I. The possibility distribution of the subjects' responses (certainty of reaction) on each element $r_j \in R$. The data Y_{ij}, which represent subjects' certainty degree of response to u_i on each $r_j \in R$, were obtained through the fuzzy statistical method. The distribution curves in Fig. 2 were obtained by plotting average subject's responses to different loads (u_i) against each grade category $r_j \in R$. From Fig. 2, it can be concluded that the plots of the possibility distribution curves were determined both by the total resistive load and the phase of breathing load.

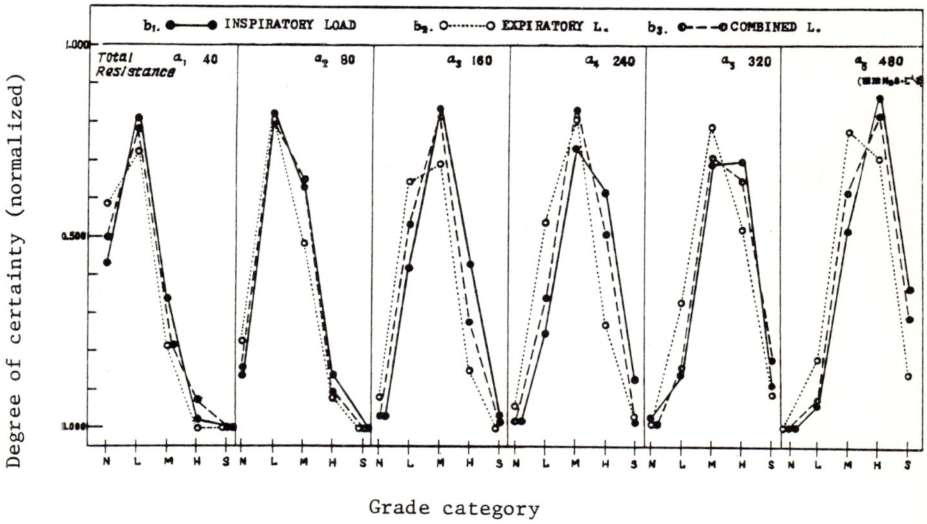

Figure 2. Category scale.

II. The distribution of certainty of each given resistive load for different grade categories. The average certainty of reaction for different grade categories was plotted against resistive loads. A family of average certainty distribution curves which belong to different grade category of the scale resulted. In Fig. 3, only the distribution curve family for the combined phase of load is shown. From the intersect of two adjacent curves, representing adjacent sensory categories, one may locate a particular critical resistance along the abscissa axis. This can be regarded as a critical resistance marking the transition from one category to the next (e.g. the critical values marking transition of sensory category from ´none´ to ´light´, ´light´ to ´moderate´ and ´moderate´ to ´heavy´ were 24, 108, 352 [$mmH_2O\ L^{-1}\ S$], respectively).

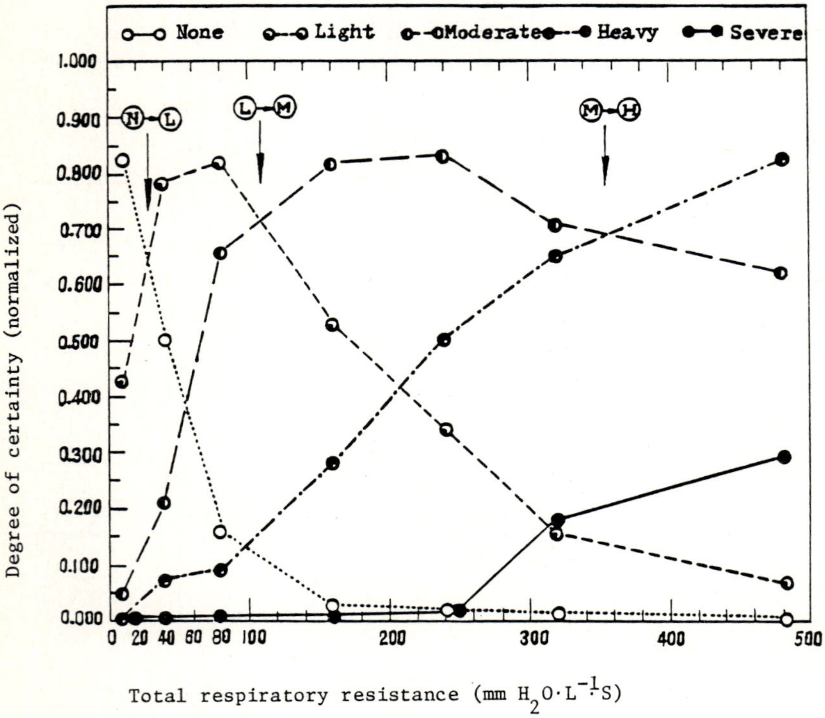

Figure 3. The curve family showing certainty distribution pattern for different grade category under the combined phase of loading.

A statistical analysis of the data associated with respiratory sensation. The perceived magnitude to each resistive load was estimated according to (ref. 6).

The ANOVA was used to test statistical significance of different levels and interaction factors. As shown in Tables 2 and 3, $F(u_i)$ in equation (1) could reflect the difference in respiratory sensation resulting from both total resistive value (resistance intensity, $p < .01$) and phase of breathing cycle impeded ($p < .01$). The interaction (the resistive value and phase of load), however, was insignificant ($p > .05$).

Relationship between calculated respiratory sensory magnitude and physical or physiological stimulus. The calculated respiratory sensory magnitude was plotted against added resistance (physical stimulus). As shown in Fig. 4, a straight line on a log-log scale resulted for each type of resistive load. The equations are also shown if Fig. 4. These results indicate that the relationship between ui and $F(u_i)$ can be expressed in the form of a power function for each added phase (i.e., inspiratory, expiratory, and combined). The paired t-test showed that the corresponding regression coefficients (exponents for the power function) of three types of loading were not significantly different. This suggest that similar mechanism are utilized in subjective estimation of respiratory resistive sensation when a different respiratory phase is loaded.

In order to explore the relationship between respiratory sensation and the physiological stimulus (the peak-to-peak mouth pressure swing, p), a plot expressing P-$F(u_i)$ relationship was drawn (see Figure 5). Figure 5 shows clearly how the sensory magnitude increased with P increasing.

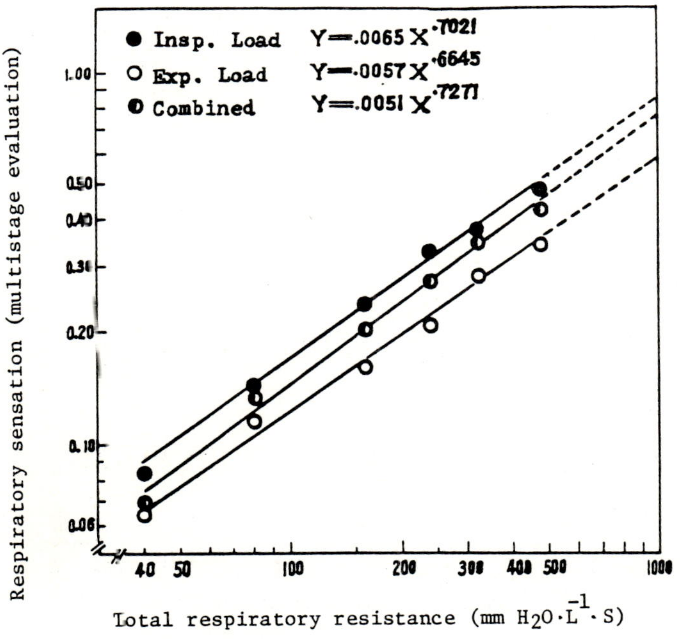

Figure 4. Relationship between perceived respiratory sensation and resistive load under three types of loading.

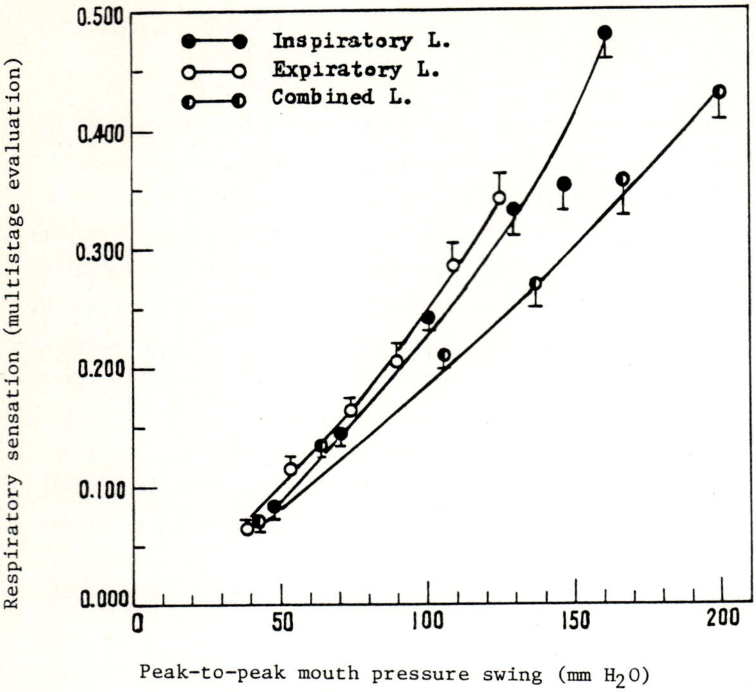

Figure 5. The resistive sensation as functions of fluctuation of the peak pressure in the mouth.

TABLE 2

Relationship between respiratory sensation and resistive load-total resistive value, phase impeded.

(B) Phase / Loaded	(A) total added respiratory resistive value*					
	(a1)40	(a2)80	(a3)160	(a4)240	(a5)320	(a6)480
(b1) inspiratory	0.0831 'light'	0.1414	0.2381 'moderate'	0.3285	0.3581	0.4754 'heavy'
(b2) expiratory	0.0639	0.1134 'light'	0.1609	0.2054 'moderate'	0.2850	0.34045
(b3) combined	0.0684	0.1335 'light'	0.2017	0.2723 'moderate'	0.3474	0.4261 'heavy'

*[mm H_2O L^{-1} S]

TABLE 3

The 6x3 ANOVA for data in table 2.

Source	SS	df	MS	F	p
total	18.5186	665			
between phases(a)	0.5968	2	0.2984	21.8882	<.01
between resistive value(b)	8.8636	5	1.7727	130.0321	<.01
interaction (a) x (b)	0.2240	10	0.0224	1.6433	<.05
error	8.8342	648	1.3633		

DISCUSSION

An empirical relationship for estimating respiratory resistive sensation is proposed (Equation 1). Let us first note two parameters, U_j and W_j there. U_j is the linguistic value of r_j, which comes from the experimental result (see Equation 3). Obviously, the distance between the linguistic value U_j and U_{j+1} which corresponds to grade categories r_j and r_{j+1} is not equal. Concretely speaking, the psychological difference between category 1 and category 2 is not the same as one between category 2 and category 3 or the distance between all other successive pairs of categories. Consequently, assignment of numbers to linguistic categories on the psychological scaling (such as category scaling) depends on the psychological attribute to be measured may or may not have a direct corresponding physical property. If it does, the linguistic value would be assigned according to Equation 3 of the paper. If the psychological attribute to be measured does not have its relevant physical property, the numbers assigned to the linguistic categories may carry out through an empirical investigation of fuzzy semantics which are related to them. Another parameter w_j in Equation 1 is a moveable weight which reflects the following empirical nature: so far as a psychological measurement is concerned, positive information is more significant that negative information. Furthermore, with the positive information, the peak value is naturally the most important one of all. Consequently, weight for the category corresponding to the peak value shall be the largest one. It is conceivable that for a category which is one further off peak category, contribution is smaller.

When much of the evaluation is conducted using the MEM, statistics associated with the frequency of the peak value can be used. These are the same for the CCS. It has further been shown that in this case the psycho-physical co-ordinates appear as a straight line in a lingle log (the physical magnitude is a log scale). Three such curves corresponding to three phases of resistive load follow the linear function, $y= a + b \log x$. However, as has been shown above, the psychophysical curve obtained with MEM could well be expressed as a power function, $y= ax^b$ or $\lg y = \lg a + b \lg x$. The fact that the results obtained with CCS differ from those obtained with MEM is probably due to the use of different methods. Firstly, because a subject makes a judgement on an object according to the given psychological dimension (e.g., his preference) generally speaking, the usual response prefers may be "It´s something between moderate and satisfactory". This seems to be a problem of possibility, rather that of probability. The fuzzy set model for category judgement has been developed to reflect the situation in which there are different degrees of reaction in more than one of the categories. The MEM is suitable for the nature only: the CCS is quite different. Here the category judgement is regarded as a random event. Using the CCS scale, a subject may be asked to determine one category from a limited number of categories. As a result, a subject usually finds it difficult to judge the category to give a definite answer. We do not think that this is a natural technique for selection. Secondly, the MEM can obtain a set of data in one run; including the main, next and negative information. Consequently, it may reveal slight differences which exist in practice, like the difference of the perceived magnitudes associated with the different resistive phases (Table 3,4). The CCS, on the other hand, only collects a simple value for each run. It is usually the most important one, however, the other values are neglected. It is difficult for the CCS to reflect slight differences (note: because there was not noticeable variety in cases of the resistive phases, the concrete curves were not shown in the article). Thirdly, the MEM is a type of evaluation conducted in two dimensions where there were only a few correlative factor for evaluation. This takes advantage of elevated reliability. The CCS, however, is evaluated in a single dimension which makes it relatively easier to not avoid one´s artificiality when making judgement. It should be noted here that the MEM could detect physical and physiological magnitudes corresponding to each category on the scale (Figures 4,5, and 6) which is useful in establishing standards regarding the use of respiratory protective equipment.

CONCLUSION

The present work based on fuzzy sets theory has resolved successfully an estimation of perceived magnitude associated with added resistive load. The obtained results suggest a possibility of the wide application of the fuzzy set theory to measure both somatic and visceral sensations.

The main findings of this study was that the loading intensity were well explained by the psychophysical power law. The severity of the sensory experience which was caused by the equal resistive intensity and loaded a different phase of breathing cycle indicated a significance. However, their speed of sensory growth did not reach a significant level. The group's mean physical and physiological magnitudes correlated to resistive sensory grade categories were determined from the Table 2 and the empirical curves shown in Fig. 3 and 4. These are useful for establishing resistive standards with respect to the use of respiratory protective equipment. Additionally the psycho-physiological function obtained by the MEM was compared with those obtained by the CCS. The possible reason between these two was discussed.

ACKNOWLEDGEMENT

We are greatly indebted to Wang Pei-zhuang and Zhao ru, Huai, Jiang Shi-zhong for their help.

REFERENCES

1 L. Silverman et al: A Study and An Evaluation of Inspiratory and Expiratory Resistance for Protective Respiratory Equipment, OSRD. No. 5339, 1945.

2 J. S. Hart, Resistance to Breathing, Appendix IV Aeromedical Laboratory, TSEAA-660-83-E, 1946.

3 E. P., D. Rubenstein, M. I. V. Jayson, E. J. M.Campbell, The Ability of Man to Detect Added Non-Elastic Loads to Breathing, Clin. Sci, 23(1962), 155-152

4 R. L. Wiley and F. W. Zechman, Jr., Perception of Added Airflow Resistance in Humans, Respir. Physiol., 2(1966), 73-87

5 S. R. Muza, S.Mcdonald. F. W. Zechman, Comparison of Subjects' Perception of Inspiratory and Expiratory Resistance, J. Appl. Physiol. Respirat. Environ. Exercise Physiol. , 56(1984). 211-216.

6 K. J. Killian D.D. Bucens, E. J. M. Campbell. Effect of Breathing Patterns on the Perceived Magnitude of Added Loads to Breathing, J. Appl. Physiol, Respirat. Environ. Exercise Physiol., 54(1983), 1216-1221.

7 R. A. Bentley, O. G. Griffin, R. G. Love, D. C. F. Muir, K. F. Sweetland, Acceptable levels for Breathing Resistance of Respiratory Apparatus, Arch. Environ. Health, 27(1973), 273-289.

8 W. P. Morgan, Psychometric Correlates of Respiration: A Review, Amer. Ind. Hyg. Assoc. J, 44(1983), 677-684.

9 Li. Bing-lin, Feng Lin-hua and Zhang Li-fan: A Study of the Acceptable Ratio of Added Inspiratory to Expiratory Resistance. Medical Journal of Chinese People's Liberation Army, 9(1984), 184-187.

10 W.P. Morgan, Psychological Problems Associated with the Wearing of Industrial Respirators: A Review, Amer. Industrial Hyg. Assoc. J, 44(9), 1983 671-676.

11 Mou-chao, Ma, and Zhi-qiang, Cao, A Fuzzy Set Model for Category Judgement and the Multistage Evaluation Method, ACTA Psychologica SINICA, 2(1983) 198-204.

12 Mou-chao, Ma and Zhi-qiang, Cao: The Multistage Evaluation Method in Psychological Measurement: An Application of Fuzzy Sets Theory to Psychology, In MM Gupta and E Sanchez (ed), Approximate Reasoning in Decision Analysis, North-Holland, 1982.

13 H. R. Pope, Holloway and E. J. M. Campbell, The Effects of Elastic and Resistive Loading of Inspiration on the Breathing of Conscious Man, Respir.-Physiol., 4(1968) 363-372.
14 Mou-chao, Ma and Pei-zhuang, Wang: An Exploratory Study on Psychological Methodology: Notes on Psychological Fuzziness and Fuzzy Statistical tests, ACTA Psychologica SINICA, 2(1985) 177-186.

Applications of Fuzzy Set Theory in Human Factors, edited by W. Karwowski and A. Mital
Elsevier Science Publishers B.V., Amsterdam, 1986 — Printed in The Netherlands

AN APPLICATION OF FUZZY SET THEORY FOR EVALUATION OF HUMAN PERFORMANCE ON AN INSPECTION TASK

M-J. WANG[1], J. SHARIT[2], and C. G. DRURY[2]
[1]Industrial Engineering Division, Alfred University, Alfred, New York 14802, U.S.A.
[2]Department of Industrial Engineering, 342 Bell Hall,
State University of New York at Buffalo, Amherst, New York 14260, U.S.A.

ABSTRACT

A frequent finding from studies in the area of inspection is the large individual differences in inspection performance, a factor that has contributed to both undermining the development of valid selection tests and complicating the training process for inspectors. In a study designed to discover cognitive factors that can potentially account for a large part of these differences, a fuzzy set approach formulated as a multi-criteria decision-making problem was utilized in order to determine whether individuals can prioritize cognitive skills considered important for inspection performance. Results indicated a close correspondence between fuzzy set and statistical approaches, suggesting the possibility for integrating the individual's subjective appraisal of the relative importance of cognitive factors into the design selection, and training process for inspection tasks.

INTRODUCTION

Over the past thirty years, an enormous amount of research has been generated on human inspection performance. Inherent to the task of inspection are errors, often classified as either type 1 or type 2. A type 1 error occurs when an acceptable item is rejected, and a type 2 errors occurs when a faulty item is accepted. When viewing the inspection task as comprised of search and decision-making components (ref. 1), inspection errors can be subdivided into (1) search error, where the inspector fails to locate a fault, and (2) decision error, where the inspector incorrectly classifies the fault as acceptable or rejectable. In addition to error data, response time variables have been frequently used to assess the inspector's performance. Two important measures are: (1) search time, consisting of the time taken to find the fault and make a decision, and (2) ending time, defined as the time taken when no fault is found.

Much of the literature in inspection pertains to models of human performance and factors affecting inspection performance. The latter can be broadly classified into task, environmental, individual, and organizational factors (ref. 2). With regard to the individual factors, a consistent finding is the large individual differences in inspection performance (refs. 3, 4, and 5). The lack of valid selection tests for inspectors, a neglected area in inspection research (refs. 5 and 6), can be at least partly attributed to these differences.

Task analysis of the inspector's job (ref. 7) clearly reveals the significant role of cognitive processes, suggesting that a large part of the individual differences could be attributed to differences in cognitive skills critical for the inspection task. To the extent that individual differences in inspection performance can be explained in terms of cognitive skills or factors, not only is the potential for the development of valid selection tests increased, but more flexible training techniques that can compensate for human limitations in these skills can also be developed.

A major objective of the study described in this paper was to determine the cognitive factors underlying inspection performance. Particular emphasis will be given towards evaluating the use of fuzzy set concepts as a method for differentiating individual inspection performance on the basis of these cognitive factors.

METHOD

Overview of the Study

To narrow down the number of potential cognitive factors contributing to inspection performance, a questionnaire survey was conducted in a New England shoe factory in which 46 experienced inspectors from five departments participated. The survey consisted of 31 questions of which the first 12 addressed the inspectors' evaluations of the importance of different cognitive factors to their jobs. Combining these findings with those derived from the more generalized task analysis of inspection performance (ref. 7) resulted in the extraction of the following four cognitive factors: judgment, attention, perception, and memory. These factors were subsequently used as the basis for selection of the pretests described below.

Six males and six females, both groups being equally distributed into three age groups (20-30, 30-40, and 40-50 years of age) to ensure a representative choice of subjects, participated in an experimental study that involved two phases. In the first phase, 11 pretests designed to measure different cognitive skills (Table 1) were administered. In the second phase, the subject performed three inspection tasks: (1) circuit pattern inspection, a pure search task; (2) computer generated symbols inspection, designed to simulate the inspection of a sheet of glass or steel; and (3) color video comparator (CVC) inspection of printed circuit boards, representing the only "real" inspection task. Results obtained from performance of CVC inspection were, relative to the other two tasks, much more reliable, so that only CVC inspection data will be considered for further evaluation. A more detailed description of this task follows.

The Experimental Task

The setup for the CVC inspection task included a color video comparator with its control panel interfaced to an HP9825A computer, HP6942A multiprogrammer, and HP9895A flexible disk memory. A six-button experimental response box was interfaced to the computer for data collection purposes. The CVC device is based on the concept of standards

TABLE 1

Summary of the 11 Pretests Designed to Measure Cognitive Skills

Cognitive Factor	Test Selected	To Measure
Attention	WAIS - Digit Span	Ability to concentrate
Attention	WAIS - Digit Symbol	Ability to attend
Recognition	Perceptual Speed	Ability to compare visual configurations and determine whether they are identical or similar
Recognition	Embedded Figure Test	Ability to recognize a small figure within a large complex figure
Detection	PSI - Basic Skills Tests: Visual Speed and accuracy	Ability to detect differences in small detail
Classification	PSI - Basic Skills Tests: Classifying	Ability to place information into predetermined categories
Memory	IPI - Aptitude Series: Memory	Ability to recall visual material
Judgment	IPI - Aptitude Series: Judgment	Ability to make logical judgments
Judgment	Circle Judgment Test	Ability to judge the difference in size between two circles
Perceptual Judgment	Line Judgment Test	Ability to discriminate the difference in length between two lines

inspection. where an item to be inspected is compared to one that is known to be free of any flaws. Video inputs from each of these items blink back and forth from the standard to the item being inspected–hence the popular name for this type of inspection: "blink inspection."

Life-sized photographs of printed circuit boards served as the stimuli, and were selected to include different combinations of flaw type and flaw location. Each board contained 298 components, with a density of 3.01 components/in^2. Of the various flaw types, three were rejectable (i.e., faults existed) and two were acceptable. A missing component, an incorrect integrated circuit, and incorrect polarity (on parts with polarity) were considered rejectable flaws. Acceptable flaws were defined for this study as wire leads from the component that were out of the solder hole, and incorrect (relative to the standard circuit board) components. A total of 24 boards were used; eight were free of any flaws, 12 contained single faults, and four had two faults. The response measures collected were search time, ending time, search error rate, decision error rate, and overall error rate. Following the completion of both phases of the experiment (the CVC inspection task was always performed last), judgments concerning the relative importance of each of the cognitive factors to CVC inspection were elicited from each subject for use in subsequent analyses involving fuzzy sets.

STATISTICAL ANALYSIS

Factor analysis was performed on all pretest and inspection performance measures (from all three tasks), resulting in the extraction of four (expected) pretest factors (Table 2) and five performance factors. Of the five performance factors, two were associated with CVC inspection: (1) search error and time on CVC inspection; and (2) decision error. Since the first of these accounted for over one-third of the total variation stemming from performance of all three tasks, this measure was considered the most reliable. It was therefore selected as the dependent variable in a mutliple linear regression model in which the four cognitive factors served as linearly independent variables. The squared semipartial correlation, defined as

$$Sr^2 = \frac{F_i}{df(res)}(1 - R^2) \tag{1}$$

where $df(res)$ = residual degrees of freedom, $F = F - ratio$ for each variable i, and R^2 = overall coefficient of determination, was utilized to measure the relative importance of each independent variable. Results are summarized in Table 3.

TABLE 2

Factor Loadings, Percent of Variance, and Labels for the Four Cognitive Factors

Variable	F_1	F_2	F_3	F_4
Classifying	.8735	-.1365	.1646	.1250
Digit Span	.7548	.1338	.0051	.2739
Digit Symbol	.7379	.1406	-.1574	-.4476
Judgment(IPI)	.7337	-.1234	-.0924	.1342
Line Judgment	-.2503	.9691	-.1923	.0882
Visual Speed and Accuracy	.4130	.8211	.1413	-.1727
Perceptual Speed	.0685	.7905	.1841	-.0859
Circle Judgment	.1275	.1127	.7910	.0242
Visual Memory	.2172	.2148	-.7971	.1264
Memory (PSI)	.3718	.0739	-.1227	.8201
Embedded Figure Test	-.0670	.4290	.0946	.0561
Percent of Variance Explained	34.7	23.5	14.1	9.6
Label	Attention	Perception	Judgment	Memory

THE USE OF FUZZY SETS

Problem Formulation

The human's ability to prioritize cognitive factors affecting inspection performance and assign weights to those priorities can serve as both a basis for selection and an index for training in inspection tasks. More generally, if the relative contribution of various cognitive skills for performance in any human-machine system is known or can be inferred, the corresponding subjective assessment of these skills can help determine the design of cognitive aids aimed at offsetting the information processing limitations of the human.

TABLE 3

Sr^2 and R^2 Associated with the Four Cognitive Factors

Cognitive Factor	Performance Measure: Time and Search Error in CVC Inspection
Attention	.373
Perception	.206
Judgment	.084
Memory	.032
R^2	.736

The problem of prioritizing cognitive factors with respect to their relative importance to inspection performance as based on the subjective assessments of a sample of individuals can be formulated as a problem of multiple objective decision-making. Since we are not concerned with whether the cognitive factor in question belongs to the set representing cognitive skills important for inspection performance but rather with how strongly it belongs to it, a fuzzy decision-making approach is more appropriate than a probabilistic approach (ref. 8). However, unlike most fuzzy set theoretic approaches to multiple objective decision making problems (refs. 9 and 10), it is the decision alternatives (i.e., the cognitive factors) that are imprecise and not the goals or constraints.

The basic problem can be formulated as follows: Let $X = \{x_1, x_2, x_3, x_4\}$ constitute the set of decision alternatives corresponding to the four cognitive factors extracted from factor analytic methods, where x_1 = memory, x_2 = attention, x_3 = perception, and x_4 = judgment. S_j is defined as the fuzzy subset of cognitive skills important for inspection performance corresponding to subject j, and characterized by a membership function $f_{S_j}(x_i)$ which associates with each of the four cognitive factors a number in the interval [0,1] indicating the grade of membership of factor i in S_j. The objective was to prioritize the weights associated with these factors and determine the correspondence between these weights and their representation in the multiple regression model as predictors of inspection performance. These weights are given by

$$f_D(x_i) = \underset{S_j}{Min}\,(f_{S_1}(x_i), \ldots, f_{S_{12}}(x_i)) \quad \text{for each } x_i \text{ in } X \tag{2}$$

where D is the fuzzy subset that results from selecting, for each x_i, the smallest membership value from any of the fuzzy subsets S_1 through S_{12}, under the assumption that the judgments of all subjects are equally reliable.

Since solutions deriving from fuzzy set operations are very sensitive to the values of the membership functions, the assignment of grade of membership is critical. Unfortunately, the subjectivity inherent in this process has represented "one of the most important drawbacks of the fuzzy set approach to decision-making" (ref. 10, p. 591). Moreover, the method by which f_D is obtained requires that human responses concerning the relative contribution of each of the factors to inspection performance be reliable. It was therefore necessary to utilize a method that not only reduced as much as possible the adverse effects of the subjectivity per se, but also enabled one to consider only those subjects whose subjective responses were reliable.

Obtaining Membership Functions

Noting that the only way to determine the intensity of belongingness of elements in a set is by relative comparison of the elements in question, Saaty (refs. 11 and 12) proposed a method for determining the grade of membership through the process of magnitude estimation derived from pairwise comparisons. Based on a scale (refs. 11, 12, and 13) that takes into account the human's ability to simultaneously compare no more than 7 ± 2 objects successfully (Table 4), a ratio scale for a set of elements can be obtained without making any assumptions concerning "cardinal" or "ordinal" consistency with respect to the judgments elicited by the individual.

Utilizing this scale, a matrix A_j of pairwise comparisons was constructed for each subject whose elements indicated the relative strength with which each cognitive factor dominated the other. The general form of this matrix is

$$A = \begin{array}{c} \\ M \\ A \\ P \\ J \end{array} \begin{array}{c} \begin{array}{cccc} M & A & P & J \end{array} \\ \begin{pmatrix} a_{11} & a_{12} & a_{13} & a_{14} \\ a_{21} & a_{22} & a_{23} & a_{24} \\ a_{31} & a_{32} & a_{33} & a_{34} \\ a_{41} & a_{42} & a_{43} & a_{44} \end{pmatrix} \end{array} \tag{3}$$

where M = memory, A = attention, P = perception, J = judgment, and $a_{ii} = 1$. If factor i is more important than factor j, a_{ij} is assigned a number from Table 4, and a_{ji} is set equal to $1/a_{ij}$. If factor j is more important than factor i, a_{ji} is assigned a number from Table 4 and a_{ij} is set equal to $1/a_{ji}$.

TABLE 4

The Rating Scale (ref.ll)

Intensity of Importance	Definition
1	Equal importance
3	Weak importance of one over the other
5	Strong importance of one over the other
7	Demonstrated importance of one over the other
9	Absolute importance of one over the other
2,4,6,8	Intermediate values between the two adjacent judgments

If the magnitudes of the weights $w' = (w_1, \ldots, w_n)$ corresponding to the strength of the n elements were known, the pairwise comparisons would result in the ratios $a_{ij} = w_i/w_j$. Although we typically have information on just the ratios, by making use of the fact that

$$Aw = nw \tag{4}$$

w can be recovered by solving the system $(A - nI)w = 0$ which ensures the existence of a largest nonzero eigenvalue equal to n whose associated eigenvector is the vector of weights. By normalizing this vector so that $\sum_{i=1}^{n} w_i = 1$, the appropriate values in terms of membership functions are obtained.

Up to this point it was assumed that judgments were consistent. In that case, the multiplication of any row i of matrix A by w would result in $(w_i, \ldots, w_i)$ due to the fact that $\sum_{j=1}^{n} a_{ij}w_j/w_i = n, i = 1, \ldots, n$, where the constant n is the largest eigenvalue of A. In reality, w contains random error. However, the largest eigenvalue would still remain near n and the remaining ones near zero. The problem therefore becomes one of finding w which satisfies $Aw = \lambda_{max}w$, where λ_{max} is the largest observed eigenvalue. The deviation of λ_{max} from n could be used as a measure of consistency since the matrix A can only be consistent if $\lambda_{max} = n$.

An advantage in using the eigenvalue procedure described above (refs. 11 and 12) for obtaining grades of membership is that it allows consideration of consistency or reliability of the person's judgments. As indicated earlier, this ability is especially critical due to the fact that the fuzzy subset D, whose membership function represents the degree of prioritization of the cognitive factors, is obtained through $S_1 \cap S_2 \cap \ldots \cap S_{12}$. The potential therefore exists for having the judgments of reliable subjects masked by one or more unreliable subjects. Saaty (ref.12) suggests using the statistic

$$m = \frac{\lambda_{max} - n}{n - 1} \tag{5}$$

for testing consistency, where λ_{max} is the largest observed eigenvalue. Based on certain properties attributed to the sampling distribution of m, it was noted that for $m > 2$ "it is possible that inconsistency is indicated" (p. 239).

Fuzzy Set Analysis

Utilizing the scale recommended in (ref. 11), the judgments elicited by the 12 subjects are summarized below:

$$A_1 = \begin{array}{c} \\ M \\ A \\ P \\ J \end{array} \begin{array}{c} \begin{array}{cccc} M & A & P & J \end{array} \\ \begin{pmatrix} 1 & 9 & 9 & 1/5 \\ 1/9 & 1 & 9 & 9 \\ 1/9 & 1/9 & 1 & 5 \\ 5 & 1/9 & 1/5 & 1 \end{pmatrix} \end{array} \qquad A_2 = \begin{array}{c} \\ M \\ A \\ P \\ J \end{array} \begin{array}{c} \begin{array}{cccc} M & A & P & J \end{array} \\ \begin{pmatrix} 1 & 1/3 & 1/2 & 1/3 \\ 3 & 1 & 2 & 1/4 \\ 2 & 1/2 & 1 & 1 \\ 3 & 4 & 1 & 1 \end{pmatrix} \end{array}$$

$$A_3 = \begin{array}{c} \\ M \\ A \\ P \\ J \end{array} \begin{array}{c} \begin{array}{cccc} M & A & P & J \end{array} \\ \begin{pmatrix} 1 & 3 & 3 & 3 \\ 1/3 & 1 & 1/3 & 1/3 \\ 1/3 & 1 & 1 & 1 \\ 1/3 & 3 & 1 & 1 \end{pmatrix} \end{array} \qquad A_4 = \begin{array}{c} \\ M \\ A \\ P \\ J \end{array} \begin{array}{c} \begin{array}{cccc} M & A & P & J \end{array} \\ \begin{pmatrix} 1 & 1 & 1/7 & 1/3 \\ 1 & 1 & 1/5 & 1/3 \\ 7 & 5 & 1 & 5 \\ 3 & 3 & 1/5 & 1 \end{pmatrix} \end{array}$$

$$A_5 = \begin{array}{c} \\ M \\ A \\ P \\ J \end{array} \begin{array}{c} \begin{array}{cccc} M & A & P & J \end{array} \\ \begin{pmatrix} 1 & 3 & 1 & 1/3 \\ 1/3 & 1 & 1/3 & 1/3 \\ 1 & 3 & 1 & 1 \\ 3 & 3 & 1 & 1 \end{pmatrix} \end{array} \qquad A_6 = \begin{array}{c} \\ M \\ A \\ P \\ J \end{array} \begin{array}{c} \begin{array}{cccc} M & A & P & J \end{array} \\ \begin{pmatrix} 1 & 1/9 & 1/7 & 1/7 \\ 9 & 1 & 1/7 & 9 \\ 7 & 7 & 1 & 7 \\ 7 & 1/9 & 1/7 & 1 \end{pmatrix} \end{array}$$

$$A_7 = \begin{array}{c} \\ M \\ A \\ P \\ J \end{array} \begin{array}{c} \begin{array}{cccc} M & A & P & J \end{array} \\ \begin{pmatrix} 1 & 1/5 & 1/7 & 1/3 \\ 5 & 1 & 3 & 9 \\ 7 & 1/3 & 1 & 3 \\ 3 & 1/9 & 1/3 & 1 \end{pmatrix} \end{array} \qquad A_8 = \begin{array}{c} \\ M \\ A \\ P \\ J \end{array} \begin{array}{c} \begin{array}{cccc} M & A & P & J \end{array} \\ \begin{pmatrix} 1 & 1/5 & 1/5 & 1/3 \\ 5 & 1 & 1/9 & 3 \\ 5 & 9 & 1 & 3 \\ 3 & 1/3 & 1/3 & 1 \end{pmatrix} \end{array}$$

$$A_9 = \begin{array}{c} \\ M \\ A \\ P \\ J \end{array} \begin{array}{c} \begin{array}{cccc} M & A & P & J \end{array} \\ \begin{pmatrix} 1 & 1 & 1/9 & 1/7 \\ 1 & 1 & 1/9 & 9 \\ 9 & 9 & 1 & 1/7 \\ 7 & 1/9 & 7 & 1 \end{pmatrix} \end{array} \qquad A_{10} = \begin{array}{c} \\ M \\ A \\ P \\ J \end{array} \begin{array}{c} \begin{array}{cccc} M & A & P & J \end{array} \\ \begin{pmatrix} 1 & 1/3 & 1/5 & 1/5 \\ 3 & 1 & 1/5 & 3 \\ 5 & 5 & 1 & 7 \\ 5 & 1/3 & 1/7 & 1 \end{pmatrix} \end{array}$$

$$A_{11} = \begin{array}{c} \\ M \\ A \\ P \\ J \end{array} \begin{array}{c} \begin{array}{cccc} M & A & P & J \end{array} \\ \begin{pmatrix} 1 & 1/5 & 6 & 1/7 \\ 5 & 1 & 5 & 3 \\ 1/6 & 1/5 & 1 & 1/7 \\ 7 & 1/3 & 7 & 1 \end{pmatrix} \end{array} \qquad A_{12} = \begin{array}{c} \\ M \\ A \\ P \\ J \end{array} \begin{array}{c} \begin{array}{cccc} M & A & P & J \end{array} \\ \begin{pmatrix} 1 & 7 & 5 & 1/3 \\ 1/7 & 1 & 1/3 & 1 \\ 1/5 & 3 & 1 & 4 \\ 3 & 1 & 1/4 & 1 \end{pmatrix} \end{array}$$

The eigenvalue problem $Aw = \lambda_{max}w$ was solved and the eigenvector corresponding to $\lambda_{max}w$ obtained for each of the 12 subjects (Tables 5 and 6).

Although the possibility of inconsistency was suggested when $m > 2$ (ref. 12), a much more conservative cutoff value was considered necessary due to the requirement of computing the intersection of fuzzy subsets S_j. When using $m > 2$ as the criterion for consistency, $\lambda_{max} = 10$ and subjects one and nine are excluded. It was also decided that subjects six and 12 be excluded from further consideration. Subject 12's judgments were suspect since that subject failed to make one of the $n(n-1)/2$ paired-comparisons, requiring that its judgment be made at a later time. The elimination of subject six, which in practice does not affect the final solution, allows for a relatively conservative consistency criterion of $\lambda_{max} = 5$ to be adopted, enabling the deviation of λ_{max} from n to be < 1 for the eight remaining subjects–an intuitively satisfying result.

TABLE 5

λ_{max} and Associated Eigenvector for Each Subject

Subject	λ_{max}	Memory	Attention	Perception	Judgment
1	10.700	.713	.552	.209	.377
2	4.434	-.177	-.442	-.396	-.785
3	4.155	-.849	-.220	.283	-.389
4	4.121	-.333	-.371	-.270	-.850
5	4.155	.410	.175	.526	.724
6	5.178	.020	.170	.407	.052
7	4.325	-.030	-.307	-.139	-.051
8	4.700	.050	-.171	-.571	-.107
9	10.148	.076	.552	.627	.545
10	4.538	-.052	-.167	-.526	-.110
11	4.697	-.888	-3.870	-.354	-.270
12	7.103	-.588	-.143	-.497	-.622

TABLE 6

Grades of Membership for Each of the Eight Consistent Subjects

Subject	Memory	Attention	Perception	Judgment
2	0.098	0.246	0.220	0.436
3	0.487	0.126	0.162	0.225
4	0.078	0.087	0.634	0.201
5	0.223	0.096	0.287	0.394
7	0.055	0.191	0.635	0.119
8	0.032	0.263	0.023	0.082
10	0.061	0.195	0.616	0.128
11	0.114	0.495	0.045	0.346

Normalization of the eigenvectors of weights derived from the eight consistent subjects resulted in the grades of membership for each of the four cognitive factors in S_j, respectively. The final decision set contains the weights reflecting the relative degree of importance of these factors for inspection performance, and is represented by the fuzzy subset

$$D = S_2 \cap S_3 \cap S_4 \cap S_5 \cap S_7 \cap S_8 \cap S_{10} \cap S_{11} \tag{6}$$

with corresponding membership function

$$f_D(x_i) = Min(f_{S_2}(x_i), f_{S_3}(x_i), f_{S_4}(x_i), f_{S_5}(x_i), f_{S_7}(x_i), f_{S_8}(x_i), f_{S_{10}}(x_i), f_{S_{11}}(x_i)) \text{ for each } x_i \text{ in } X. \tag{7}$$

Following fuzzy set operations, the normalized membership for each factor in D was found to be:

$$D = \left\{ \frac{0.130}{Memory}, \frac{0.0354}{Attention}, \frac{0.183}{Perception}, \frac{0.333}{Judgment} \right\} \tag{8}$$

indicating the degree of membership of each of the four cognitive factors in the fuzzy subset of cognitive skills important for CVC inspection performance. The priority of importance is seen to be attention, judgment, perception, and memory.

Comparisons Between Fuzzy Set and Statistical Approaches

Comparisons between fuzzy set and statistical (nonsubjective) approaches were based on the normalized membership function values derived from fuzzy set operations and the normalized squared semipartial correlations (Sr^2) derived from multiple regression analysis. As indicated in Table 7, both approaches resulted in the same order of priority ($r = .924$, $p < .05$) with respect to the importance of the cognitive factors for CVC inspection performance.

TABLE 7

Comparison of the Normalized Grades of Membership with the Normalized Sr^2

Cognitive Factor	Normalized Membership Function	Normalized Sr^2
Memory	.130	.046
Attention	.354	.537
Perception	.183	.084
Judgment	.333	.296

Some Comments on the Use of the Eigenvalue Procedure

The eigenvalue procedure described above had been utilized by Yager (ref. 13) for evaluating the relative importance of fuzzy goals and constraints in a multiple objective decision-making example that demonstrated the method by which a selection is made from a set of alternative candidates for a job. As he pointed out, one would naturally want to avoid selecting an alternative from the decision set D if it has a low membership in an objective whose importance exceeds that of the others. Since the membership function for each alternative in D is determined by the lowest membership in all objectives, the likelihood of selecting a weak alternative will be less if the grade of membership of weak alternatives in important objectives is made even lower. Letting B be a fuzzy subset over X and $\alpha \geq 0$ a scalar, then raising B to the α power is defined as a fuzzy subset over X with membership function $f_B{}^{\alpha}(x_i) = (f_B(x_i))^{\alpha}$ for all x_i in X (ref. 13). By assigning to each objective a number α, one could increase the membership function of less important objectives ($\alpha < 1$) and decrease the membership function of the important objectives ($\alpha > 1$). These operations increase the grade of membership of those elements with low values more than those with high values, and decrease the grade of membership of elements with large values less than those with low values, respectively.

The usefulness of the eigenvalue procedure for the multiple objective decision-making example in (ref. 13) was that by obtaining a scale on which the importance of each objective can be measured, solution of the eigenvalue problem results in the eigenvector of weights (the α's) that allows trade-offs in the priorities among objectives to be taken into account. In our study, the coordinate system, so to speak, of the decision problem was reversed. Yager (ref. 13) focused on the ramifications of having objectives with varying degrees of importance; values for the membership function were assumed. Here, the intention was to obtain grades of membership for objectives (or more correctly, for criteria) that were represented by a set of subjects within which each individual element (i.e., each subject) was treated as equally important. To ensure that this assumption was met, it was necessary that the grades of membership for the subjects be as reliable as possible. The usefulness of Saaty's methodology (refs. 11 and 12) for this study was thus based not only on its ability to obtain grades of membership but also on its ability to provide the mechanisms by which persons whose judgments were inconsistent and thereby unreliable could be determined.

CONCLUSIONS

Evidence exists (refs. 14 and 15) indicating that people are adept at thinking in fuzzy terms, at least to the extent that their ratings of the degree to which objects belong to a particular class are consistent. This result implies that ratings can potentially reflect the fuzziness underlying human cognitive processes (ref. 16), a particularly important finding considering that the assignment of grades of membership remains the primary drawback in the use of fuzzy sets.

In this study, an eigenvalue procedure described by Saaty (refs. 11 and 12) was used to directly obtain grades of membership representing the importance of each of four cognitive factors for inspection performance. Other approaches for obtaining grades of membership have been suggested (refs. 16 and 17). The approach selected, however, does not assume cardinal or ordinal consistency in human judgments, and is based on theoretical foundations which allow consistency in human judgment to be estimated.

The results obtained from this study suggest that the human's perception of relative importance of various cognitive processes can serve as a selection device for inspection tasks which have already been validated with respect to the cognitive skills necessary for their satisfactory performance. In addition, it might be worthwhile to consider integrating this subjective information into a training program that would attempt to either redistribute the degree to which the human depends on particular cognitive skills or compensate for possible differences in these skills through appropriate decision support. Finally, since the cognitive factors extracted in this study also underlie human performance in many other types of tasks, the possibility exists for considering the human's assessment of the relative importance of cognitive skills in the design, selection, and training process of various human-machine systems.

REFERENCES

1. C. G. Drury, Human Factors, 17 (1975) 257-265.
2. E. D. Megaw, Applied Ergonomics, 10.1 (1979) 27-32.
3. D.E. Chapman and M.A. Sinclair, in C. G. Drury and J. G. Fox (Eds.), Human reliability in quality control, Taylor and Francis, London, 1975, pp 231-251.
4. T. J. Gallwey, Visual Inspection for Several Fault Types and its Predictors, Unpublished Dissertation, SUNY Buffalo, N.Y., 1980.
5. T. J. Gallwey and C. G. Drury, Human Factors, (1985), in press.
6. T. J. Gallwey and S. J. Czaja, An Evaluation of the Harris Inspection Test, Proceedings of the Human Factors Society 125th Annual Meeting, 1981, 634pp.
7. C. G. Drury, in S. Gael (Ed.), Job Analysis Handbook, J. Wiley, N.Y. 1984.
8. L.A. Zadeh. Information and Control, 8 (1965) 338-353.
9. R. E. Bellman and L. A. Zadeh, Management Science, 6 (1970) 141-164.
10. R. R. Yager and D. Basson, Decision Sciences, 6 (1975) 590- 600.
11. T. L. Saaty. Journal of Cybernetics. 4 (1974) 53-61.
12. T. L. Saaty, Journal of Mathematical Psychology, 15 (1977) 234-280.
13. R. R. Yager, Int. J. Man-Mach. Stud. 9 (1977) 375-382.
14. E. H. Rosch, in T. E. Moore (Ed.), Cognitive development and the acquistion of language, Academic Press, N.Y., 1973.
15. E. H. Rosch, Journal of Experimental Psychology: General, 104 (1975) 192-233.
16. G. C. Oden, Memory and Cognition, 5 (1977) 198-204.
17. M. Kochen and A. N. Bodre, Journal of Cybernetics, 4 (1976) 49-59.

Part V.

MENTAL AND PHYSICAL WORKLOAD AND STRESS EVALUATION

Applications of Fuzzy Set Theory in Human Factors, edited by W. Karwowski and A. Mital
Elsevier Science Publishers B.V., Amsterdam, 1986 — Printed in The Netherlands

COMPARISON OF MENTAL WORKLOAD AND AVAILABLE CAPACITY IN COMPLEX PERSON-MACHINE SYSTEMS

Mark H. Chignell[1] and P.A. Hancock[2]

[1] Department of Industrial & Systems Engineering

[2] Safety Science and Human Factors Departments

University of Southern California, Los Angeles, CA 90089

ABSTRACT

A method is described for building intelligent person-system interfaces which are responsive to changes in task demand. This approach uses on-line mental workload (MWL) assessment procedures which are comprised of both physiological and performance measures. Comparison of available perceptual and cognitive capacity with the current level of MWL demanded of the operator by the task activates a compensatory procedure that shifts the processing load toward or away from the human in order to optimize overall system performance. This procedure is dependent on the derivation of the error signal which is input into the compensatory system element. The overall success of the intelligent interface relies largely on the validity and accuracy of this error signal. The error signal is based on a fuzzy set-theoretic formulation. A multivariate profile of physiological responses and measures of primary task performance are mapped onto two fuzzy sets. A fuzzy discriminating criterion, related to the fuzzy cluster-analytic model, is used in assigning the membership functions for each of two MWL categories. The resources required by the task are specified in terms of set membership functions for each attentional resource. This information is combined into an error signal used in subsequent dynamic task reallocation.

INTRODUCTION

In this chapter we present a brief view of an adaptive human-machine system and the components which constitute such an assembly. The goal of an adaptive human-machine system is to perform a defined task optimally, within the constraints of the environment and the processing resources available to the human and the machine, both singly and in combination. This goal can be achieved only if there is a cooperative relationship between the human and the machine which allows precise orchestration of the tasks performed by each of these two components of the system. In the past, such a cooperative relationship was unachievable as input to the machine was constrained by a complex command syntax which was difficult to learn, and translational interfaces were unavailable. As a result of recent technological advances, however, the development of intelligent interfaces which can support effective human-machine communication and cooperation is now possible [1].

A knowledge-based adaptive mechanism (KBAM) is capable of controlling the task allocation and load leveling functions within a human-machine system [2]. Load leveling is required because of fluctuations in the perceived or actual difficulty of the task over time. Figure 1 shows how the mismatch between task demands and available resources can vary over time. Load leveling is particularly important when the mismatch is sufficient to produce inadmissible overload or underload as illustrated by the shaded areas within the Figure. Any KBAM which acts as a buffer, task generator or load leveler between the task and the operator, must be capable of assessing the MWL imposed within a short time frame and in a sufficiently accurate fashion. Task reallocation can be accomplished through user-generated commands instructing the machine to assume control of certain tasks, or by the intervention of a KBAM system. In the latter mode, upon detection of an abnormal level of MWL, the adaptive mechanism redefines and reallocates the task or subcomponents of an overall task, unless specifically directed to do otherwise by the user.

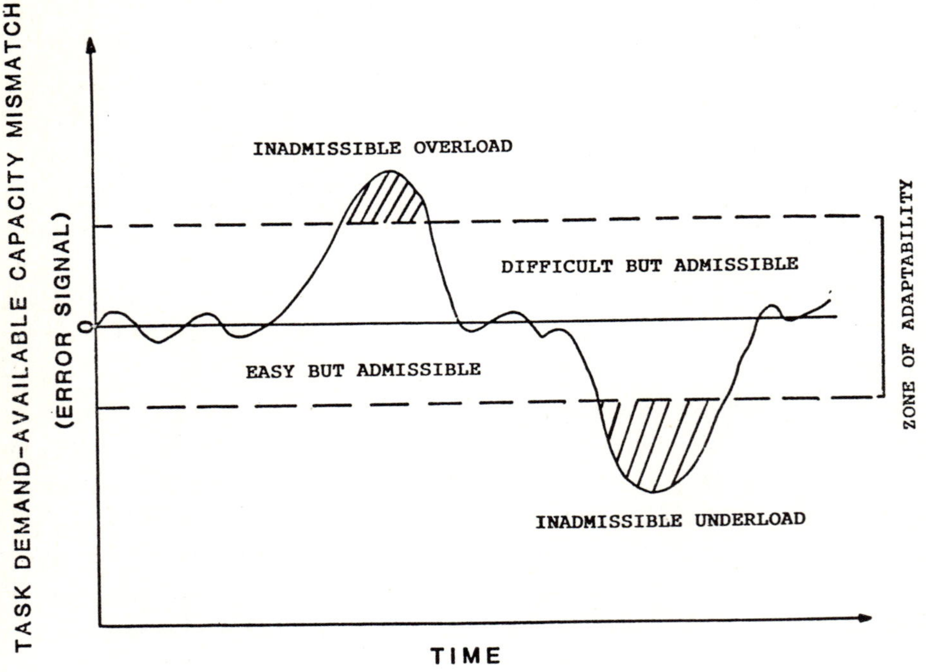

Figure 1. Schematic Representation of the Time-Varying Mismatch between Task Demands and Available Capacity.

The major purpose of the current chapter is to show how the error signal required by a KBAM system may be derived. This signal, through which subsequent task reallocation and redefinition is achieved, is produced through the combination of fuzzy set theory, attentional resource theory, and contemporary MWL assessment methods.

MENTAL WORKLOAD ASSESSMENT

There are a variety of methods by which MWL may be measured [3, 4] including 1) subjective ratings, 2) primary task performance measures, 3) dual task assessment, and 4) physiological assessment techniques.

Subjective Rating

Subjective ratings are used as a measure of the behaviorally perceived difficulty of a task. This is a potentially useful measure since the goal of a KBAM is to relieve the operator of excessive MWL, whether the overloading is objectively or subjectively defined. However, rating of perceived difficulty is itself a task and it may result in an additional loading for the operator. This is particularly true if the rating responses are required during actual performance. Furthermore, the use of subjective ratings may be confounded by variations in the motivation of the performer which are not directly related to task difficulty. In addition, it is possible that subjective reactions do not map directly to the level of performance on the primary task at hand. This dissociation [5] has important implications for the derivation of the KBAM error signal.

Attempts to improve the validity of subjective ratings have led to the use of multiple ratings and averaging techniques. These advanced methods for collecting subjective ratings are, however, inappropriate for monitoring MWL on-line as they are almost exclusively post hoc and static rather than predictive and dynamic. Also, multiple rating judgments and subsequent interpretation require information integration over relatively long periods of time. In spite of the attractiveness and face validity of subjective ratings, as currently formulated they are not well suited to continuous measurement of MWL, are somewhat intrusive into the primary task, and if and when used in a KBAM they would require supplementation by other assessment methods.

Performance Measures

Measures based on primary task performance may be the most obvious method of workload assessment. There are several methodological approaches to the measurement of performance or system output measures, and these have been discussed by Chiles and Alluisi [6]. Welford's [7] concept of the analytical approach focuses on details of the performance of the task itself and examines not only overall achievement but also the way in which the task goals are attained. The advantage of this particular method is that the various decisions and responses that make up performance are considered in the context in which they normally occur. As a result the full complexities of any interaction between different elements in the task may be observed. This approach requires that several performance scores are taken of repetitive performance events, and consequently this analytical approach requires extensive data collection.

Welford [7] has suggested that the analytical approach allows the more subtle effects of workload to be examined by making transparent the strategies employed by the individual performer. Examples of analyses which might be made include assessment of the tradeoff between speed and accuracy, and between errors of omission and commission, and methods of operation which in various ways seek to increase efficiency and to reduce excessive load. The approach has two major problems. First, that the detailed scores required can be difficult to obtain for tasks such as process monitoring in which most of the decisions which are made do not result in any overt action. Second, that even where there are sufficient observable actions, recordings are often elaborate and analysis of results laborious. Consequently, its use in real-time systems is problematic and awaits methodological refinement at the present time.

Synthetic methods require a task analysis of the system which divides the overall task into operational phases. Performance times and operator reliabilities are assigned to the individual tasks and sub-tasks on the basis of the availability of derived data. The information on performance times is then accumulated and the total compared with the predicted duration of the task. The ratio of time required to time predicted can be employed as an index of workload. This approach is similar to the calculation of standard times in work measurement, where performance rating is used to adjust for the effects of skill and motivation in assessing task-related workload on the basis of performance measures [8].

MWL assessment through performance analysis is also possible using multiple measures of primary task performance. This has the beneficial effect of reducing the likelihood that important strategy changes will go unnoticed [3]. One area of concern, however, is that individual measures are differentially sensitive to the various aspects of the task, which may make scaling of workload difficult for specific tasks. Performance measures typically are integrated over time and are often incapable of reflecting sudden changes in MWL and consequently do not provide a satisfactory predictive capability. It has been argued [3] that the measures of primary task performance are task specific. Each time a new situation is examined, new measures must be developed and tested. In many situations, increasing workload may have little effect on performance as the operator has sufficient reserve capability to adapt to the increased load. Performance may not drop off appreciably until a point of critical overload is reached [9] by which time compensatory actions may be too late to recapture or recover a stable performance state.

Secondary Task Methods

Secondary task measures are based upon the rationale that the human operator has only a limited amount of resources to devote to one or more tasks. It is assumed that the imposition of a second task in addition to the major or primary task can reach or potentially exceed an operator's capacity. The MWL of the primary task is then calculated as being inversely proportional to the level of secondary task performance. This method suffers from a number of problems, such as individual load shedding strategies, impingement of one task on the other, and competition between the tasks for access to a single sensory or output channel.

Recently, secondary task methods of MWL assessment have assumed a multiple resource model of attentional capacity [10]. In this view, separate resource pools are used for different aspects of the task. Initially, it was postulated that these pools were functionally separated and could, in principle, become completely drained. Later experimental results have implied that task integrality may sometimes occur, with the addition of an extra task improving, rather than hurting performance on the tasks already being performed, even though the new task is addressing the same resource pool as the previous tasks [11]. The existence of task integrality makes the use of secondary task methodology even more problematic with respect to workload measurement.

Although secondary task methods have been used to measure MWL in the laboratory, they appear to be unsuited for application in realistic complex work environments which typically consist of more than two tasks, often with a high loading on attentional resources. In such environments it is often infeasible to add a further task purely for MWL assessment purposes. It should also be noted that the use of one or more tasks to indirectly assess MWL as a performance decrement conflicts with the major goal of a KBAM, which is to optimize the overall performance of a person-machine system.

Physiological Measures of Mental Workload

Since individuals who are subjected to some degree of MWL commonly exhibit changes in a variety of physiological functions, researchers have advocated the measurement of these changes to provide an estimate of the level of MWL experienced [12, 13]. In dealing with physiological measures we should recognize that many aspects of operator behavior other than MWL can have an effect on the physiological measures [14]. In order to reduce the confounding effects of extraneous factors such as emotional responses which will impact on some measures more than others, a multivariate model is required for relating profiles of physiological response to level of MWL. To the extent that such a model can only be approximate, and levels of MWL can not be identified with complete accuracy, the problem becomes one of fuzzy classification on the basis of noisy multivariate data. Before considering a fuzzy set-theoretic formulation of this problem however, we shall summarize the types of physiological measures which may be appropriate and the ways in which they may be related to MWL.

The interaction between the individual and the environment is reflected in physiological response. As MWL presumably affects the activity of the CNS, measures may variously reflect processes such as demand for increased energy, progessive degradation of the system or homeostatic action of mechanisms designed to restore system equilibrium disturbed by the requirements of cognitive tasks. Currently available physiological measures appear to be similar in some respects to subjective ratings in that they reflect a subjective physiological response to task difficulty. As Ursin and Ursin [12] suggested, physiological methods do not measure the imposed load directly, but instead give information concerning how the individuals themselves estimate the load and their ability to cope with it. Thus physiological measures appear to reflect a subjective response to the task, in the form of perceived task difficulty and anxiety associated with that perceived difficulty. The great advantage that physiological measures have over other methods is that they can be measured continuously and, with careful application, they are not intrusive with respect to performance on the primary task. Furthermore, to the extent that perceived loading is predictive of performance breakdown, the individual's perception of the workload may be precisely the criterion required for task reallocation using a KBAM.

THE MISMATCH BETWEEN TASK DEMANDS AND AVAILABLE RESOURCES

Physiological measures appear to be the most useful form of MWL assessment for a KBAM system. Not only do they supply a continuous stream of data which is amenable to statistical analysis, but they also reflect the human's physiological perception of the current workload and they do not alter the existing task structure substantively. Performance measures also provide important clues about the operator's ability to perform the task as currently defined. The problem of converting a matrix of physiological and performance measures into a signal which provides the necessary information for task redefinition and reallocation within a KBAM can only be solved if a complete specification of the information required by the KBAM is given. The informational requirements of a KBAM result from its purpose, which is to optimize system performance, and from the processes used to achieve that purpose.

In order to maintain satisfactory levels of both MWL and system performance, a KBAM assesses the mismatch between task demands and available capacity and redefines the task in order to reduce the detected mismatch. The input to the KBAM is an error signal representing the mismatch between current task demands and the available capacity of the human operator (Figure 2).

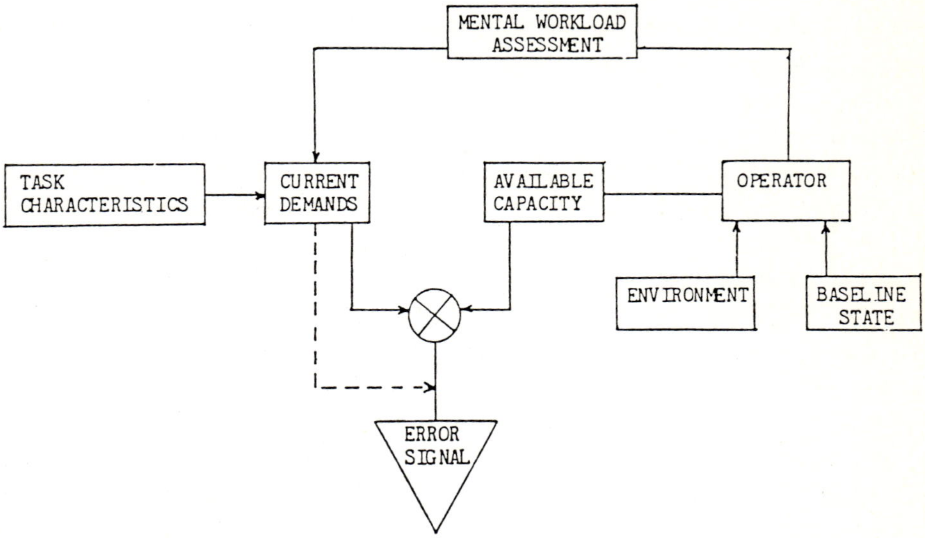

Figure 2. The Formation of an Error Signal as the Mismatch between Current Demands and Available Capacity.

The key component of the KBAM is a reasoning process which selects a task allocation policy that changes the loading on the human in such a way as to improve overall system performance. This process must have access to both overall system goals (a model of the task) and information about what the person and machine components of the system are capable of accomplishing (person and system models). In addition, it is required to decide which loading strategies and task allocation policies are available for selection, and what the implications of the current error signal(s) are.

We can use a process of elimination to determine what information should be supplied by the error signal. Models of the task, person, system, and available loading strategies are already available in the other inputs to the loading strategy reasoner, but none of these inputs give information about the current status of the person. Thus the error signal is a dynamically changing input which has to be combined with the static information provided by the other informational sources available to the loading strategy reasoner.

Not only must the error signal provide information about how well the person is performing particular subtasks, but it must also indicate any discrepancies between overall system performance and system goals in order that the available capacity of the person can be integrated with the needs of the entire system. It would be inappropriate to optimize the workload experienced by the operator if the overall system failed or suffered as a consequence. In this chapter we focus on the component of the error signal which provides information about the status of the human operator. This information will include the amount and direction of MWL (whether increasing or decreasing) experienced by the human.

MWL indicates the direction in which task-induced loading on the person should be altered, but not how the task loading should be achieved. An error signal that was based solely on MWL and performance measures would not provide information about the particular aspects of the person's task or subtasks that were producing over- or under-loading. Thus the input to a KBAM load leveling system also needs to provide a third type of information, namely the extent to which different aspects of the task are using up available resources. We shall now consider the types of resource that are available to the person before considering the error signal in fuzzy set-theoretic terms.

ATTENTIONAL RESOURCES

Early research demonstrated that auditory processing depends on limited resources [15, 16, 17]. More recent research has has found that a concept of capacity as a single and global resource which can be allocated to all tasks [18] is insufficient to explain the relationship between task difficulty and human performance. Some tasks will be difficult because they require a lot of resources in general, while other tasks may have components which interfere with each other. Tracking, recognition, and decision making tasks have been used to identify the functional relationship between multiple resources and input/output limitations of the human operator [19].

The inteference between tasks can be of two types, which are referred to as <u>structural limitation</u> and <u>structural interference</u>. Structural limitation occurs in situations for which dual-task performance deteriorates as a result of the physical constraints on the processing system. Thus, the eye cannot view two separated locations at once, the mouth utter two words at once, nor a specific limb be in two places at the one time. The concept of structural interference can account for such instances as the difficulty in simultaneously performing two independent motor acts such as rubbing the head and patting the stomach. In this example, excessive demands are made upon the processing resource responsible for response output. Although also related to the similarity of demands on the motor system, structural interference is not due to the physical constraints of the limbs and therefore such limitation may be overcome with practice. Instead, structural interference is a curtailment of central nervous system capacity responsible for conjoint action of sensors and/or effectors.

Wickens [19] suggested a three-dimensional model of resources where the dimensions were defined by stages of processing (perceptual-central versus response), codes of perceptual and central processing (verbal versus spatial), modalitites of input (visual versus auditory), and response (manual versus vocal). This model has proved useful in accounting for empirical results. It is the foundation of considerable applied research in areas such as the design and operation of complex systems and is used in the error signal definition below.

A FUZZY ERROR SIGNAL

As discussed above, the error signal which is input to a KBAM should have two major components, one dealing with the induced workload and resources utilized by the human, the other more system-related, dealing with the discrepancy between system goals and performance, and the extent to which different aspects of the task are using up the resources available to the system. The formulation of the first (human) component of the error signal based on a fuzzy subset mapping will now be outlined.

Mental Workload

A fuzzy subset A of a universe of discourse U is defined by a membership function:

$$f_A : U \rightarrow [0,1] \tag{1}$$

which associates with each element u of U a number $f_A(u)$ in the interval [0,1], where $f_A(u)$ represents the grade of membership of u in A [20]. For the purposes of MWL assessment, let the universe of discourse M, say, be a vector comprised of a concatenation of physiological and performance measures. The inclusion of performance measures makes the MWL estimate more task specific, since tasks will differ markedly in terms of the MWL implications of different levels of performance. The selection of physiological measures to be monitored will be based on the practicality of the required measurement method (in terms of cost, intrusiveness and reliability) and the strength of the arguments (presumed causal connections) relating the measure to task-related CNS activities.

In developing a unitary measure of MWL on the basis of a set of physiological and performance measures we will adapt a method suggested by Mital and Karwowski [21]. Let us define two fuzzy subsets labeled "excessive mental workload" and "insufficient mental workload" respectively. We then have two corresponding membership functions, m_{EMWL} and m_{IMWL} which map the vector M of physiological and performance measures into the two fuzzy subsets. The membership function m_{EMWL} for instance, associates with each measure M_i in M a number m_{EMWL} (i) in the interval [0,1] which represents the grade of membership of M_i in the fuzzy subset "excessive mental workload". The closer this number is to 1, the stronger the indication that the individual is under excessive mental workload. The numbers $m(M_1)$, $m(M_2)$,... can be assigned for each of the measures in M and this vector of membership functions (corresponding to the original vector of measures) can then be condensed into an overall estimate of the degree of excessive MWL present. One method of combining the measures is to use a weighted average, thereby allowing for differences in the relative importance of each measure. It is suggested [21] that expert opinion be used to derive these weightings, but we consider an empirically-based approach for obtaining appropriate weightings in the example below. We have chosen to use two fuzzy subsets to assess MWL, rather than one, to allow for more sensitivity in measurement. In process control tasks, for instance, where long periods of insufficient MWL may be interspersed with short bursts of excessive MWL, it may be reasonable to assess separately the two extremes of MWL level, rather than to combine them on a single scale (fuzzy subset). As an example, one might have two situations both representing .5 membership in the fuzzy subset "high mental workload". In the first situation, the .5 represents an averaged response to excessive MWL followed closely by insufficient MWL. The second situation represents a case of a constant and fairly high MWL. Clearly, a KBAM would need to be able to distinguish between these two situations when making corrective responses.

Resources Required by the Task

Mental Workload is experienced in response to the stress imposed by the task. Using the multiple resource view of Wickens [10] we can describe a task in terms of the resources that it requires for adequate performance. At present, we do not have adequate methods for assessing the resource requirements of tasks. One can, however, use expert opinion, possibly supplemented with experimental analyses, to estimate the resource requirements of a task.

Using similar notation to that given in the previous section, we can define a vector R of resource requirements R_i which is mapped onto a fuzzy subset labeled "excessive resource requirements". The components of R are obtained using the classification outlined by Wickens [19]. The union of these resource requirements, i.e., MAX_i (R_i), defines a second global measure of MWL which should covary with the estimate based on the vector M of physiological and performance measures.

Spare Resource Capacity

The resources required by a task can be represented as a vector R, as described above. A similar vector U showing the utilization of those same resources can be constructed. In a given situation, this vector would be specified by instantiated membership functions for each of the relevant resource components (which would include functions for visual input, auditory input, spatial code, verbal code, encoding, central processing, manual response, and vocal response) into the fuzzy subset "complete resource utilization". In practice, however, it is easier to measure spare capacity using probe secondary tasks and similar techniques than it is to measure resource utilization directly.

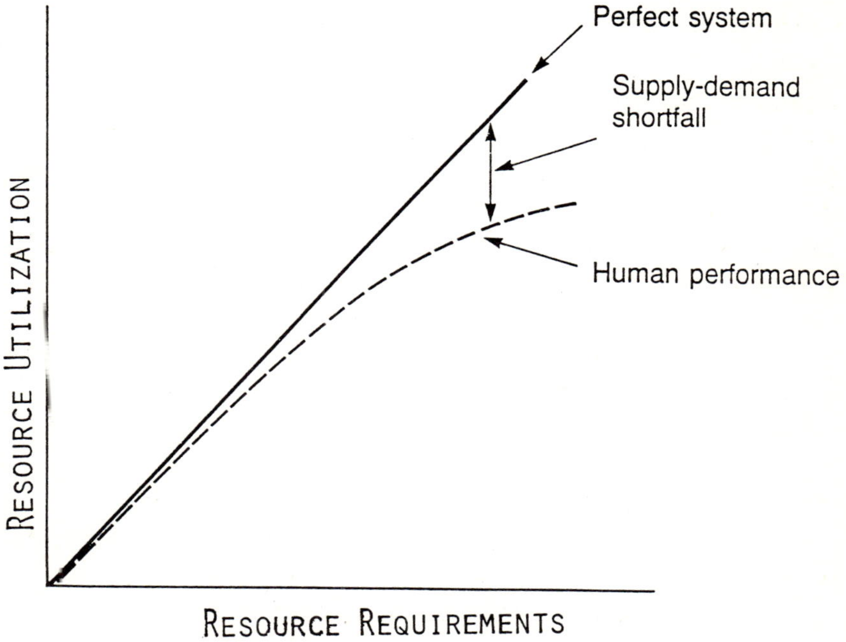

Figure 3. Hypothetical Relationship between Requirements for, and Utilization of, Human Attentional Resources.

In manufacturing engineering, production planning is based on the availability of different resources. In any system, a certain amount of spare resource capacity should be available at all times in case of unexpected breakdowns or increases in demand [22]. Figure 3 gives a schematic view of the relationship between task demands and resources supplied, and the resulting effect on task performance. To the extent that spare resource capacity indicates whether task demands can be met by current resources it serves as a predictor of performance. It is the job of a task allocation process to ensure that, under all but emergency conditions, sufficient amounts of spare capacity are available for all the task-relevant resources.

The tasks performed by the system define the implied resource requirements vector R as described above. Using probe task techniques it should then be possible to identify an additional vector C specifying the degrees of membership for each of the resource components in the fuzzy subset "full resource capacity available."

Given the vector C, the vector U corresponding to utilization of resources can be estimated as the inverse of the spare capacity membership functions, i.e.,

$$\forall_i U_i = 1 - C_i \tag{2}$$

Global measures of MWL tell a KBAM system what change in MWL level is needed, but not how to implement that change. The current mismatch between resources required by the task and resources utilized by the operator will provide the information needed to implement a more favorable set of task requirements. Thus the vector difference between R, the resource requirement vector, and U, the resource utilization vector, defines A, the vector on which task reallocation is to be made, i.e.,

$$A = [(R_1 - U_1), (R_2 - U_2), \ldots] \tag{3}$$

The size of this mismatch, $R_{deviation}$ can be derived as the Hamming distance between the vectors R and C, i.e.,

$$R_{deviation} = Sum_i[ABS\ (R_i - C_i)] \tag{4}$$

$R_{deviation}$ can be used as a third estimate of global MWL.

<u>Overview of the Error Signal</u>

The purpose of the error signal is to provide dynamically changing information about the human operator both in terms of overall MWL being experienced and mismatches between resource requirements and resources utilized. The utilization of this error signal in task allocation is then a two-stage process. First, a global assessment of MWL is made, i.e., is MWL too high or too low and by how much? Second, the resource mismatch vector is used to implement a task reallocation policy designed to return the operator to a more desirable level of MWL. Although the fuzzy logic formalism for combining the various membership functions into an error signal is straightforward, the problem of how to derive the membership functions and associated numbers in the first place is not. It is likely that existing techniques of work measurement and task analysis can be extended to provide estimates of the vector R. It is also likely that a variety of methods, including evoked potential measurement (the P300 component in particular), analysis of eye movement patterns, and the like will lead to reasonable estimates of resource utilization. It is assumed here that the scaling of physiological and performance measures as membership functions on the fuzzy subsets "excessive mental workload" and "insufficient mental workload" will be based on experimental functions derived from existing experimental data. The final measurement problem consists of assigning membership functions in the fuzzy subsets "excessive mental workload" and "insufficient mental workload" on the basis of the vector M. This problem is considered from a fuzzy classification perspective in the next section.

FUZZY CLASSIFICATION

The KBAM method of adaptive control requires a fuzzy error signal that includes the information on MWL and performance which is vital for task rescheduling and reallocation. As defined earlier, the fuzzy error signal will consist of three global measures of MWL, the first based on a weighted average of M, the second based on the maximum resource requirement (Max_i R_i) of the task and the third based on the measure $R_{deviation}$ defined above. The error signal will also include the vector A, specifying the difference between requirements and utilization for each type of human attentional resource. The problem of deriving the first global measure of MWL on the basis of multivariate profiles of physiological and task performance measures is one of fuzzy classification. Methods for making this classification will now be considered.

If the categories are well defined <u>a priori</u>, one can use a discriminating function to allocate a given profile to a particular category. This procedure assumes that the categories can be represented as a partitioning of the space of multivariate profiles. In fuzzy classification, profiles are not exclusively inside or outside a single category, but have grades of membership in one or more categories. Since appropriate discriminating functions for MWL scaling on the basis of physiological and performance measure are currently unavailable, techniques which require fewer a priori assumptions will now be considered.

Cluster analysis is a second type of classification which assumes no prior information about the categories to be formed. It is essentially an unsupervised pattern recognition technique [23] which uses measures of similarity or distance between the profiles to develop a classification solely on the basis of the data input [24]. While this method is appealing because it requires no prior assumptions about the categories to be formed, it is difficult to assess the validity of the obtained clusters [25]. Since cluster validity is difficulty to assess, an unsupervised technique such as fuzzy cluster analysis will not be appropriate for assigning MWL because one cannot ensure that MWL, rather than another criterion, such as motivation or fatigue, is being used as the basis for classification.

A third method of classification uses supervised training [23]. Calibration of MWL can be achieved using the following supervised training method. MWL and performance are observed under a wide range of conditions, with global MWL being assigned as grade of membership in the fuzzy subsets "excessive MWL" and "insufficient MWL" on the basis of subjective ratings by the person doing the task and ratings made by the experimenter on the basis of observation and task analysis. A large collection of these observations, with the associated fuzzy classifications, would then form a training sample. New samples could then be classified without further rating of global MWL using a nearest neighbour allocation rule. Each new profile would be matched against the training sample. It would then be assigned the fuzzy classification given to the most similar profile in the training sample. Euclidean distance could be used in assessing the nearness of profiles, although a number of other measures are available [26].

ERROR SIGNAL DERIVATION: AN EXAMPLE

The derivation of the error signal begins with the acquisition of the data required for the vectors M, R, and C. Table 1 shows a set of hypothetical data along with associated measures of global MWL (M_{EMWL}, R_{Max}, $R_{deviation}$) and the resource requirements-utilization mismatch (A). The weighted average of M using an abritrary vector of weights (W) is .39. This is then normalized by multiplying the obtained average by the number of weights used divided by their sum. In this example, the normalized weighted average is .65, which is the same as the unweighted average. It remains to be seen whether the additional effort of weighting is justified by the increased sensitivity of the M_{EMWL} measure. If an appropriate training sample were available, nearest neighbour allocation could be used instead of the weighted average approach, as outlined in the previous section, to derive an estimate of M_{EMWL}.

$M = [\ .7,\ .5,\ .6,\ .8,\ .9,\ .4\ ]$

$W = [\ .8,\ .8,\ .4,\ .3,\ .8,\ .5\ ]$

$R = [\ .9,\ .1,\ .5,\ .8,\ .8,\ .7,\ .9,\ .1\ ]$

$U = [\ .9,\ .3,\ .4,\ .9,\ .8,\ .6,\ .7,\ .3\ ]$

$M_{EMWL} = .39 \times 6/3.6 = .65$

$R_{Max} = .9$

$R_{deviation} = .9$

$A = [\ 0,\ -.2,\ .1,\ -.1,\ 0,\ .1,\ .2,\ -.2\ ]$

TABLE 1

Hypothetical data to illustrate the error signal derivation method. The four physiological measures M_1 -> M_4 are auditory canal temperature, heart rate variability, blood pressure, and galvanic skin response (GSR), respectively. M_5 and M_6 are measures reflecting the speed, and accuracy, of task performance. W is a vector of weightings assigned to each of the components of M. For the purposes of this example we only consider the subset "excessive mental workload." A separate vector of weights would be required for the fuzzy subset "insufficient mental workload." The resources used in R and U are ordered as follows. 1 = visual input, 2 = auditory input, 3 = spatial code, 4 = verbal code, 5 = encoding, 6 = central processing, 7 = manual output, 8 = vocal output. This data was generated on the basis of a conceptual analysis of the same-different matching task [27].

Thus the nearest M_j in the training sample might be [.7, .4, .7, .8, .9, .3] with an associated M_{EMWL} rating of .7. The rating of .7 assigned to the nearest neighbour in the training sample could then be used as the estimate of M_{EMWL} for the current M vector.

R_{Max} is .9, reflecting the high resource requirements for visual input and manual output which is characteristic of the matching task considered in this example. $R_{deviation}$ is .9 (out of a maximum value of 8.0) indicating that resource utilization matches resource requirements fairly closely, despite the high level of MWL implied by the first two global measures. The vector A implies desirable changes in the requirements for each resource. In the example, inspection of A indicates that it might be desirable to decrease the requirement for manual output, while small increases in the requirements for auditory input and vocal output should be possible without adverse effect on global MWL.

SUMMARY

In order to produce systems that take greatest advantage of the unique capabilities of humans and machines and to provide such systems with the facility to operate in real time in work environments where demands change frequently, it is necessary to extract information on how hard the human operator is working. As a biological entity, the human is not amenable to several of the techniques which can be applied to the assessment of machine workload. The lack of physical activity, with concomitant increase in cognitive activity, further compounds the difficulty of the measurement problem in many tasks [13].

The derivation of a valid error signal, especially in relation to the limits of the human operator in general, and specific operators in particular, is the major barrier to the implementation of knowledge-based adaptive mechanisms. In the present chapter we advocate the use of fuzzy set-theoretic measures to assess human MWL and the resource requirements-utilization mismatch. This has the advantage of providing specific information concerning what is, at present, an indeterminant phenomenon, and allows the use of fuzzy classification techniques to synthesize the major forms of current MWL information into a viable error signal. Implementaion of this procedure is the subject of current investigation.

The assessment of MWL and the derivation of an error signal for a KBAM is well suited to the fuzzy set-theoretic approach. It would be precipate, however, to implement an associated calculus of operations such as fuzzy set union and set intersection at this time. Instead, we have developed the rationale for a fuzzy error signal in the KBAM methodology based on simple averaging, differencing and rating techniques. Since there is uncertainty over both the assessment of MWL and its incorporation within the error signal we have a case of Uncertainty squared where calibration based on careful experimental work is required. An alternative view on fuzzy controllers is given in the work of Mamdani [28]. We prefer the fuzzy classification approach to alternative methods which use concatenation of fuzzy set operators because fuzzy classification assumes less about the ways in which physiological responses and performance measures interact both with each other and amongst themselves.

For the current functioning of complex systems we continue to rely upon the adaptability of the humans and their unique and often poorly specified capabilities. Trends toward rapidity and complexity (e.g., increase in system degrees of freedom) in system structure no longer permit the inclusion of such a poorly understood and unquantified component. While automation has had a major impact on the workplace, it cannot yet replace those unique capabilities of the human which are not as yet implementable in terms of computational models. This chapter presents an approach to the instantiation of cooperative action, in which adaptation is now a machine function. It is our view that such adaptive action is not only desirable, but with current developments, vital to continued harmonious human-machine interaction. While initial application is directed toward systems which operate at the extremes of performance, we view the development of intelligent interfaces in general, and knowledge-based adaptive mechanisms in particular as central issues in the Human Factors endeavor.

REFERENCES

[1] Chignell, M.H., Hancock, P.A. & Loewenthal, A. (Eds.) (1986) Intelligent Interfaces: Theory, Research and Design. Amsterdam: North-Holland.

[2] Chignell, M.H. & Hancock, P.A. (1985) Knowledge-based load leveling and task allocation in human-machine systems. Proceedings of the Twenty-First Conference on Manual Control, Columbus, Ohio, June 1985.

[3] Williges, R.C. & Wierwille, W.W. (1979) Behavioral measures of air crew mental workload. Human Factors, 21, 549-574.

[4] Moray, N., (Ed.)., (1979) Mental Workload, its Theory and Measurement. N.Y.: Plenum Press.

[5] Derrick, W.L. (1983) Examination of workload measures with subjective task clusters. Proceedings of the Human Factors Society, 27, 134-138

[6] Chiles, W.D. & Alluisi, E.A. (1979) On the specification of operator or occupational workload with performance - measurement methods. Human Factors, 21, 515-528.

[7] Welford, A.T. (1978) Mental workload as a function of demand, capacity, strategy and skill. Ergonomics, 21, 157-167.

[8] Barnes, R.M. (1980) Motion and Time Study: Design and Measurement of Work, 7th edn. N.Y.: Wiley

[9] Hancock, P.A. & Chignell, M.H. (1985) The principal of maximal adaptability in setting stress tolerance standards. In R.E. Eberts and C.E. Eberts (Eds.), Trends in Ergonomics/Human Factors II. Amsterdam: North-Holland.

[10] Wickens, C.D. (1980) The structure of attentional resources. In Nickerson, R. & Pew, R. Attention and Performance VIII. Hillsdale, N.J.: Erlbaum.

[11] Kramer, A.F. & Wickens, C.D. (1985) Event-related brain potentials and resource allocation: From dual-task decrements to dual-task integrality. Proceedings of the Human Factors Society, 29, 966-970.

[12] Ursin, H. & Ursin, R. (1979) Physiological indicators of mental workload. In Moray, N., (Ed.)., Mental Workload, its Theory and Measurement. N.Y.: Plenum Press.

[13] Hancock, P.A., Meshkati, N. & Robertson, M.M (1985) Physiological reflections of mental workload. Aviation and Space Environmental Medicine, 56 1110-1114.
[14] Kalsbeek, J.W.H. (1971) Standards of acceptable load in ATC tasks. Ergonomics, 14, 641-650.
[15] Broadbent, D.E. (1958) Perception and Communication. London: Pergamon Press.
[16] Treisman, A.M. (1964) Monitoring and storage of irrelevant messages in selective attention. Journal of Verbal Learning and Verbal Behavior, 3, 449-459.
[17] Neisser, U. (1967) Cognitive Psychology. N.Y.: Appleton-Century-Crofts.
[18] Kahneman, D. (1973) Attention and Effort. Englewood Cliffs, N.J.: Prentice-Hall.
[19] Wickens, C.D. (1984) Engineering Psychology and Human Performance. Columbus, OH: Charles Merrill.
[20] Dubois, D. & Prade, H. (1980) Fuzzy Sets and Systems. Theory and Applications. N.Y.: Academic Press.
[21] Mital and Karwowski, 1986, Towards the Development of Human Work Performance Standards in Futuristic Man-Machine Systems: A Fuzzy Modeling Approach. Fuzzy Sets and Systems, 19 (in press)
[22] Dilworth, J.B. (1983) Production and Operations Management: Manufacturing and Non-Manufacturing, 2nd Edn. N.Y.: Random House.
[23] Duda, P. & Hart, P. (1973) Pattern Classification and Scene Analysis. N.Y.: Wiley.
[24] Everitt, B.S. (1974) Cluster Analysis. London: Heinemann.
[25] Chignell, M.H. & Stacey, B.G. (1980) Practical problems associated with the use of cluster analysis. Psychological Reports, 46, 131-134.
[26] Gregson, R.A.M. (1975) Psychometrics of Similarity. N.Y.: Academic Press.
[27] Chignell, M.H. & Krueger, L.E. (1984) Further evidence for priming in perceptual matching. Temporal, not spatial, separation enhances the fast-same effect. Perception & Psychophysics, 36, 257-265.
[28] Mamdani, E.H. (1976) Advances in the linguistic synthesis of fuzzy controls. International Journal of Man-Machine Studies, 8, 669-678.

Applications of Fuzzy Set Theory in Human Factors, edited by W. Karwowski and A. Mital
Elsevier Science Publishers B.V., Amsterdam, 1986 — Printed in The Netherlands

THE POSSIBILITY FUNCTION OF FUZZY CRITICAL FLICKER FREQUENCY - CHANGES UNDER MENTAL LOAD AND FATIGUE

T. MAREK and Cz. NOWOROL

Jagiellonian University, Institute of Psychology, Group of Mathematical Psychology and Statistics, Cracow, (Poland)

ABSTRACT

Changes in threshold values of critical flicker frequency (CFF) are widely used as a measure of mental work load and mental fatigue. A new concept of threshold measure based on the theory of fuzzy sets is proposed. Twenty four video display terminal operators were investigated three times: before, after three and six hours of mental work load. A new procedure of threshold measure based on fuzzy critical flicker frequency (FCFF) was used. The possibility function (Poss) of FCFF was obtained for each investigated person. It was stated that changes in possibility function of FCFF are due to mental work load and fatigue.

BACKGROUND

The theory of fuzzy sets proposed by Professor Zadeh (refs.15,16,17,18,19) offers the perception theory new concepts to use as powerful tools for building the real pivot of many complicated problems of human behaviour. The theory generates many interesting ideas and provides a natural way to reduce the gap between reality and its theoretical representation. It is an especially effective approach in the most important basic areas of the perception theory which are psychophysical problems, laws and associated phenomena (eg. thresholds of various kinds). The modern perception theory is built on ground of neurophysiological and neuropsychological data as well as on the ground of the modern psychophysical investigations which are based on mathematical theories and models like Fourier methods, network theory, probability theory, structural information theory, sequential analysis, cognitive algebra, theory of signal detection, taxonomy methods, coding theory and so on (refs.1,2,3,4,5,6,7).

The theory of fuzzy sets when applied to the psychophysical concepts of thresholds can constitute fuzzy measures regarded to as subjective scales by which the grade of fuzziness is measured. Terano and Sugeno (ref.12) pointed out that "a fuzzy measure is a means for measuring "fuzziness" while a probability measure is one for measuring "randomness"". The nature of threshold (in psychophysics) seems to be more fuzzy than random. Thus, the fuzzy modeling can be very useful in investigations of threshold.

A flicker is a common term applied to the sensation experienced when the eye is stimulated intermittently. The flicker stimulus consists of two phases: flash and die-away in turn. According to the classical definition the threshold is the border line dividing the stimulus which causes one type of reaction from an other stimulus which causes a different reaction. In the light of this definition the threshold of licker frequency (critical flicker frequency - CFF) is the subjective border line dividing a flicker from continuous light.

Changes of the CFF value are used as an indicator of the functional state of the nervous system in neuropsychological and neuroophtalmological diagnoses. It is also frequently used as the indicator of fatigue resulting from mental load. At present it is one of the most powerful indices used to evaluate that particular type of fatigue. Recently, clear evidence that mental fatigue and CFF value depend on the functional state of the nonspecific brainstem and thalamic projection systems has been reported. Thus, fatigue and CFF values are connected with the level of arousal (ref.10).

It is also known that the changes of CFF are correlated with the sucjective assessment of fatigue. Weber and his co-workers (ref.13) showed that flicker frequency threshold (CFF) was correlated especially with the self-scaling of fatigue based on the scale stretched between eight opposite designations ($p<0.01$). These designations were: refreshed - tired ($r = 0.809$), strong - weak ($r = 0.833$), awake - sleepy ($r = 0.976$), vigorous - exhausted ($r = 0.833$). Many authors point out that decrease in CFF value is connected with increase in subjective feeling of fatigue. Changes in CFF value are widely used as the measure of mental work load and fatigue bacause of the relationships described above.

One of the important problems in the research area of CFF is its measurement and its interpretation as the indicator of mental fatigue (ref.11). A new approach based on the theory of fuzzy sets has been proposed by applying possibility function (Poss). The aim of the present study was to describe the CFF in terms of the fuzzy sets theory where the possibility of several flicker values was considered for threshold. In previous investigations of mental fatigue, based on the flicker - test a decrease of CFF value has been treated as the only indicator of fatigue. It is characteristic of this classical approach that one and only one value of flicker was treated as the flicker frequency threshold value.

In contrast with the classical term CFF we will use fuzzy critical flicker frequency (FCFF). Similarly, instead of threshold of flicker frequency the term fuzzy threshold of flicker frequency is used.

In our approach we assume:

$U = \{u;\ u$ - flicker value$\}$ = universum

FT = u is a threshold value = flicker threshold fuzzy function

$\pi_{FT}(u)$ = Poss(FT = u) = possibility function

Consider the following example which illustrates our fuzzy approach to the CFF. A person examined observes the flicker stimuli of different given frequencies displayed for fixed duration with three seconds' exposure of each stimulus and two seconds' break between them. The accuracy of exposure is 0.1 Hz. Each stimulus is randomly presented twenty times. The persons task is to classify each stimulus as a flicker or a constant light. This procedure deals with the critical interval of the flicker values (high frequences). Applying the fuzzy sets theory the interval consists of values which may possibly be the threshold values. Thus the interval contains values of the flicker threshold fuzzy function. For each flicker value (u) the odds ratio p(u) is calculated:

$$p(u) = \begin{cases} \dfrac{n_f(u)}{n_c(u)} & \text{if } n_f(u)<n_c(u) \\ \\ \dfrac{n_c(u)}{n_f(u)} & \text{if } n_c(u)<n_f(u) \end{cases} \qquad (1)$$

where: n_f = number of psychophysical responses where u is flicker light;
n_c = number of psychophysical responses where u is constant light.

As shown in Formula 1 the odds ratio p(u) is the subjective psychophysical measure of the possibility where a flicker value (u) is the threshold. This empirical possibility value belongs to a closed interval from zero to one. As the odds ratio increases to one so the possibility that the value u is the threshold increases.

Table 1 contains four values of u from the critical interval and their odds ratios (p) (see Formula 1) for one subject examined in a pilot investigation.

TABLE 1

Four values and their odds ratios for one subject (example).

Flicker value u (Hz)	Number of psychophysical responses where u is flicker light n_f	Number of psychophysical responses where u is constant light n_c	Odds ratio p
40.1	7	13	0.53846
40.2	10	10	1.0
40.3	12	8	0.66667
40.4	18	2	0.11111

Figure 1 shows values of p for four given u's from Table 1.

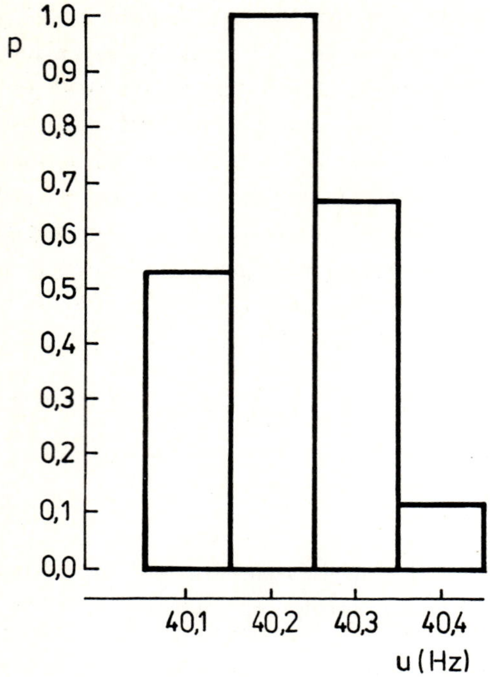

Fig. 1. Histogram of p-values for four given u points (see Table 1) Data obtained for one subject in pilot investigation.

It can be seen from Table 1 and Figure 1 that u = 40.2 Hz is threshold value (p = 1.0) while u = 40.3 Hz has a greater possibility of being the threshold value (p = 0.66667) than u = 40.1 Hz (p = 0.53846). The value that has the lowest possibility of being the threshold out of all four values is u = 40.4 Hz (p = 0.11111).

This paper presents the results of an empirical investigation which was concluded to build a fuzzy model for threshold of flicker frequency and also attempts to apply the resultant model for an indication of mental fatigue.

METHOD

Subjects

Twenty four video display terminal operators (all female) whose ages ranged from 22 to 26 were subjects in the study. Their task was to enter data consisting of alphanumeric sentences. This task is thought to cause mental work load.

Procedure and technique

The operators were examined three times during the morning shift. These times were: I - before work, II - after 3 hours of work, III - after 6 hours of work.

The flicker - test was used as the main testing tool. One eye flicker equipment was utilized. A red luminescent diode served as the source of stimuli and was presented on a dark background. The diode was characterized by a luminescence level of 35 cd/m^2. The ratio of flash and interval was constant (1:1 = 1) and the visual angle was about one degree. The frequency range was from 30 Hz to 50 Hz and the measuring accuracy was 0.1 Hz. The procedure described in our discussion of the background was applied. In our preliminary investigation it was stated that the critical interval of the flicker values is equal to 0.6 ± 0.1 Hz. Thus, the interval of length 0.7 Hz was taken into account in every case.

Additionally, the self-scaling of mental fatigue, based on four pairs of opposite designations which was inscribed in the background was used. A ten centimeter scale was stretched between the designations.

MODEL OF FUZZY FLICKER FREQUENCY THRESHOLD

Mean values were calculated for seven u points of the critical intervals which were collected in three investigative sessions. Formula 1 was then applied. Figures 2, 3 and 4 show the value of p obtained for seven u fuzzy critical flicker frequency points in the Ist, IInd and IIIrd investigative sessions respectively.

It can be easily seen that flicker u-values decrease in successive investigative sessions: I, II and III (see Figs. 2,3 and 4). This finding is important in the classical approach for measurement of the CFF and also its application to a fatigue diagnosis. That finding however is insignificant for our fuzzy set point of view. In the fuzzy approach it is highly important that the possibility of being a threshold for particular u points (belonging to the critical intervals) increases in the three successive sessions. Thus, without loss of generality we can assume that the critical interval ranges from 0.1 to 0.7 Hz.

Table 2 presents the p-values related to the common critical interval ranging from 0.1 to 0.7 Hz.

On the basis of the empirical data (see Figs.2,3,4 and Table 2) obtained the possibility function of fuzzy flicker frequency for each investigative session can be ascertained using Formula 2.

$$\pi_{FT}(u) = \int_{0.1}^{0.7} \exp(\alpha + \beta u + \gamma u^2)/\pi \text{ , where: } \alpha,\beta,\gamma = \text{coefficients.} \qquad (2)$$

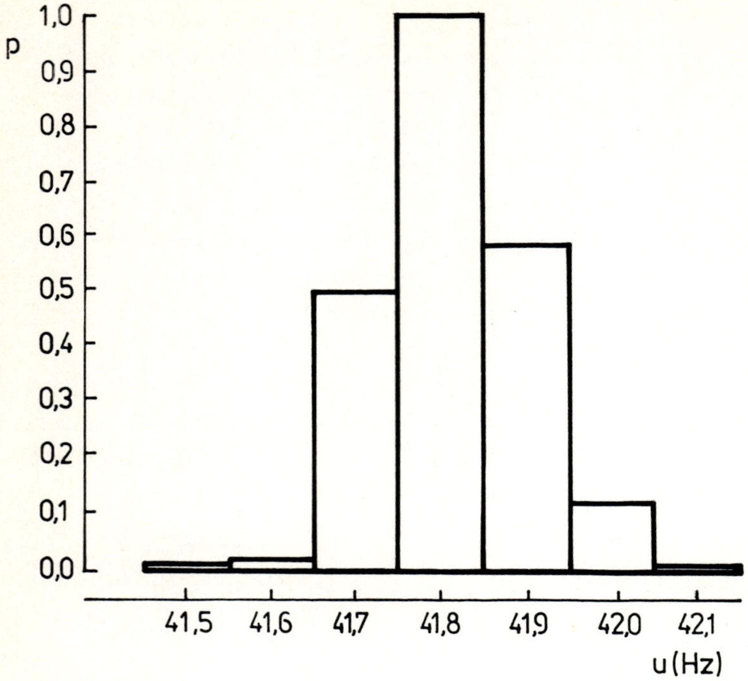

Fig. 2. Histogram of possibility mean values p obtained for seven u fuzzy critical flicker frequency points in the Ist investigative session - before mental work load.

TABLE 2

The p-values related to the common critical interval form 0.1 Hz to 0.7 Hz for three investigative sessions: the Ist, the IInd and the IIIrd.

Value of u	Investigative sessions - p-values		
	I	II	III
0.1	0.00200	0.04420	0.30770
0.2	0.00610	0.09290	0.75287
0.3	0.49333	0.73350	0.85292
0.4	0.99999	0.97000	0.93175
0.5	0.57832	0.55555	0.55555
0.6	0.12321	0.26600	0.68998
0.7	0.00351	0.08990	0.25000

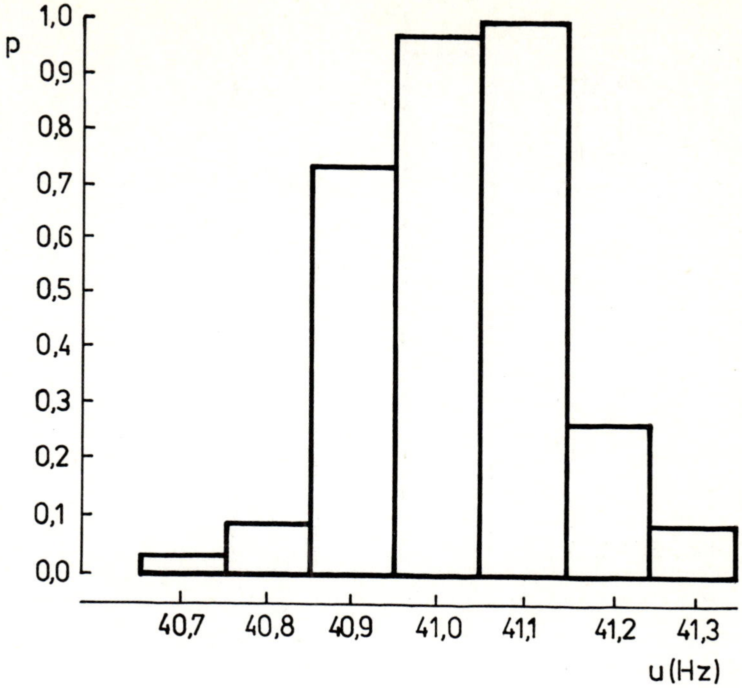

Fig. 3. Histogram of possibility mean values p obtained for seven u fuzzy critical flicker frequency points in the IInd investigative session - after 3 hours of mental work load.

Table 3 contains coefficients α,β,γ of the possibility function for the three investigative sessions. As shown all the coefficients decrease (in regard to absolute values) with the work time - investigative sessions: after three and six hours of mental work load. This indicates that following curve under formula 2 (for the Ist, the IInd and the IIIrd investigative session) lies over the previous one (see Fig. 5).

Comparing the appropriate values from Table 2 and 4, it can easily be seen that the possibility π-values lie closely to p-values. This means that the possibility fuzzy function of critical flicker frequency describes precisely the quantity of threshold for each velue from the critical interval. That is confirmed by values of F-test ($p<0.05$). The values are: $F_I=1570.0$, $F_{II}=116.248$, $F_{III}=34.123$.

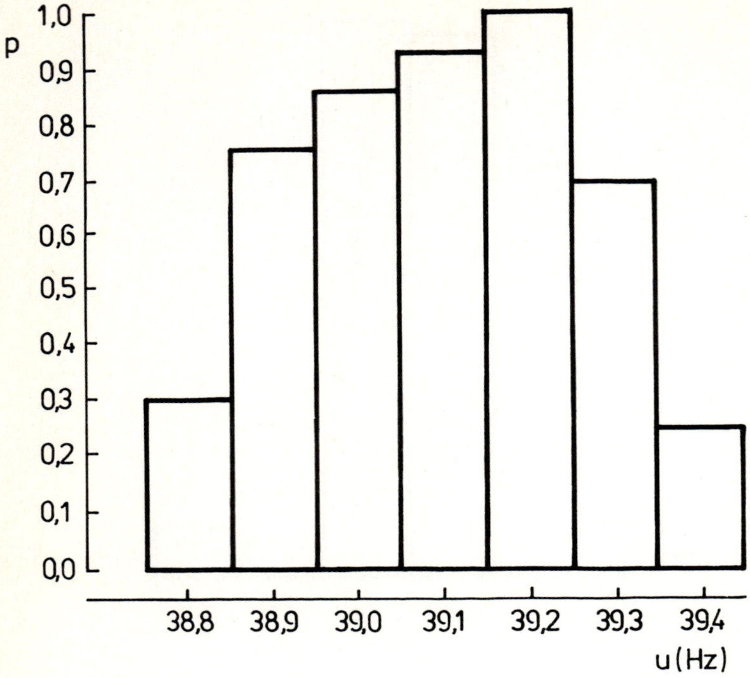

Fig. 4. Histogram of possibility mean values p obtained for seven u fuzzy critical flicker frequency points in the IIIrd investigative session - after 6 hours of mental work load.

TABLE 3

The coefficients α,β,γ of possibility function for three investigative sessions: the Ist, the IInd, and the IIIrd.

Coefficients	Investigative sessions - values of coefficients		
	I	II	III
α	-10.72450	- 5.68031	- 2.12691
β	52.63617	26.90920	10.82037
γ	-64.55387	-31.39395	-13.36573

Table 4 contains the π-values calculated for the critical intervals under Formula 2, using the appropriate coefficients from Table 3, respectively.

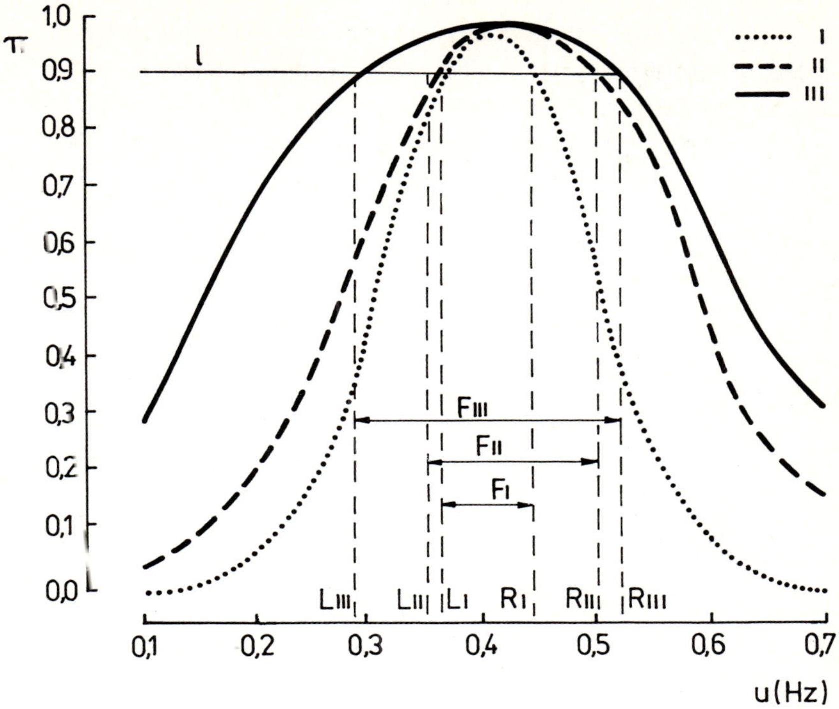

Fig. 5. Graphs of the three possibility curves. On the horizontal coordinate u-values from 0.1 Hz to 0.7 Hz - critical interval are designated. On the vertical coordinate π-values are designated. Curves I, II and III represent the possibility functions - models of possibility that u is the threshold in the conditions: before mental work load, after three hours and six hours of mental work load, respectively. : $\pi(u) = 0.9$. F_I, F_{II} and F_{III} mean threshold-level sets for the Ist, the IInd and the IIIrd investigative sessions, respectively. $|F|$ = length of interval F: $|F_I| = 0.0827571$, $|F_{II}| = 0.156132$, $|F_{III}| = 0.2244848$. L and R mean left and right end-points of a given interval, a given threshold level set: $L_I = 0.3663133$, $R_I = 0.4490704$, $L_{II} = 0.350507$, $R_{II} = 0.506639$, $L_{III} = 0.292538$, $R_{III} = 0.5170228$.

Let us define α-level set F_α according to Zadeh's idea (ref.18)

$$F_\alpha = \{u : \pi(u) \geq \alpha\} \qquad (3)$$

Let threshold-level set means 0.9-level set.

Figure 5 presents the three possibility curves drawn in using common Cartesian coordinate system.

In Figure 5 there are designated threshold-level sets for each curve (in each condition: before, after three and six hours of mental work load). These sets constitute the real pivot of our fuzzy approach to flicker frequency threshold as the indicator of mental fatique. The lengths of 0.9-level sets can be treated as a measure of the fuzzy flicker frequency threshold. It is easily seen from Figure 5 that it increases as mental work load is prolonged.

TABLE 4
The π-values calculated under Formula 2 for three investigative sessions: the Ist, the IInd and the IIIrd.

Value of u	Investigative sessions - π-values		
	I	II	III
0.1	0.00223	0.03676	0.30773
0.2	0.06207	0.21135	0.60807
0.3	0.47545	0.64855	0.91969
0.4	1.00000	1.00000	1.00000
0.5	0.57991	0.92848	0.94350
0.6	0.09235	0.43317	0.63996
0.7	0.00404	0.10786	0.33225

The model built of fuzzy flicker frequency threshold (model of the FCFF) and the measure connected with it seem to be quite useful for the evaluation of mental load and fatigue.

LENGTH OF THRESHOLD-LEVEL SET AS AN INDICATOR OF FATIGUE

For each three sessions possibility functions were calculated for each subject. These we accomplished using Formula 2. The threshold-level set (0.9-level set) was calculated for building the models. Table 5 contains the mean values and standard deviations of the threshold-level set lengths.

A sequential analysis of variance was conducted to detect the significant differences between means (refs.14,8). It was stated ($n = 20$, $\alpha = 0.05$ and $\beta = 0.05$) that all means differed significantly. Thus, it was found that threshold-level sets length increased under mental work load.

TABLE 5
Means and standard deviations of threshold-level set length for each of the three investigative sessions.

Investigative session	Mean	Standard deviation
I	0.0819786	0.0015031
II	0.1570410	0.0029107
III	0.2393875	0.0035012

A further analysis of the data obtained using self-scaling of fatigue was conducted. Table 6 contains these data.

TABLE 6
Mean values of self-scaling of fatigue for the three investigative sessions: the Ist, the IInd and the IIIrd.

Opposite designations	Mean values of self-scaling for investigative sessions		
	I	II	III
refreshed - tired	1.3	3.1	9.2
strong - weak	0.6	2.8	7.6
awake - sleepy	1.2	3.4	8.7
vigorous - exhausted	3.0	4.5	8.8

The correlation coefficients were calculated between self-scaling of fatigue and the threshold-level set length. Table 7 contains these coefficients. As shown the obtained correlation coefficients were statistically significant ($p<0.01$).

TABLE 7
The correlation coefficients between the self-scaling of fatigue and the threshold-level set length for the three investigative sessions ($p<0.01$).

Opposite designation self-scaling of fatigue	Threshold-level set length $F_{0.9}$
refreshed - tired	r = 0.878
strong - weak	r = 0.823
awake - sleepy	r = 0.820
vigorous - exhausted	r = 0.851

These high correlations between self-scaling of fatigue and the threshold--level set length indicate that the increase of the latter can be a useful indicator of mental fatigue.

CONCLUSION

As pointed out in the background there are a variety of other models of flicker frequency threshold, including for example the sequential model (ref.9) and a model based on the Fourier methods (ref.2). However, the current state of these models does not seem to offer much to the problems of the mental load and mental fatigue measurement. It would appear that the FCFF model based on the fuzzy set theory seems to be more powerful in that area. The sequential and Fourier models are analytical. The first describes the decision process and the latter the physiological processes. The fuzzy model synthetically combines both describing the threshold as the psychophysical phenomenon.

REFERENCES

1 L. Burigana, W. Gerbino, R. Luccio, S.C. Masin, C.V. Remondino, G. Vidotto, L. Zanuttini, Convegno di Psicofisica, Report 88, Universita di Padova, Instituto di Psicologia, 1983.

2 T. Caelli, Visual Perception. Theory and practice, Pergamon Press, Oxford-New York-Toronto-Sydney-Paris-Frankfurt, 1981.

3 H. Davson (Ed.), The eye. Academic Press, New York-London-San Francisco, 1976.

4 S.J. Dimond, Neuropsychology, Butterworths, London-Boston-Sydney-Wellington -Durban-Toronto, 1980.

5 J.P. Frisby, Seeing. Illusion, Brain and Mind, Oxford University Press, Oxford-New-York-Toronto-Melbourne, 1979.

6 K.S. Fu, Sequential Methods in Pattern Recognition and Machine Learning Academic Press, New York-London, 1968.

7 H.G. Geissler (Ed.), Modern Issues in Perception, VEB Deutscher Verlag der Wissenschaften, Berlin, 1983.

8 T. Marek, Cz. Noworol, The Sequential Approach to Analysis of Variance. Application in Psychological Research, Polish Psychological Bulletin 3 (1984) 195-200.

9 T. Marek, Cz. Noworol, Sequential Model of Critical Flicker Frequency Threshold, Polish Psychological Bulletin 4 (1985) 66-88.

10 T. Marek, Cz. Noworol, Bipoint Flicker Research of VDT Operators, in F. Klix (Ed.), Proceedings of the First Seminar Main-Computer Interaction Research, Elsewier, Amsterdam, 1985.

11 T. Marek, W. Pieczonka-Blaszczyk, Method of Critical Flicker Frequency Measurement and Its Accuracy, Acta Scientiarum Literarumque DXIX, Universitas Iagiellonica, 29 (1979) 17-24.

12 T. Terano, M. Sugeno, Conditional Fuzzy Measures and Their Applications, in L.A. Zadeh, K.S. Fu, K. Tanaka, M. Shimura (Eds.), Fuzzy Sets and Their Applications to Cognitive and Decision Processes, Academic Press, New York-San Francisco-London, 1975, pp. 151-170.

13 A. Weber, C. Jermini, E.P. Grandjean, Relationship between Objective and Subjective Assesment of Experimentally Induced Fatigue, Ergonomics 18 (1975) 151-164.

14 G.B. Wetherill, Sequential Methods in Statistics, John Wiley, New York, 1968.

15 L.A. Zadeh, Fuzzy Sets, Information and Control 8 (1965) 338-353.

16 L.A. Zadeh, Fuzzy Algorithms, Information and Control 12 (1968) 94-102.

17 L.A. Zadeh, Probability Measures of Fuzzy Ivents, Journal of Mathematical Analysis and Applications, 23 (1972) 421-427.

18 L.A. Zadeh, Calculus of Fuzzy Restrictions, in L.A. Zadeh, K.S. Fu, K. Tanaka, M. Shimura (Eds.), Fuzzy Sets and Their Applications to Cognitive and Decision Processes, Academic Press, New York-San Francisco-London, 1975, pp. 1-39.

19 L.A. Zadeh, Fuzzy Sets as a Basis for a Theory of Probability, Fuzzy Sets and Systems 1 (1978) 3-28.

Applications of Fuzzy Set Theory in Human Factors, edited by W. Karwowski and A. Mital
Elsevier Science Publishers B.V., Amsterdam, 1986 — Printed in The Netherlands

DEVELOPMENT OF ACCEPTABLE WORK LIMITS FOR COGNITIVE TASKS

ANIL MITAL[1] and W. KARWOWSKI[2]

[1]Ergonomics Research Laboratory, University of Cincinnati, Cincinnati, Ohio 45221-0072 (U.S.A.)

[2]Waldemar Karwowski, Department of Industrial Engineering, University of Louisville, Louisville, Kentucky 40202 (U.S.A.)

ABSTRACT

Recent shifts in the nature of industrial work force and increased level of automation have led to abundance of jobs for which there are no established work limits or task loading limits. While work performance standards exist for physical tasks, limits on acceptable levels of work are yet to be developed for cognitive tasks. This paper describes an approach to quantify qualitative human responses and describes a procedure to integrate them with other quantitative responses resulting from physical, cognitive, behavioral, or psycho-social factors. The theory of fuzzy sets is utilized in developing the modeling concept which can be used in developing work-performance standards or optimizing overall man-equipment system effectiveness.

INTRODUCTION

The development and implementation of new technology in the late seventies and early eighties is causing a tremendous change in the nature of the industrial base in the United States. The employment has been gradually and steadily shifting from industrial manufacturing to service industries. In the 1980s proportionately more people have found employment in service industries than traditional manufacturing industries. It is projected that by 1990 approximately 114 million people will be employed in service industries compared to 25 million in the industrial manufacturing sector (ref. 1).

The above changes, coupled with increasing levels of automation in traditional industries, are altering the character of traditional jobs. The emphasis is shifting from physical to cognitive tasks; the majority of jobs now require, or will require, extensive cognitive skills. This is primarily the result of high technology implementation. The need for overall economic growth and desire for continued higher living standards, however, has forced retention of physical components (ref. 2).

Since the work has become either completely or predominently of information processing nature, there is a critical need to develop work performance standards which take into account both cognitive and physical aspects of it. Traditionally, the emphasis has been on accounting for physical elements only.

Also, at the present time there are no procedures for developing standards that account for the presence of both physical and cognitive elements. In order to develop such performance standards, one must consider all types of internal and external stressors (factors which stimulate or cause changes in behavioral or physical responses) and their corresponding stresses (responses itself). The stressors may be physical or cognitive and their responses, qualitative or quantitative. All stressors, and their responses, must be considered if the effectiveness of humans is to be optimized.

Such comprehensive standards can be addressed with a comprehensive model that includes both physical and mental processing of human tasks. Such a comprehensive approach may be found in the General Living Systems Model of Miller (ref. 3) who concentrates on critical subsystems of a living system. This model has been further developed and applied by a number of investigators to a wide variety of situations, ranging from medical to health care administration (ref. 4). An early application at the conceptual level has been suggested by Ragade (ref. 5) at the level of man-equipment systems design.

While the General Living Systems Model is a top-down approach, work in human factors research largely stems from a bottom-up approach as very specific detail task level measurements are envisioned. The two can be matched in a framework of inquiry which correlates General Living System Process Measures with the ergonomic concepts of task level effectiveness. It is in this framework of matching that fuzzy set theory (ref. 6) can possibly make a significant contribution in terms of the modeling effort.

PROBLEM DESCRIPTION

The quantitative measure of various stressors, which influence performance and human responses, is called work-load (ref. 7,8). The levels of work-load are acceptable if the stress (levels of responses) they cause does not interfere with a worker's capability to operate safely and efficiently. In other words, the external work-load is acceptable if homeostasis, or steady-state, can be maintained. The goal of ergonomists or human engineers is to determine maximum levels of work-load that do not violate the steady state. While work-loads below this 'optimum' level will be inefficient, higher work-load levels could lead to destablization of the steady state.

The task of determining 'optimum' work-loads for tasks which are cognitive, or predominantly cognitive, is a very difficult one. Unlike physical tasks (stressors), which have well defined stresses (responses) and work-load limits, based on the acceptable levels of these stresses, cognitive tasks do not have any established work-load limits. The difficulty in developing such limits (standards) arises primarily from the fact that not every human responds to a given cognitive stressor; when there is a response, there is either no uniform pattern or it is qualitative.

There may be as many as one hundred, or more, different responses, some qualitative and some quantitative, to cognitive stressors as opposed to four or five for physical stressors. Due to these reasons, there is a need to develop optimum levels of acceptable work-loads, based on responses to cognitive and physical stressors that are unique to an individual. In a comprehensive approach of considering both the physical and the mental processing of tasks, one can define stresses in two contexts. The first is purely a mental context (cognitive context) while the second is in a physical or physiological context.

This paper presents a modeling approach to quantify cogntive stresses, if they are qualitative, and to integrate them with physical stresses in a manner suitable for developing human work-performance standards defined as the upper limits on acceptable work-load levels. The concepts of fuzzy set theory were utilized in developing the model.

There have been a number of applications of fuzzy sets in man-machine systems. These include visual inspection tasks (ref. 9), analysis of memorizing and forgetting (ref. 10), modeling of man-equipment systems (ref. 11), mental stress quantification (ref. 12), and the study of manual lifting stresses (ref. 13,14). These studies indicate that many problems of complex man-equipment systems can be solved or better understood if other methodologies, such as fuzzy set theory, are applied.

EFFECTS OF COGNITIVE/LIFE STRESSORS

In a man-equipment system, the system effectiveness is largely dependent upon the individual's well being (ref. 15). Since there are numerous stressors influencing a worker's behaviour and responses, a brief review of relevant literature is pertinent. Appelbaum (ref. 16), Sleight and Cook (ref. 17) and Eaton (ref. 18) have listed several mental stressors and situations which cause them. Table 1 lists the various researchers and the responses to various mental stressors reported by them.

TABLE 1

Investigations and responses to stressors reported by various researchers

Responses	Investigations
Increased heart rate	Mital and Mital (ref. 19); Mital et al. (ref. 20); Stromme et al. (ref. 21); Barnard and Duncan (ref. 22);

	McLean (ref. 23); Zwaga (ref. 24); Grandjean et al. (ref. 25);
Changes in blood pressure	Mital and Mital (ref. 19); Mital et al. (ref. 20); McLean (ref. 23); Kasl and Cobb (ref. 26);
Body Temperature	Mital and Mital (ref. 19); Mital et al (ref. 20); Renbourn (ref. 27);
Changes in blood glucose levels	Burrows et al (ref. 28);
Utilization of 17-hydroxycorticosteroid level in Parotid fluid	Street et al. (ref. 29);
Excretion of adrenaline ad noradrenaline	Levi (ref. 30);
Changes in serum uric acid and cholesterol levels	Kasl et al. (ref. 31);
Disruption of circadian rhythm	Caplan et al. (ref. 32);
Increased cathecholamine excretion in urine	Grandjean et al. (ref. 25); Rubin et al. (ref. 33);
Significant cortisol and/or growth hormone respones	Miller et al. (ref. 34); Abplanalp et al. (ref. 35);
Prevalence of thyroid gland hot spots	Voth et al. ref. 36);
Increase in galvanic skin response	Raouf and Manney (ref. 37);
Changes in speech signal patterns	Hecker et al. (ref. 38);
Absenteeism	Hinkle and Plummer (ref. 39);
Adaptive dysfunctions	Selye (ref. 40);
Distraction and accident proneness	Henderson (ref. 41); Viney (ref. 42); Felton (ref. 43); Sachs (ref. 44); Rogg (ref. 45);
Fatigue	Grandjean et al. (ref. 25);
Performance decrement	Basila et al. (ref. 46); Manenica and Corlett (ref. 47); Margolis and Kroes (ref. 48); Berkum (ref. 49); Korchin and Basowitz (ref. 50);
Prevalence of worry	Margolis (ref. 51); Margolis and Kroes (ref. 48); Grandjean et al. (ref. 25); Rubin et al. (ref. 33); Aitken (ref. 52);
Psychiatric hospitalization	Brenner (ref. 53);

Mental/physical ill health	Buell and Eliot (ref. 54); Cooper et al. (ref. 55); Cooper and Marshall (ref. 56); Barnard et al. (ref. 57); Benson (ref.58); House (ref. 59); Rosenman et al. (ref. 60);

From Table 1 it is clear that mental stressors cause a variety of stresses. Some of the resulting stresses may be of short term nature (e.g. anxiety, tension, elevated heart rate, blood pressure, etc.) while others may be chronic (e.g. feelings of fatigue, depression, etc.). Since no single stress (response) represents all stressors, it is critical, therefore, to consider the composite effect of all stresses in some form. To achieve this, a method which allows integration of all responses, must be used. To date, no method which accomplishes this has been proposed in the published literature. The major objective of this work was to develop such a method and to demonstrate how it can be used in the develpment of optimum levels (standards) of work performance, or standards, in a man-equipment system.

THE MODELING TECHNIQUE

Various factors (stressors) may cause different responses in the human component of a man-equipment system. As long as these responses do not cause destabilization of the steady state conditions, an individual can operate effectively and the negative or detrimental effects of stressors are fully compensated for. Once the response levels exceed these limits, the state of homeostasis cannot be maintained and the individual is said to be under strain (ref. 7). Fig. 1 shows the order of this progression.

Work performance standards based only on the existence (occurence) or non-occurence of negative effects of stresfull responses (resulting in decrements of performance) would be inadequate and limited. Since the nature of the stress responses to the task and associated strain of the human operator are ill-defined, their quantification requires the use of an approach which would allow to account for vagueness and imprecision intrinsic to human behaviour. It is this gradual nature of different stress responses, with a soft transition from being bearly noticable to being distractive, that calls for the application of the concept of fuzziness in such analysis. Therefore, the theory of fuzzy sets is used here to model the natural imprecision associated with a degree of disturbances to the homeostasis of an individual (human operator's strain).

Numerous stressors, physical and cognitive, cause different quantitative and qualitative responses (say <u>n</u>). Our objective is to determine the maximum acceptable level of these <u>n</u> responses (as the upper limit of compensatory zone) integrated together.

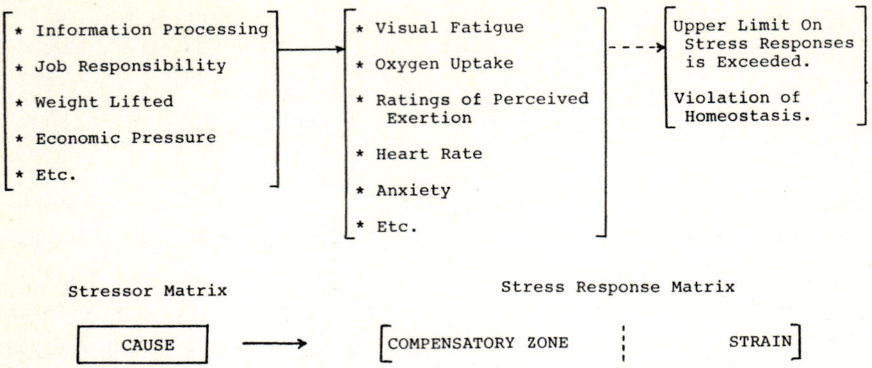

Fig. 1. Progression of human responses (stresses).

Let us define two sets - yes and no, such that these indicate the state of a person who is either under strain or is within the compensatory zone, respectively. An individual, with completely destabilized steady state, will be identified by a response membership value equal to 1 in the set yes or a response membership value equal to zero in the set no. A person whose responses are at the lower end of compensatory zone will be identified by a response membership value equal to 1 in the set no or a response membership value equal to 0 in the set yes (see Table 2). The n different responses, mentioned above, form the elements of these sets yes and no. It must be kept in mind here that sets yes and no form fuzzy sub-sets since the transition from membership to non-membership in either set is gradual, rather than sharp (ref. 6).

TABLE 2.
Transition from steady state to strain.

Level of Compensatory Zone.	At the lower end of the compensatory zone no	Under strain-beyond the upper end of compensatory zone yes
Response Membership Value	0 if under no strain	1, if under strain.

A fraction between 0 and 1 would then represent the grade of membership of each of the n responses in either yes or no. For example, a membership grade of 0.9 represents a closer association with the fuzzy set yes. Similarly, a membership grade of 0.1 represents a closer association with the fuzzy set no. A membership grade of 0.5 represents a crossover point.

MODEL DESCRIPTION

Let us define a universe of discourse:

$$U = [Y_1, Y_2, Y_3, \ldots, Y_n]$$

where $Y_1, Y_2, \ldots, Y_n$ are the n responses which constitute the elements of this universe. Element Y_1 might represent the heart rate; element Y_2, the blood pressure; ....; and element Y_n, the level of mental concentration, for example.

Let us also define a fuzzy set yes of a universe of discourse U, characterized by a membership function $f_{yes}: U \rightarrow [0,1]$. This membership function associates with each element Y of U a number $f_{yes}(Y)$ in the interval [0,1] which represents the grade of membership of Y in yes. The closer this number is to 1, the stronger is the grade of membership of Y in yes and stronger the indication that the individual is closer to destabilization of homeostasis.

The set yes is fuzzy since many aspects of possible indicators of strain (Y's) cannot be represented as being included in or excluded from the set and thus making the set boundaries imprecise. The membership of any element Y is then determined by a degree of membership [f(Y)] of Y in yes $[f_{yes}(Y)]$. The set yes may be now represented by:

$$\underline{yes} = [f(Y_1)/Y_1, f(Y_2)/Y_2, \ldots, f(Y_n)/Y_n] = \int f_{yes}(Y)/Y \qquad (1)$$

where the integral sign stands up for the union of n fuzzy singletons.

At this point, it is arbitrarily decided that if the fuzzy set yes yields a value greater than or equal to 0.5, the operator is under strain. The value, 0.5, corresponds to that level of response which is unacceptable (just above the upper limit of the compensatory zone).

As mentioned before, many responses are quantitative. For instance, the heart rate (Y_1-measured in beats/minute) which is a number. In such cases, the degree of membership f(Y) can be modeled easily. Fig. 2 shows a plot between some kind of mental activation (e.g. information processing) and incremental heart rate due to that activation (heart rate above the level predicted by oxygen consumption). The degree of membership can be determined as follows. Let the incremental heart rate (IHR) = Y1, then:

$$f(Y_1) = \begin{cases} 0, \text{ if incremental heart rate} \leq 5 \\ \dfrac{\text{I.H.R.} - 25}{5}, \text{ if } 5 < \text{incremental heart rate} < 30 \\ 1, \text{ if incremental heart rate} \geq 30 \end{cases} \qquad (2)$$

where the degree of membership is a linear function of a single variable. In the above example, an increment of 5 in the heart rate represents 'no strain'.

An increment of 30 or more indicates that the individual is definitely under 'strain'.

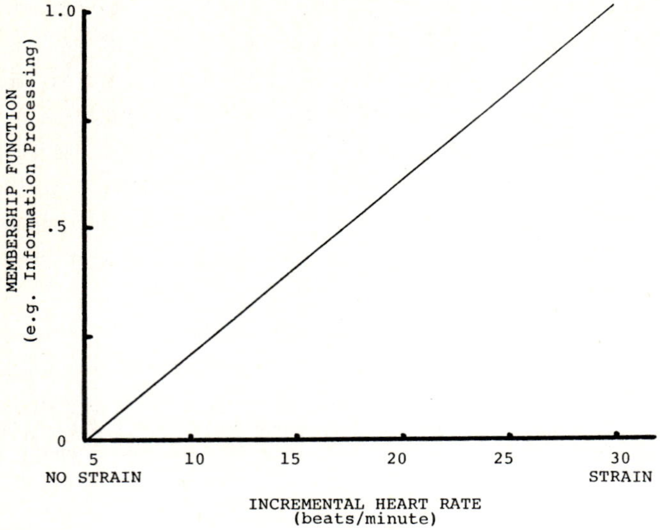

Fig. 2. Membership function for the amount of information processed (indicated by the incremental heart rate).

It should be noticed that the degree of membership could be represented by a non-linear function if the trend is non-linear. The degree of membership for other quantitative parameters may similarly be determined.

The determination of membership function value for qualitative responses requires a different approach. For example, take the level of mental concentration (Y_n). Qualitatively, this could be <u>low</u>, <u>normal</u>, or <u>high</u> using a broad classification. These categories (<u>low</u>, <u>normal</u>, <u>high</u>) are regarded as primary terms (values) of the linguistic variable "mental concentration", and in the context of this response (Y_n - level of mental concentration) are defined by giving each a fuzzy representation (Fig. 3).

Let these primary terms (fuzzy subsets) be defined by a universe of discourse V:

$$V = [0, 0.1, 0.2, \ldots, 0.9, 1] \tag{3}$$

with the membership intervals [0,1] as follows:

$$\underline{\text{low}} = [\frac{0}{0}, \frac{0}{.1}, \frac{0}{.2}, \frac{0}{.3}, \frac{0}{.4}, \frac{.1}{.5}, \frac{.3}{.6}, \frac{.7}{.7}, \frac{.9}{.8}, \frac{1}{.9}, \frac{1}{1}] \tag{4}$$

$$\underline{\text{normal}} = [\frac{0}{0}, \frac{.5}{.1}, \frac{.9}{.2}, \frac{1}{.3}, \frac{.9}{.4}, \frac{.5}{.5}, \frac{.2}{.6}, \frac{.05}{.7}, \frac{0}{.8}, \frac{0}{.9}, \frac{0}{1}] \tag{5}$$

$$\underline{high} = [\frac{1}{0}, \frac{1}{.1}, \frac{.9}{.2}, \frac{.7}{.3}, \frac{.3}{.4}, \frac{.1}{.5}, \frac{0}{.6}, \frac{0}{.7}, \frac{0}{.8}, \frac{0}{.9}, \frac{0}{1}] \quad (6)$$

It should be pointed here that the definitions of these primary terms are subjective. If the level of mental concentration is low, the individual is very likely under strain and consequently the membership function value should be closer to 1 (as is given by the definition of low). In a similar manner normal and high are defined. In equations (4) - (6), the denominator indicates the level of strain while the numerator, the degree of membership.

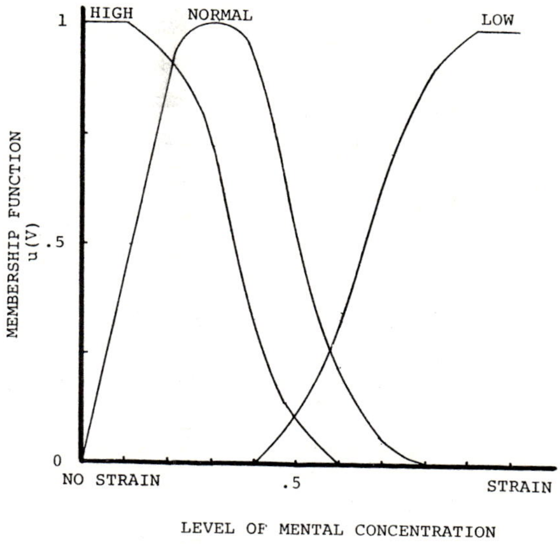

Fig. 3. Membership function for various levels of mental concentration.

In order to modify the meaning of these primary linguistic terms (which express the qualitative output or influence) and to increase the sensitivity and accuracy of quantification hedges such as very, very very, etc. may be used For example, if low is defined by (4), then:

$$\underline{very\ low} = [\underline{low}]^2 = [\frac{0}{0}, \frac{0}{.1}, \ldots \frac{.01}{.5}, \frac{.09}{.6}, \frac{.49}{.7}, \frac{.81}{.8}, \frac{1}{1}] \quad (7)$$

Similarly other appropriate hedges can be used to modify terms like normal and high.

As in the case of 'level of mental concentration', primary linguistic terms can be defined for other qualitative responses. The fuzzy sets low, normal (or medium), and high might have to be defined slightly differently depending upon how they relate to yes.

Once the outcomes of all the elements have been quantified, yes can be expressed as a "mixed" fuzzy numerical set in which representation of the degree of membership for all involved elements is not unique:

$$\underline{yes} = [\frac{.45}{Y_1}, \frac{.7}{Y_2}, \ldots, \frac{[0/0, 0/.1, 0/.3, 0/.4, .1/.5, .3/.6, .7/.7, .9/.8, 1/.9, 1/1]}{Y_n}] \tag{8}$$

When the degree of membership itself is a fuzzy set, as is the case with Y_n, it is reasonable to select the value of strain (denominator) corresponding to the highest grade of membership (numerator), i.e., 1.0. This would reduce yes (when level of mental concentration is "low") to:

$$\underline{yes} = [\frac{.45}{Y_1}, \frac{.7}{Y_2}, \ldots, \frac{1}{Y_n}] \tag{9}$$

Notice that if the level of mental concentration was "high", meaning low or no strain, we would choose the value 0 corresponding to the highest grade of membership of "high" as given in equation (6). In this case yes would be given by:

$$\underline{yes} = [\frac{.45}{Y_1}, \frac{.7}{Y_2} \ldots, \frac{0}{Y_n}] \tag{10}$$

For "normal", yes would be:

$$\underline{yes} = [\frac{.45}{Y_1}, \frac{.7}{Y_2}, \ldots, \frac{.3}{Y_n}] \tag{11}$$

The determination that an operator is under strain, must be based upon all the information collected. This means the whole fuzzy set yes (9) must be utilized. To make the appearance of yes simpler, membership grades below 0.5 may be replaced by 0 and membership grades 0.5 or above may be replaced by 1. This yields:

$$\underline{yes} = [\frac{0}{Y_1}, \frac{1}{Y_2}, \ldots, \frac{1}{Y_n}] \longrightarrow [\frac{f(Y)i}{Y_i}] \tag{12}$$

A value of 1 represents an individual under strain while 0 stands for steady-state condition. More the number of grades with value 1, stronger the indication of strain. If the outcome shown by equation (10) is unacceptable, then the levels of stressor, or stressors, that has caused it is unacceptable.

Instead of using equation (12), it is also possible to convert equation (9) into a fuzzy singleton, using a weighted average, indicating the overall level of the response due to stressor(s). Equation (9) thus becomes:

$$\underline{yes} = [\frac{W_1 \cdot f(Y_1) + W_2 \cdot f(Y_2) + \ldots + W_n \cdot f(Y_n)}{W_1 + W_2 + \ldots + W_n}] \tag{13}$$

where W_1, W_2, ... W_n are the respective weighting factors associated with n responses. Different weighting factors are necessary since, for a particular individual, all responses are not equally important. The same response also may not appear in all individuals.

The value of equation (13) can be determined for a wide variety and levels of stressors and the appropriate performance standards can be based on these values.

CONCLUDING REMARKS

A method to develop work performance standards, in man-equipment systems dominated by cognitive tasks, is proposed. The approach discussed attempts to integrate all responses to stressors instead of looking only for one good response. A review of the literature shows that consideration of a few simple responses, as in the case of physical stressors, can not lead to the determination of acceptable or unacceptable levels of performance. Since different individuals respond differently to different stressors, all responses must be considered.

REFERENCES

1 Bureau of Labor Statistics, U.S. Department of Labor, Washington, D.C., 1981.

2 A. Mital, S.S. Asfour, and R.L. Shell, Future applications of ergonomics, Society of Automotive Engineers, SAE Publication No. 840039, 1984.

3 J.G. Miller, Living Systems, McGraw-Hill, New York, 1978.

4 S.L. Merker, Application of Living Systems Process Analysis to Hospital Systems, Invited Symposium, Abstracts of the Work Conference on Systems, Caracas, July 11-14, 1983, pp. I-4.

5 R.K. Ragade, Androids and Living Systems Theory: A Framework for Evaluation at the Man-Machine Interface, In A.W. Smith (Ed.), Systems Methodologies and Isomorphies: Vol. I, Intersystems Publication, New York, 1984, pp. 135-139.

6 L.A. Zadeh, Outline of new approach to the analysis of complex systems and decision process, Institute of Electrical and Electronics Engineers Transaction on Systems, Man, and Cybernetics, SMC-3 (1978) 28.

7 J.S. Weiner, The measurement of human workload, Ergonomics, 25 (1982) 953.

8 J. Sharit and G. Salvendy, Occupational stress: review and reappraisal, Human Factors, 24 (1982) 129-162.

9 N. Malvache and D. Willayes, Contribution of the fuzzy sets theory to man-machine systems, In M.M. Gupta, M.M. Ragade and R.R. Yager (Eds.) Advances in Fuzzy Sets Theory and Applications, North-Holland, Amsterdam, 1979.

10 M. Kokawa, K. Nakamura, and M. Oda, Experimental approach to fuzzy simulation of memorizing, forgetting, and interference process, In L.A. Zadeh, K.S. Fu, K. Tanaka, and M. Shimura (Eds.), Fuzzy Sets and Their Applications to Cognitive and Decision Processes, Academic Press, New York, 1975.

11 J. Grobelny and M. Nowakowski, Applications of the fuzzy set theory to description and modeling of man-machine systems (in Polish), Instytut Organizacji i Zarzadzania, Politechnika Wroclawska, Wroclaw, Pre (118) (unpublished), 1980.

12 A. Mital and O.M. Ulgen, Mental stress quantification and identification decision modeling, In Proceedings of the Human Factors Society 26th Annual Meeting, 1982, pp. 474-478.

13 W. Karwowski, A fuzzy sets based model of the interaction between stresses involved in manual lifting activities, Ph.D. dissertation, Texas Tech University, Lubbock, Texas, USA, 1982.

14 W. Karwowski, M.M. Ayoub, L.R. Aley, and J.L. Smith, Fuzzy approach in psychophysical modeling of human operator-manual lifting systems, Fuzzy Sets and Systems, 14 (1984) 65.
15 J.R. Dejong, Measurement of Man at Work, paper 3:1, Taylor & Francis Ltd, London, 1971.
16 S.H. Appelbaum, Managerial/organizational stress: identification of factors and symptoms, Health Care Management, Winter, 1980.
17 R.B. Sleight and K.G. Cook, Problems in occupational safety and health: a critical review of select worker physical and psychological factors, HEW (NIOSH) report, contract no. HSM-99-72-043, Cincinnati, Ohio, USA, 1974.
18 M.T. Eaton, Detecting executive stress in time, Industrial Medicine and Surgery, 35 (1967) 115.
19 A. Mital and C. Mital, Mental stress and physiological responses, In A. Mital (Ed.) Trends in Ergonomics/Human Factors I, North-Holland, Amsterdam, 1984.
20 A. Mital, C.L. Burford and S.S. Asfour, Effect of mental stress and performance on the physiological responses, Paper presented at the American Industrial Hygiene Association Conference, Chicago, USA, 1979.
21 S.B. Stromme, P.C. Wikeby, A.S. Blix, H. Ursin, E. Baade and S. Levine, Psychobiology of stress:A Study of Coping Men, Academic Press, New York, 1978.
22 R.J. Barnard, and H.W. Duncan, Heart rate and ecg responses of firefighters Journal of Occupational Medicine, 17 (1975) 247-250.
23 A. McLean, Occupational Stress, Charles C. Thomas Publishers, Springfield, 1974.
24 H.J.G. Zwaga, Psychophysiological reactions to mental tasks: effort or stress, Ergonomics, 16 (1973) 61-67.
25 E.P. Grandjean, G. Wotzka, R. Schaad and A. Gilgen, Fatigue and stress in air traffic controllers, Ergonomics, 14 (1971) 159-165.
26 S.V. Kasl and S. Cobb, Blood pressure changes in men undergoing job loss: a preliminary report, Psychosomatic Medicine, 32 (1970) 19-38.
27 E.T. Renbourn, Psycho-physiological stress and the soldier, with a review of the literature, Report No. 118, Clothing and equipment, Physiological Research Establishment, Great Britain, 1961.
28 G.C. Burrows, T. Cox and G.C. Simpson, The measurement of stress in a sales training situation, Journal of Occupational Psychology, 50, (1977) 45-51.
29 R.L. Street, R. Singh and P.N. Hale Jr., The evaluation of mental stress through the analysis of parotid fluid, Human Factors, 12 (1970) 453-455.
30 L. Levi, Physiological methods of measuring industrial stress, Occupational Mental Health Notes, April, 1970.
31 S.V. Kasl, S. Cobb and G.W. Brooks, Changes in serum uric acid and cholesterol levels in men undergoing job loss, Journal of American Medical Association, 206 (1968) 1500-1507.
32 R.D. Caplan, S. Cobb and J.D.G. French Jr., White collar work-load and cortisol: disruption of a circadian rhythm by job stress, Journal of Psychosometic Research, 23 (1979) 181-192.
33 R.T. Rubin, R.G. Miller, B.R. Clark, R.E. Poland and R.J. Arthur, The stress of aircraft carrier landings II.3-Methoxy-4-hydroxyphenylglycal excretion in naval aviators, Psychosomatic Medicine, 32 (1970) 589-597.
34 R.G. Miller, R.T. Rubin, B.R. Clark, W.R. Crawford and R.J. Arthur, The stress of aircraft carrier landings I. corticosteriod responses in naval aviators, Psychosomatic Medicine, 32 (1970) 581-588.
35 J.M. Abplanalp, L. Livingston, R.M. Rose, and D. Sandwisch, Cortisol and growth hormone responses to psychological stress during the menstrual cycle, Psychosomatic Medicine, 39 (1977) 158-177.
36 H.M. Voth, P.S. Holzman, J.B. Katz and R.S. Wallerstain, Thyroid "hot spots": their relationship to life stress, Psychosomatic Medicine, 32 (1970) 561-568.
37 A. Raouf and W. Manney, Variations in cycle time and certain physiological measures of workers performing a paced assembly task, International

Journal of Production Research, 16 (1978) 409-426.
38 M.H.L. Hecker, K.N. Stevens, V.B. Gottfried and C.E. Williams, Manifestations of task-induced stress in the acoutsic speech signal, The Journal of Accoustical Society of America, 44 (1968) 993-1001.
39 L.E. Hinkle and N. Plummer, Life stress and industrial absenteeism, Industrial Medicine and Surgery, 21 (1952) 363-375.
40 H. Selye, The stress of life-new focal point for understanding accidents, Industrial Medicine and Surgery, 33 (1964) 621-625.
41 M. Henderson, The accident-prone car driver - does he exist?, The Medical Journal of Australia, 2 (1971) 909-912.
42 L. Viney, Accident proneness: some psychological research, Medical Journal of Australia, 2 (1971) 916-918.
43 J.S. Felton, Sensitivity to emotional ill health, Proceedings of the Annual Conference of NASA Clinic Directors,Environmental Health Officials, and Medical Advisors, New Orleans, U.S.A., 1969.
44 B.C. Sachs, Psychosomatic aspects of accidents, Industrial Medicine and Surgery, 31 (1962) 525-532.
45 S.G. Rogg, The role of emotions in industrial accidents, Archives of Environmental Health, 3 (1961) 519-522.
46 B. Basila, S. Suominen, G. Salvendy and G.P. McCabe, Non-work related movements in machine-paced and in self-paced work, Proceedings of the Human Factors Society 23rd Meeting, 1979, pp. 149-153.
47 I. Manenica and E.N. Corlett, A study of light repetitive task, Applied Ergonomics, 8 (1977) 103-109.
48 B.L. Margolis and W.H. Kroes, Occupational stress and strain, Occupational Mental Health, 2 (1972) 4-6.
49 M.M. Berkum, Performance decrement under psychological stress, Human Factors, 6 (1964), 21-30.
50 S.J. Korchin and H. Barowitz, Perceptual adequacy in a life stress, Journal of Psychology, 38 (1954) 495-502.
51 B.L. Margolis, Stress is a work hazard too, Industrial Medicine and Surgery, 42 (1973) 20-23.
52 R.C.B. Aitken, Prevalence of worry in normal aircrew, British Journal of Medicine and Psychology, 42 (1969), 283-286.
53 M.H. Brenner, Patterns of psychiatric hospitalization among different socioeconomic groups in response to economic stress, Journal of Nervous and Mental Disease, 148 (1969), 31-37.
54 J.C. Buell and R.S. Eliot, The role of emotional stress in the development of heart disease, Journal of American Medical Association, 242, (1979) 365-36
55 C.L. Cooper, M. Mallinger and R. Kahn, Identifying sources of occupational stress among dentists, Journal of Occupational Medicine, 51, (1978) 227-234.
56 C.L. Cooper and J. Marshall, Occupational sources of stress: a review of literature relating to coronary heart disease and mental ill health, Journal of Occupational Psychology, 49 (1976) 11-28.
57 R.J. Barnard, G.W. Gardner, N.V. Diaco and A.A. Kattus, Near maximal ecg stress testing and coronary artery disease risk factor analysis in los angeles city firefighters, Journal of Occupational Medicine, 17 (1975), 693-695.
58 H. Benson, Your innate asset for combating stress, Harvard Business Review, 52 (1974), 49-60.
59 J.S. House, Occupational stress and coronary heart disease: a review and theoretical integration, Journal of Health and Social Behavior, 15 (1974) 12-27.
60 R.H. Rosenman, W. Hahn, H. Werthessen, C. Jenkins, H. Messinger, R. Kositchek, M. Wurm, M. Friedman and R. Straus, Coronary heart disease in the western collaborative group study, Journal of the American Medical Association, 195 (1966) 86-92.

Applications of Fuzzy Set Theory in Human Factors, edited by W. Karwowski and A. Mital
Elsevier Science Publishers B.V., Amsterdam, 1986 — Printed in The Netherlands

FUZZY LOGIC CONTROLLER AS A MODEL OF THE THERMAL REGULATION PROCESSES IN MAN

JERZY GROBELNY

Institute of Prod. Eng. and Management, Tech. University of Wrocław, Poland

ABSTRACT

A linguistic description of the "active subsystem" in the thermoregulation system in man is presented. A representation of the linguistic variables in terms of fuzzy set theory makes the model's operationalization and simulation research possible. The example given shows a way of behaviour of the linguistic model in comparison with the "traditional" model and the real laboratory experiment.

INTRODUCTION

The modelling of the thermoregulation system in man has been a subject, of intensive research since the early sixties. The development of the calculation technique created possibilities for constructing the complex, quantitative thermoregulation models which include numerous variables. The quantitative, computer method of representation is a convenient instrument for designers of work-stand environments, physiologists and ergonomists. Therefore there is of great interest in such models. Since they are still not perfected, continuing efforts are being made to improve them. It seems that there are two main reasons of these imperfections:

a) incompleteness of empirical research on thermoregulation;
b) application of the modelling methods which are not adequate to a situation (there is lack of others).

This work mainly concentrates on the latter. Namely, a new approach which describes the active subsystem of a thermoregulation system modelled on the notions of fuzzy set theory (Zadeh, ref. 9) and Fuzzy Logic Controller concepts (Mamdani, ref. 3; Tong ref. 6, 7) is presented. Such an approach may be regarded as small out significant step to fulfill a demand for an application of adequate descriptional methods in the processes of formal modelling of the biological regulation systems.

A MAN - "SOFT" SYSTEM

Formal methods applied in modelling of systems concerning man and processes taking place in a human body were set up in substance, for a description of systems and phenomena having technical nature. Such systems, because of their nature, may be described by exact quantitative relations. The usability of classical (standard) mathematical methods has been confirmed many times, in descriptions of "hard" technical systems. However an application of such "hard" descriptive methods in the fields which examine different spheres of human functioning is increasingly criticized (Zadeh, ref. 9). This fact is emphasized because the fundamental features of the systems concerning a man or processes taking place in a human body are completely different from the features distinctively used for technical systems. Human systems have very complex relationships, many variables and these variables and relations are very inexact. It is often said that human systems are "non-sharp" because, with reference to human body reactions (of psychological and physiological nature), it is very difficult to define the exact limits or exact relations describing these reactions. Moreover the variables which compose the given relations are often difficult or impossible to measure. Therefore, Zadeh (ref. 9) proposed that these systems, referred to by him as "soft-systems", were not to be described formally by classical (standard) methods. He advises other methods to avoid inadequacy: "hard" methods - "soft" described object. One of the possible ways of building new, "soft" methods for the description of "soft" systems may be basing these methods on the fuzzy set theory whose foundations were suggested by Zadeh (ref. 9). The approach to modelling of thermoregulation systems presented in this paper is based on this theory.

A LINGUISTIC SYNTHESIS OF REGULATION ALGORITHMS

The foundations of the fuzzy set theory (Zadeh, ref. 9) allows for the formulation of the rules of building the control algorithms which are based on the "compositional rule of inference"*. A basic condition which makes the use of this rule possible is knowledge of the relation (between the spaces, eg: X and Y).

On these grounds and also knowing one value (or a set of values) regarded as INPUT it is possible to find a value (or a set of values) which is regarded as OUTPUT using the compositional rule of inference

---

*Composition of two relationships (in the sense of Zadeh) it is the max-min product of matrixes representing this relationships.

(Tong, ref. 6; Grobelny, ref. 2). Generally, there are two methods of formulating the R relation:

Method I - consists of subjective "direct" defining of a membership function for this relation, Zadeh (ref. 9) presents many examples of the representation of the linguistic expressions such as: "more or less equal", "a little greater than", "proportional", etc. However there is a considerable limitation of this method of synthesis. Namely, a man is not able "to call" or interpret logically (by means of memberships function) the relation which "connects" more than two different spaces.

Method II - consists of the synthesis of R relation by the appropriate connection of a set of logical expressions such as IF "A" THEN "B" which may be formulated on the basis of knowledge of a given process. A and B are the linguistic expressions defined in the adequate spaces and are represented by means of fuzzy sets. The connections of the set of such expressions (and, as a metter of fact, the expressions representations set) in the matrix of R relation may be performed in the following two steps:

a) Every expression such as: IF "A_i" THAN "B_i" is represented as a fuzzy relation. At the same time if A_i is defined in the space X, and B_i in Y, then:

$$R_i(x,y) = \min (f_{A_i}(x), f_{B_i}(y)) \quad \forall x \in X, \quad \forall y \in Y \tag{1}$$

(f - denotes the membership function).

b) The general R relation is obtained as the union of R_i relations:

$$R = \sum_i R_i \tag{2}$$

(Σ - denotes the union of matrix R_i).

Formula (2) corresponds with the logical structure (Tong, ref. 6): IF "A_1" THEN "B_1" ELSE IF "A_2" THEN "B_2" ELSE...ELSE IF "A_n" THEN "B_n".

R relation constructed in the above presented fashion represents a "non-sharp sum" of knowledge on a given system (phenomenon). If X is interpreted as a space of "error signal", and Y as a space of "actions" undertaken to decrease the error, then the R relation can be a formal representation of "fuzzy" knowledge on the regulation process. Together with the "compositional rule of inference", the R relation may be regarded as a "fuzzy regulation algorithm".

The necessity of computer application to testing of the regulation models and the fact that modelled systems (objects of regulation) are of non-fuzzy nature cause problems connected with implementation and interpretation of the fuzzy regulators. The expressions which form a regulator structure are interpreted in terms of fuzzy sets. In spite of that, the real object of regulation-INPUT (signal of error) as well as OUTPUT (action of the regulator) is the definite, single values (or the vectors of a fuzzy singleton by Tong (ref. 6), is defined as a special fuzzy set having only one element. A value of a membership function which is equal to one correspond to this element. In such a situation the "compositional rule of inference" is reduced to the following form:

$$f_B(y) = f_R(x_m, y) \tag{3}$$

(x_m - is the actually measured input signal value).

This formula was used to determine the output signal in the proposed model.

One step of the example control process based on the two expressions can nowe be showed as follow.

Let $X_1 = X_2 = 1,2,\ldots,5$ denoting "error" and "action" spaces respectively, and assume that a given process is described by two following expressions:

1° IF "value level in X_1 is much greater than 1 "THEN" the value increment in X_2 is negative big".

2° IF "value level in X_2 is medium "THEN" the value increment in X_2 is negative small".

The fuzzy sets representing respective linguistic expressions can be defined as in table 1.

TABLE 1

Definitions of fuzzy sets representing linguistic expressions

$X_1 = X_2$	1	2	3	4	5	EXPRESSION
$f_A(x_1)$	0.0	0.2	0.5	0.8	1.0	A = value level "much greater then 1"
$f_B(x_1)$	0.1	0.6	1.0	0.6	0.1	B = value level "medium"
$f_{A'}(x_2)$	1.0	0.8	0.4	0.1	0.0	A' = value increment "negative big"
$f_{B'}(x_2)$	0.0	0.1	0.5	1.0	0.6	B' = value increment "negative small"

Taking the above into account we can constitute (due to (1))

$$R_1 = \begin{matrix} 0.0 & 0.0 & 0.0 & 0.0 & 0.0 \\ 0.2 & 0.2 & 0.2 & 0.1 & 0.0 \\ 0.5 & 0.5 & 0.4 & 0.1 & 0.0 \\ 0.8 & 0.8 & 0.4 & 0.1 & 0.0 \\ 0.1 & 0.8 & 0.4 & 0.1 & 0.0 \end{matrix} \quad R_2 = \begin{matrix} 0.0 & 0.1 & 0.1 & 0.1 & 0.1 \\ 0.0 & 0.1 & 0.5 & 0.6 & 0.6 \\ 0.0 & 0.1 & 0.5 & 1.0 & 0.6 \\ 0.0 & 0.1 & 0.5 & 0.6 & 0.6 \\ 0.0 & 0.1 & 0.1 & 0.1 & 0.1 \end{matrix} \qquad (4)$$

The union of these matrices (formula (2)) which represents the logical structure "1° ELSE 2°" follows:

$$R = R_1 + R_2 = \begin{matrix} 0.0 & 0.1 & 0.1 & 0.1 & 0.1 \\ 0.2 & 0.2 & 0.5 & 0.6 & 0.6 \\ 0.5 & 0.5 & 0.5 & 1.0 & 0.6 \\ 0.8 & 0.8 & 0.5 & 0.6 & 0.6 \\ 1.0 & 0.8 & 0.4 & 0.2 & 0.1 \end{matrix} \qquad (5)$$

Assuming that in X_1 the resulting value level $x_1 = 4$, then using (3) we obtain:

$$f_B(x_2) = 0.8 \quad 0.8 \quad 0.5 \quad 0.6 \quad 0.6 \qquad (6)$$

The fuzzy set B defined by its membership function (6) in X_2 corresponds to the measurement of $x_1 = 4$ in the space X_1. B also represents the "fuzzy control action" which should be translated (in real world applications) into a single numerical value. In other words the choice of one element in the control action space should be made. Following Tong's (ref. 6) algorithm which represents the human thermal controller we adopted the method of choice based on the mean value of elements with maximal membership function values. For our simple example we obtain $x_2 = \frac{1+2}{2} = 1,5$. This is our control action since $x_2 = 1$ and $x_2 = 2$ are the elements with the maximal (0.8) values of the membership function (6).

A SYNTHESIS OF THE ACTIVE SUBSYSTEM OF THERMOREGULATION IN MAN

Here are the specific features of the thermoregulatory system in man, which are of great importance in the modelling of this system (Stolwijk and Hardy, ref. 4)

a) The signals of errors in core temperature (Σ_{cr}) and of the skin (Σ_{sk}) may be assumed INPUT variables of the regulator. They have the following definitions:

$$\Sigma_{cr} = T_{cr} - 36.6 \qquad (7)$$

$$\Sigma_{sk} = T_{sk} - 34.1 \qquad (8)$$

T_{cr} and T_{sk} are, respectively, temperatures of body core and the skin. The values 36.6° and 34.1°C are the observed levels of T_{cr} and T_{sk} in a state of "thermophysiological neutrality".

b) $\Delta\Sigma^{-}_{sk}$ is a signal of negative value (exactly: negative increase of this signal) which causes a reduction of a blood flow to a skin from the rest areas of circulatory systems.

c) $\Delta\Sigma^{+}_{sk}$ is a signal of positive value (the increase) which causes an increase of blood flow in a skin and moreover, activities the work of the sweat glands.

d) $\Delta\Sigma^{-}_{cr}$ is a signal of negative value which causes a decrease of a blood flow in a skin, but this process is less effective than the similar one caused by the negative $_{sk}$ signal.

e) $\Delta\Sigma^{+}_{cr}$ is a signal of positive value which causes the increase of secretion of sweat and an increase in the rate of blood flow in the skin.

f) The simultaneous appearance of negative Σ_{cr} and Σ_{sk} signals causes an increase of metabolic processes and a decrease of blood flow rate in a skin.

The presented general features of the individual influence of the respective signals may be extended to the effect of their combinations. One of the signals is regarded as a modifier which intensifies or reduces the reaction caused by a single signal. However, it should be emphasized that the nature of such a modification need not be "linear" (e.g.: the positive $\Delta\Sigma^{+}_{sk}$ signal is progressively more important with the increase of the positive $\Delta\Sigma^{+}_{cr}$ signal).

The experimentally supported finding of the increased metabolism level, only in the case of negative Σ_{sk} and Σ_{cr} signals (point f), allows one to accept the idea initially proposed by Stolwijk and Hardy (ref. 5). It consists of separating two thermoregulation mechanisms in the model - a mechanism of the heat loss and a mechanism of heat production and preservation.

To take into account the natural limits of the regulator potentiality the functional division of the proposed model was assumed after Tong (ref. 6). The model consists of the "absolute" regulation part and the incremental regulation part. This means that according to the quantity followed (observed) as the error, there is an adequate action formulated on OUTPUT: the increase action when the error is not big and the absolute action (i.e.: limit for a regulator) when the error is big. The general representation of the "absolute regulation" rules is the following:

IF "the error is bigger than e "THEN
"the regulator action equals u", (9)

where "e" and "u" are the singular numerical values which correspond to the limit error and the action of the real human thermoregulation system. A general form of sentences which make a structure of the thermal regulator in man may be presented as follows:

IF "$\Sigma_{cr} = E_1 \cdot \Delta\Sigma_{cr} = E_2 \cdot \Sigma_{sk} = E_3 \cdot \Delta\Sigma_{sk} = E_4$"THEN" $\Delta V_{bl} = E_5$
$\cdot \Delta m_{rsw} = E_6$" for heat loss and (10)

IF "$\Sigma_{cr} = E_1 \cdot \Delta\Sigma_{cr} = E_2 \cdot \Sigma_{sk} = E_3 \cdot \Delta\Sigma_{sk} = E_4$ "THEN" $\Delta V_{bl} =$
$= E_5 \cdot \Delta M = E_6$" for production and preservation of a heat (11)

where

Σ_{cr} - the difference between reference temperature of the body core and the current core temperature (a signal from the body core),

$\Delta\Sigma_{cr}$ - the increase of the above difference in the examined time interval,

Σ_{sk} - the difference between reference temperature of the skin and the current temperature of the skin,

ΔV_{bl} - the increase of the skin blood flow rate,

Δm_{rsw} - the increase of the sweat secretion rate,

E_i - the linguistic expression which defines the level of a given quantity (ex.: POSITIVE BIG, NEGATIVE MEDIUM, POSITIVE MEAN, ZERO, etc.),

"$\cdot$" - the conjunction "AND".

The linguistic expressions which were used to construct a proposed model and their fuzzy representations in seven elements artificial space are presented in TABLE 1A of the APPENDIX.

In TABLE 2A of the APPENDIX the list of linguistic expressions applied in the performed experiments is presented. These expressions make a structure of the heat loss regulator. The list was formulated on the basis of general regularities in functioning of the thermoregulation system (a - f points) and the analysis of the detailed data presented in literature (Gagge et al, ref. 1, Stolwijk and Hardy, ref. 4,5).

THE "PASSIVE" SYSTEM

Following Stolwijk and Hardy (ref. 4) we assumed the simple system of cylinders as physical representation of the extremities and the trunk of human body. The sphere represented the head. These are caused by necessary simplifications of physical processes to be modelled mathematically.

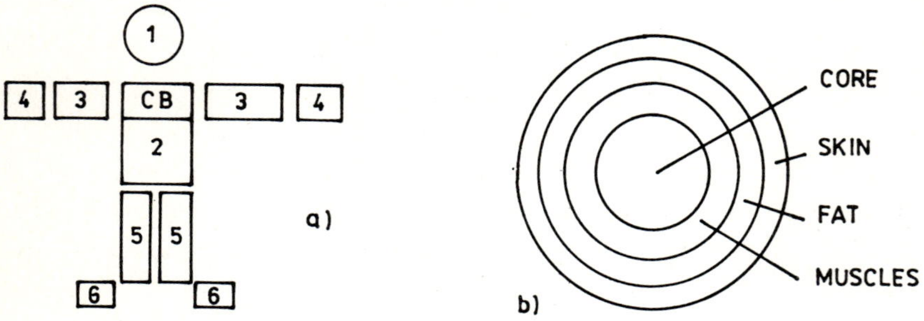

Fig. 1a. Schematic representation of the "passive system". 1 - head, 2 - trunk, 3 - arms, 4 - hands, 5 - legs, 6 - foots, CB - central blood, b - layers in the cylinder.

Three general assumptions should be accepted to a mathematical model of such a "physical" representation of human body:

1) In each compartment, parameters determining thermophysiological state are concentrated (i.e. each compartment is assumed to be homogeneous).
2) Heat exchange by conduction between cylinders is ommited.
3) Counter-current heat flows between adjacent arteries and veins are not taken into account.

Such an assumptions allows the modelling of the thermophysiological state of a given layer of a casual cylinder (Fig. 1) according to following general equations:

$$Q_t = Q_{t-1} + DT(HF_{t-1}^{(IN)} - HF_{t-1}^{(OUT)}) \tag{12}$$

in which Q_t - heat level in the discrete moment t

D_t - duration time (the time between t-1 and t).

$HF^{(IN)}$ and $HF^{(OUT)}$ - heat flow values coming into the given layer and going out of the layer respectively. The value of HF in the discrete moment t can be determined generally as:

$$HF_t^{(ij)} = f(T(i) - T(j), R) \tag{13}$$

$HF_t^{(ij)}$ - designates the heat flow between ith and jth adjacent layer and $T(i)$ is temperature level of ith layer. R is the function describing the regulatory response of the human body in a given state. Generally:

$$R = f(T_{sk}, T_{cr}) \tag{14}$$

which means that regulatory "action" is dependent upon the mean values of skin temperature of the whole body (T_{sk}) and body core temperature (T_{cr}).

The data given in Stolwijk and Hardy (ref. 4,5) and Gagge et al. (ref. 1) allows the application the PASSIVE SYSTEM explained above. To use our Fuzzy Logic Controller approach we constitute some additional assumptions in the following section.

"PASSIVE" - "ACTIVE" SYSTEMS COOPERATION - SIMULATION EXPERIMENT

The Fuzzy thermal controller was constituted by using fuzzy sets representation of the "natural" expressions in 7 elements artificial discrete space (TABLE 1A, APPENDIX).

To make it useful for thermal processes in man control we determined a way of "translate" artificial discrete points into real, physical values which appear as "error values" and "control action" in the temperature regulation process. In a following table the "means of translation" used in our experiments is shown.

TABLE 2

Translation system for the Fuzzy Temperature Controller

ERROR	$\Sigma_{sk}(\Delta\Sigma_{sk})$	-14	-11	-7.0	-3.0	1.0	2.5	4.0	6.0
	$\Sigma_{cr}(\Delta\Sigma_{cr})$	-0.3	-0.25	-0.15	-0.05	0.05	0.15	0.25	0.3
	INTERVALS INTERPRETATION								
ARTIFICIAL SPACE		1	2	3	4	5	6	7	
ACTION	V_{bl}	-3	-2	-1	0	1	2	3	
	m_{rsw}	50	25	10	0	10	25	50	

The bounds of intervals for Σ_{sk} and Σ_{cr} (and its increments) were estimated from Stolwijk and Hardy (ref. 5) data and Amosov et al. (ref. 2). The values of ΔV_{bl} and Δm_{rsw} were found in a series of simulation experiments for "hot environments". The "e" and "n" values for the absolute regulation were adopted from Gagge et al. (ref. 1) and Amosov et al. (ref. 2) as limits of the possibilities for the "average" thermophysiological system in man (max m_{rsw} = 950 g/min, max V_{bl} = 160 l/h).

General regulatory "actions" are "distributed" on all cylinders in accordance with the distributions of sweat glands, vasodilators, and vasoconstrictors estimated by Stolwijk and Hardy (ref. 5).

The above defined model was tested on a computer. The results of the typical thermophysiological experiment are depicted in Figure 1.

These results are of interest. Model construction presented in this way does not impose the strict constraints concerning the number of the expressions which make a structure; the way of defining the linguistic expressions (and their number); and the size of discrete space of representation of these expressions. All of these areas are connected with accuracy and the quality of the real system mapping.

The presented "linguistic" method of temperature control model construction seems to be relatively easy for an expert in the area. It can also be fruitful to use presented the suggested approach in the modelling of the other physiological regulatory systems which involve expert knowledge as proposed by Tong (ref. 7). It is also to note the comparison between the suggested approach and the "new fuzzy controller" concept of Tsukamoto (ref. 8) which is based on multivalued logic. The present method, however, needs more intensive study.

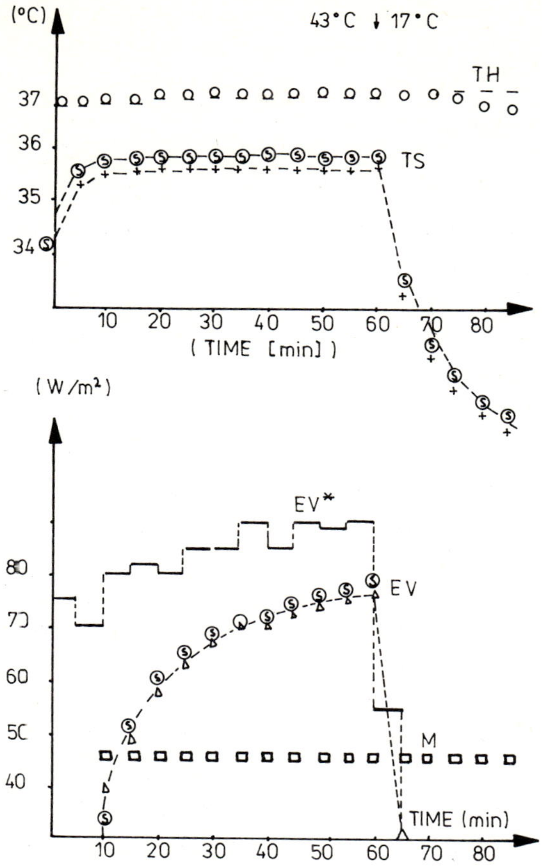

Fig. 1. Experiment in which a subject clad in shorts sat for 1 hr in a 43°C environment followed by 30 min at 17°C. TH - temperature of the head (dashed line - real experiment O - computer simulation) TS-skin temperature, Ⓢ - in Stolwijk and Hardy's model, + - proposed fuzzy model, EV - evaporative heat loss in real experiment, Ⓢ - in Stolwijk and Hardy's model, △ - in fuzzy model, M - metabolism rate □ - fuzzy model and Stolwijk's.

APPENDIX

TABLE 1A

Subjective representations of the linguistic expressions in the discrete space

LINGUISTIC EXPRESSION	SYMBOL	ELEMENT OF THE REPRESENTATION SPACE						
		1	2	3	4	5	6	7
Negative big	NB	1.0	0.6	0.0	0.0	0.0	0.0	0.0
Negative medium	NM	0.6	1.0	0.5	0.0	0.0	0.0	0.0
Negative small	NS	0.0	0.5	1.0	0.5	0.0	0.0	0.0
Zero	ZE	0.0	0.0	0.5	1.0	0.5	0.0	0.0
Positive small	PS	0.0	0.0	0.0	0.5	1.0	0.5	0.0
Positive medium	PM	0.0	0.0	0.0	0.0	0.5	1.0	0.6
Positive big	PB	0.0	0.0	0.0	0.0	0.0	0.6	1.0
Any positive	ANYP	0.0	0.0	0.0	1.0	1.0	1.0	1.0
Any negative	ANYN	1.0	1.0	1.0	1.0	0.0	0.0	0.0
ANY	ANY	1.0	1.0	1.0	1.0	1.0	1.0	1.0

TABLE 2A

Expressions used in the experiment (the structure of fuzzy controller of body temperature)

n	E_1	E_2	E_3	E_4	E_5	E_6
1.	ZE	ZE	ZE	PS	PS	ZE
2.	ZE	PS	ZE	PS	PS	ZE
3.	ZE	ZE	ZE	PM	PM	PS
4.	ZE	PM	ZE	ZE	PM	PS
5.	ZE	PS	ZE	PS	PS	PS
6.	ZE	PM	ZE	PS	PM	PM
7.	ZE	PB	ZE	ANYP+ZE	PB	PB
8.	ZE	PM+PS	ZE	PB	PM	PB
9.	ZE	ZE	ZE	PB	PB	PM
10.	PS+PM	ZE	ANYP	PS	ZE	PS
11.	PB	ZE	ANYP	PS	PS	PM
12.	PS+PM	PS	ANYP	PS	PS	PS
13.	PS+PM	ZE	ANYP	PM	PS	PM
14.	PB	ZE	ANYP	PM	PM	PB
15.	PB+PM+PS	PS	ANYP	ZE	PS	PS
16.	PB	ZE	ANYP	PB	PB	PB
17.	PB+PM+PS	PM	ANYP	ZE	PM	PM
18.	PB+PM+PS	PB	ANYP	ZE	PB	PB
19.	ANYP	ZE	ANYP	ZE	ZE	ZE
20.	PS+PM	PM	ANYP	PS	PM	PM
21.	PS+PM	PS+PM	ANYP	PM+PB	PM	PB
22.	ANYP	ZE	ANYP	NS	NS	UM
23.	ANYP	ZE	ANYP	NM	NM	NB
24.	ANYP	ZE	ANYP	NB	NB	NB
25.	ANYP	PB	ANYP	NM+NB	PM	ZE
26.	ANYP	PB	ANYP	NS	PM	PS
27.	ANYP	PM	ANYP	NS	PS	ZE
28.	ANYP	PS	ANYP	NB	NM	NB
29.	ANYP	NB	ANYP	NB	NB	NB
30.	ANYP	NS	ANYP	NS+ZE	NS	NS
31.	ANYP	NM	ANYP	NS+NM+ZE	NM	NM
32.	ANYP	NB	ANYP	ANYN	NB	NB
33.	ANYP	NS+NM	ANYP	NB+NM	NM	NB
34.	ANYP	ANYP	ANYP	ZE	ZE	ZE
35.	ANYP+ZE	ZE	ANYP+ZE	PM	PM	PS
36.	ANYP+ZE	ZE	ANYP+ZE	PB	PB	PM
37.	ANYP+ZE	PS+PM+PB	ANYP+ZE	PB	PB	PB
38.	ANYP+ZE	PB	ANYP+ZE	ANYP+ZE	PB	PB
39.	ANYP+ZE	PS+PM	ANYP+ZE	PS+PM	PM	PM

REFERENCES

1 Gagge A.P., Stolwijk J.A.J., Nishi Y., An Effective Temperature Scale Based on a Simple Model of Human Physiological Regulatory Response, ASHRAE Trans. 77/I/ (1971), 247-262.

2 Grobelny J., The Model of the Impact of the Microclimate on Man, PhD thesis, Techn. Univ. of Wrocław, (1980), (in Polish).

3 Mamdani E.H., Advances in the Linguistic Synthesis of Fuzzy Controllers, Int. J. Man-Machine Studies 8, (1976), 669-678.
4 Stolwijk J.A.J., Hardy J.D., Control of Body Temperature, in: Handbook of Physiology, Am. Physiol. Soc. (1975), 45-68.
5 Stolwijk J.A.J., Hardy J.D., Temperature Regulation a Man - A Theoretical Study, Pflügers Archiv. 191, (1966), 129-162.
6 Tong R.M., Some Problems with the Design and Implementation of Fuzzy Controlles, Hill Lane, (1976), Cambridge.
7 Tong R.M., A Retrospective View of Fuzzy Control Systems, Fuzzy Sets and Systems 14, (1984), 199-210.
8 Tsukamoto Y., An Approach to Fuzzy Reasoning Method, in: Advances in Fuzzy Sets Theory and Applications, North-Holland, Amsterdam, 1979.
9 Zadeh L.A., The Concept of Linguistic Variable and its Application to Aproximate, Reasoning, Am. Elsev. Pub. Comp. N.Y., 1973.

Applications of Fuzzy Set Theory in Human Factors, edited by W. Karwowski and A. Mital
Elsevier Science Publishers B.V., Amsterdam, 1986 — Printed in The Netherlands

DEFINITION OF ENVIRONMENTAL HEAT STRESS AS A FUZZY SET

HALA A. HAFEZ[1], M.M. AYOUB[2], and JERRY D. RAMSEY[2]

[1]Dept. of Design & Production Engr., Ain Shams University, Cairo, Egypt

[2]Dept. of Industrial Engr., Texas Tech University, Lubbock, Texas

ABSTRACT

The interaction between man and the thermal environment while performing any kind or level of activity involves physiological and psychological aspects and may form a rather complete system. Considering the uncertainties relating to the complexity of the system, and with the recent successful applications of fuzzy sets theory to analyze such complex ergonomic systems, this work was conducted to define the environmental heat stress in a form so that the thermal environment can be analyzed using fuzzy sets theory.

INTRODUCTION

Imprecision and indeterminacy are ordinarily considered to be statistical random characteristics and are taken into account by classical methods of probability theory. In real situations, a frequent source of imprecision is not only the presence of random variables, but the impossibility of operating with exact data as a result of the complexity of the system and imprecision of the constraints and objectives. At the same time, classes of objects appear in the problems that do not have clear boundaries; the imprecision of such classes is expressed in the possibility that an element does not only belong or not belong to a certain class, but intermediate grades of membership are also possible (ref.1).

The theory of fuzzy sets, introduced by Zadeh (ref.2), may be regarded as a body of concepts and techniques for dealing with imprecision and uncertainties which are associated with classes in which the transition from membership to nonmembership is gradual rather than abrupt. Abrupt membership is when an element is either a member or is not a member in a subset of a reference set. The notion of gradual or weighted membership is when an element may belong totally, may belong more or less, or is not a member of the subset.

The fuzzy sets theory has been used in applied sciences and various branches of engineering as well as other fields where the situation of mutually interacting complex systems (natural and artificial), with human beings is encountered. Kandel (ref.1) pointed out that problems involving subjective evaluations seem to be better stated in terms of fuzzy models rather than probabilistic or deterministic ones. The usefulness of fuzzy sets application in human factors research has been demonstrated by Karwowski (ref.3) in which the author listed examples of recent applications and suggested areas for future research.

According to Fanger (ref.4), wherever artificial climates are created for human occupation, the aim is that the thermal environment be adapted so that each individual is in thermal comfort. McIntyre (ref.5) argues that thermal comfort is not an exact concept, nor does it occur at an exact temperature. The author explains that a person may be comfortable over a range of temperatures and if the temperature is altered so that it moves outside this range, the onset of discomfort is not sudden.

The first condition necessary for thermal comfort for a person under long exposure to a given environment is the existence of heat balance. However, man's thermoregulatory system is quite efficient and will therefore create heat balance within wide limits of the environmental conditions, even if comfort does not exist (ref.4).

Being uncomfortable is not the only problem with working in hot conditions. For the same amount of work, man experiences increasing physiological strain as the temperature of the environment rises. Body temperature increases during the first thirty to forty minutes of any physical exertion (ref.6) thereafter stabilizing at some level, when and if thermal balance is attained. The level of body temperature is governed by the intensity of work performed and by the individual's physical work capacity and heat acclimatization in such a manner that a person's body temperature levels off at about 38°C when working at an intensity corresponding to half his maximal oxygen uptake.

For different work loads, the body temperature reaches steady state at levels dependent on the intensity of the work.

Hence, the measure of the acceptability of a given heat stress is imprecise in nature. Thermal conditions are not always totally acceptable or unacceptable to human beings, but some intermediate grades of acceptability exist in which the environment may be more or less acceptable, slightly acceptable, intolerable, etc. The range of environmental heat stress conditions that can be considered acceptable represent a class with no sharp transition from membership to nonmembership, and hence could be associated with fuzziness.

The purpose of this paper is to define the environmental heat stress in a form so that the thermal environment can be analyzed using fuzzy sets.

THE CONCEPT OF A FUZZY SET

A fuzzy subset A of a reference set E, in which x is an element of E is a set of ordered pairs

$$A = \{(x, f_A(x)) \mid x \varepsilon E\}$$

where $f_A(x)$ is a membership characteristic function that takes its values in a totally ordered set M, and which indicates the degree or level of membership. M is called a membership set. It is assumed that M is in the closed interval [0,1], where 0 and 1 represent the lowest and highest grades of membership, respectively. Thus a fuzzy subset A of E can be defined precisely by associating with each value of x a number between 0 and 1, which represents its grade of membership in A. If M contains only two points 0 and 1, the "fuzzy set" A is understood as a "nonfuzzy set" or ordinary set.

One method for estimating a membership function is by defining an ideal point or reference point from which the degree of similarity of a given object to that ideal element is assessed.

METHOD

Let E = {x} represent the environment. Also, let an element x of the environment be represented by the environmental temperature as an indication of the environmental heat stress. In order to define environmental heat stress as a fuzzy set, a membership function has to be established.

A fuzzy subset A of E is the set of acceptable environmental temperatures. The membership function $f_A(x)$ represents the measure of acceptability or grade of membership of the environmental temperature in the subset A.

$A = \{(x, f_A(x)) \mid x \varepsilon E\}$

and $f_A(x)$ taking values between 0 and 1 such that

$f_A(x) = 1$ for $t \leq R$

< 1 for $t > R$

where R denotes a reference value for the environmental temperature represented by the optimal air temperature, and t is any temperature.

To select a reference temperature, it is necessary to determine from what standpoint that temperature is to be considered ideal. Therefore, the reference temperature may differ in accordance with the criterion of selection, be it comfort, mental performance, physical performance, or any other criterion. This temperature may then be used to develop the acceptability measures of the different temperatures, forming the membership function.

For this purpose, the preferred temperature range determined by subjective estimates of comfort was examined. ASHRAE (ref.7) defined thermal comfort as "that condition of mind which expresses satisfaction with the thermal environment." Ramsey, et al. (ref.8) indicated that some investigators employ the phase "thermally neutral" as synonymous to thermal comfort. The authors explained that comfort is judged by a person exposed to a particular thermal environment, and as it is based on personal feelings of comfort, the estimate of thermal comfort is subjective in nature.

Ramsey, et al. (ref.8) concluded that there is a general agreement on the preferred temperature between the studies that have used subjective estimates of thermal sensation. They reported that Fanger (ref.4) and Fanger, et al. (ref.9) suggested preferred air temperatures for seated persons dressed in cotton twill shirts and trousers of 25.7°C and 24.5°C, respectively, and that

ASHRAE (ref.10) recommended a comfort zone of 22.75°C-24.95°C for office work. The comfort curve as voted upon by a group of seated adults wearing normal indoor clothing was centered with a mean, mode and median all near an air temperature of 22.9°C (ref.11), while the range reported by Humphreys (ref.12), based on a large number of independent studies, was between 17°C and 26°C.

The relationship between subjective ratings and temperature was also investigated by Griffiths and Boyce (ref.13). The assessment yielded optimal values between air temperatures of 18.3°C and 21.1°C. The comfort temperature for sedentary people reported by McIntyre (ref.5) based on a study by Rohles and Nevins (ref.14) using a large subject sample was 25.6°C.

From the above reported studies, a neutral temperature of 25°C was chosen to represent the reference temperature based on the comfort criterion since this value is common to most of the comfort studies.

In order to find the acceptability measures of the higher temperatures based on subjective estimates of comfort, the Predicted Percentage of Dissatisfied (PPD) method developed by Fanger (ref.4) was used. Using this method, the author calculated PPD values of 1300 subjects. Those values represent the percentage of persons expected to be dissatisfied with a given thermal environment, and hence indicate the thermal unacceptability and consequently the thermal acceptability of the environment.

Fanger (ref.4) reported that the PPD at 24.4°C and 25.6°C was 3.8 and 5.6 percent, respectively. By linear interpolation, in the present study, the PPD at 25°C was found to be 4.7 percent. Fanger (ref.4) also reported that the PPD at 28.9°C and 32.2°C was 41.2 and 78.7 percent, respectively. By subtracting 4.7 percent from all the previous PPD values in order to make the PPD at 25°C equal to zero, according to the choice of this temperature as the reference point in the present work, PPD values of 36.5 and 74 percent were obtained for 28.9°C and 32.2°C, respectively.

Since PPD represents thermal unacceptability, then by subtracting from 100 and dividing the result by 100, thermal acceptability of 28.9°C and 32.2°C was found to be 0.635 and 0.26, respectively. By plotting the values thus obtained and assuming a linear relationship, the membership function in Figure 1 is obtained. Therefore, the set of environmental temperatures that can be considered acceptable from the comfort standpoint might be represented by a fuzzy set whose membership function is given by

$f_{A1}(x) = 1$ for $t \leq R$, and

$= 3.5 - 0.1t$ for $t > R$

It is worth noting that although the comfort range below 25°C was not addressed in this study, it is assumed that temperatures up to 25°C are considered acceptable. And since the purpose of this work was to define hot climatic conditions only, therefore temperatures much lower than 25°C are out of the scope of this study.

The preferred temperatures reported in the literature have been based on men at or near the sedentary activity level. According to ASHRAE (ref.10), during physical activity a change occurs in man's physiology, and physiological thermal neutrality in the sedentary sense does not exist. Some form of thermal regulation always occurs within man's body during exercise.

It was therefore decided to consider a physiological criterion related to the environment when the acceptability of heat stress during physical activity

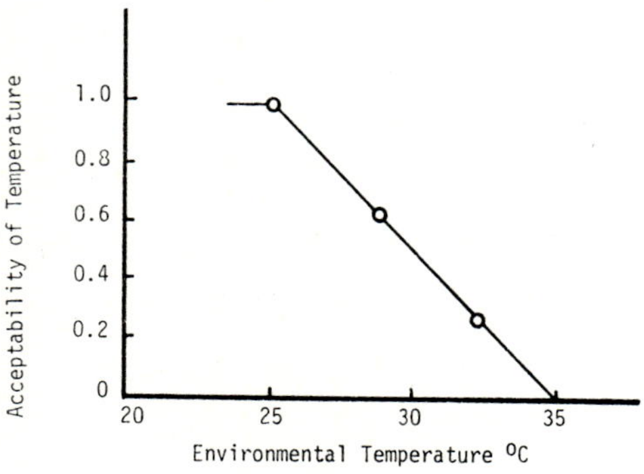

Figure 1. Acceptability of environmental temperature based on comfort

is of interest. The most appropriate criterion in heat stress studies is heat storage expressed in terms of deep body temperature.

The results of Lind (ref.15), in which Dukes-Dobos and Henschel (ref.16) corrected the original levels of the upper limit of the Prescriptive Zone (PZ) so as to make it safe for 95 percent of the worker population, are shown in Figure 2. For the three work rates examined: 180, 300 and 420 kcal/hr, these limits were 30, 26.94 and 25°C Effective Temperature (ET), respectively. These temperatures represent the critical levels of climatic stress above which there is an abrupt increase in body thermal equilibrium above the levels determined solely by the metabolic rate.

In order to estimate the acceptability of the climatic stress for each work load, the rectal temperature corresponding to the critical limit at each load was used as a reference point. From Figure 2 these rectal temperatures are 37.6, 37.78 and 37.98°C for 180, 300 and 420 kcal/hr, respectively. The upper limit of rectal temperature at which the acceptability of the climatic stress is zero, was considered 39°C in all cases. By linear proportionality, the acceptability of two values of rectal temperature above the reference point for each work load was calculated, and hence the acceptability of the corresponding climatic temperatures was derived. For example, for the 180 kcal/hr work load, the acceptability of 37.6°C and 39°C rectal temperatures is 1 and 0, respectively, therefore by linear proportionality the acceptability of the two intermediate rectal temperatures 38°C and 38.24°C is 0.71 and 0.54, respectively, which are the same acceptability values for the corresponding climatic temperatures 32.93°C and 34.4°C ET, respectively.

Assuming that ET may be considered approximately equal to air temperature, which is the case when relative humidity is equal to 100 percent and air velocity is equal to zero, and assuming a linear relationship, the acceptability of climatic temperatures at each work load was plotted in Figure 3. Therefore, the set of environmental temperatures that can be considered acceptable from the phsiological standpoint during physical activity might by represented by a fuzzy set whose membership function is given by

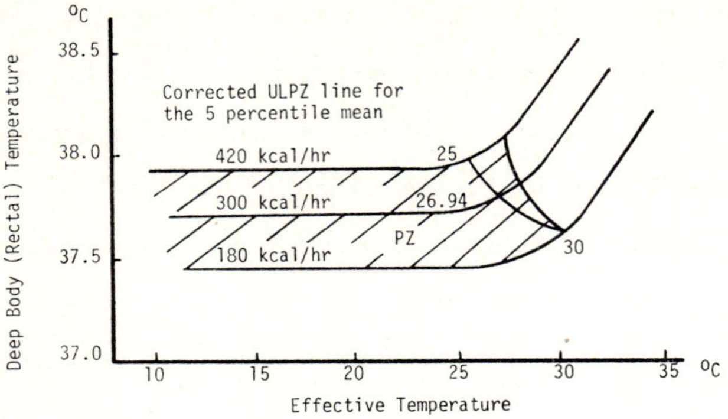

Figure 2. Deep body temperature versus climatic temperature at different work loads (adapted from Dukes-Dobos and Henschel, 1973)

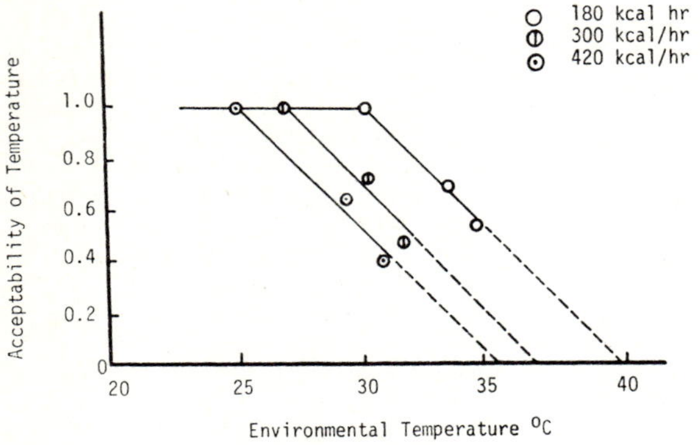

Figure 3. Acceptability of environmental temperature based on heat storage

$f_{A2}(x) = 1$ for $t \leq R$, and

$= b - mt$ for $t > R$

where b and m are the intercept and slope of the straight line representing the membership function for each work load, respectively, and R is the reference point at each work load.

Having established two membership functions, one for the acceptability of environmental temperature based on comfort and the other for the acceptability of environmental temperature based on heat storage, one can select either function for representing the thermal environment depending upon which criterion is of concern. If both criteria are of interest, i.e., if it is required that the environment be comfortable and at the same time acceptable from the physiological standpoint, then the two fuzzy sets may be combined.

DISCUSSION

The thermal environment to which an individual is exposed during work may influence to varying degrees his physiological behavior and performance. To assess the effects of such environments using the theory of fuzzy sets, the fuzziness associated with the thermal environment as a system has to be defined. The formulation of a membership function to define heat stress was carried out in this study. The environmental heat stress as defined by the established membership function may then be used in fuzzy sets applications for representing the interaction between man and his working environment. One such application was demonstrated by Hafez (ref.17) and reported by Ayoub and Hafez (ref.18) in which the authors investigated the combined effects of the different stresses, including environmental heat stress, associated with manual lifting on the lifting capabilities of individuals.

The membership functions in the present investigation were established for environmental heat stress, as represented by air temperature, for only three levels of activity, and linear relationships were assumed. Provisions must be made, however, to modify the functions such that they include the other environmental factors, namely relative humidity, air velocity, and radiant heat. To accomplish this, a heat stress index such as Wet Bulb Globe Temperature (WBGT) may be used. Other factors which may influence the acceptability of heat stress such as acclimatization, clothing, sex, physical fitness, etc. also need to be taken into consideration. Membership functions may also be found for other activity levels. In addition, other forms of the membership function, such as a parabolic function, need to be investigated.

REFERENCES

1 Kandel, A., Fuzzy statistics and policy analysis, in: Wang, P.P. and Chang, S.K. (eds.), Fuzzy Sets: Theory and Applications to Policy Analysis and Information Systems (Plenum Press, New York, 1980).

2 Zadeh, L.A. Fuzzy sets, Information and Control 8 (1965) 338-353.

3 Karwowski, W., Why do ergonomist need fuzzy sets, Proceedings of the Ninth Congress of the International Ergonomics Association (Taylor & Francis, London and Philadelphia, 1985).

4 Fanger, P.O.; Thermal Comfort (McGraw-Hill, New York, 1972).

5 McIntyre, D.A., Design requirements for a comfortable environment, in: Cena, K. and Clark, J.A. (eds.), Bioengineering, Thermal Physiology and Comfort (Elsevier Scientific, New York, 1981).

6 Astrand, I.; Axelson, O.; Eriksson, U. and Olander, L., Heat stress in occupational work, AMBIO 4 (1985) 37-42.

7 ASHRAE, Thermal comfort conditions, ASHRAE Standard 55-74, (New York, 1974).

8 Ramsey, J.D.; Burford, C.L. and Beshir, M.Y., Effects of heat on safe work behavior, Final Report, NIOSH Contract No. 210-79-0021 (March, 1982).

9 Fanger, P.O.; Hojbjerre, J. and Thomsen, J.O.B., Man's preferred ambient temperature during the day, Arch. Sci. Physiol. 27 (1973) A395-A402.

10 ASHRAE, Handbook of Fundamentals, Chapter 8, Physiological principles, comfort and health (ASHRAE, New York, 1981).

11 Hindmarsh, M.E. and Macpherson, R.K., Thermal comfort in Australia, Australian Journal of Science 32 (1962) 335-339.

12 Humphreys, M.A., The dependence of comfortable temperature upon indoor and

13 Griffiths, I.D. and Boyce, P.R., Performance and thermal comfort, Ergonomics 14 (1971) 475-468.

14 Rohles, F.H. and Nevins, R.G., The nature of thermal comfort for sedentary man, ASHRAE Transactions 77 (1981) 239-246.

15 Lind, A.R., A physiological criterion for setting thermal environmental limits for everyday work, Journal of Applied Physiology 18 (1963).

16 Dukes-Dobos, F. and Henschel, A., Development of Permissible heat exposure limits for occupational work, ASHRAE Journal 15 (1973) 57-62.

17 Hafez, H.A., An Investigation of Biomechanical, Physiological and Environmental Heat Stresses Associated with Manual Lifting in Hot Environments, unpublished Ph.D. dissertation, Texas Tech University (May, 1984).

18 Ayoub, M.M. and Hafez, H.A., Combining stresses for manual materials handling, Proceedings of the Ninth Cóngress of the International Ergonomics Association (Taylor & Francis, London and Philadelphia, 1985).

Applications of Fuzzy Set Theory in Human Factors, edited by W. Karwowski and A. Mital
Elsevier Science Publishers B.V., Amsterdam, 1986 — Printed in The Netherlands

EVALUATION OF LIFTING CAPACITY DETERMINATION METHODOLOGIES: A FUZZY SET APPROACH

S.S. Asfour, T.M. Khalil, A.M. Genaidy, and S.M. Waly

Department of Industrial Engineering, University of Miami, Coral Gables, Florida 33124, U.S.A.

ABSTRACT

The main objective of this study is to demonstrate, using fuzzy modeling techniques, that the acceptable maximum effort (AME) is a better predictor of lifting capacity of workers than the maximum voluntary contraction (MVC) utilized in the literature. Hamming distance between AME and lifting capacity was always lower than that between MVC and lifting capacity. Also, the similarity index was higher between AME and lifting capacity, than between MVC and lifting capacity. Thus, it is suggested that if a static muscle testing regimen is selected in job design tests to assess the lifting capacity of workers, then AME not MVC should be the measure used.

INTRODUCTION

Musculoskeletal injuries resulting from manual materials handling activities are a major source of lost time and compensation claims. Reduction in the number of these injuries has therefore become a major concern of different researchers and practitioners. Recent studies showed that most of these injuries result from over exertion.

The psychophysical criteria have been adopted by several researchers in the field of manual materials handling to predict the lifting capacity of individuals (ref.1 and ref.2). Lifting capacity, which will be referred to in this paper as the maximum acceptable dynamic effort (MADE), is defined as the amount of load an indvidual can lift, under a given set of conditions, for an extended period of time.

Static strength has been recommended to assess the lifting capacity of workers (ref.3 and ref.4). Static strength has been defined as the capacity to produce torque or force by a single isometric maximum voluntary contraction (MVC) (ref.5). It has been shown that static strength measurement techniques are safe with an injury risk smaller than 1:1000 (ref.6). The time needed to administer this test is minimal. Five different measurements can be obtained in less than 15 minutes by a trained technician. Keyserling (ref.7) found that static strength is a reliable measure with coefficient of variation less than 10% and that test/retest values are highly correlated (r = 0.9). Also, static strength measuring devices are available on the market at values much lower than dynamic strength measuring devices (below $2000 in the U.S.). However studies by Garg et al. (ref.4), Khalil et al. (ref.8), and Mital et al. (ref.9) found poor correlation between the maximum isometric voluntary contraction (MVC) and dynamic lifting capability.

This investigation is undertaken to introduce an alternative psychophysical submaximal isometric test to MVC which requires the subject to exert himself/herself all the way to his/her maximum. In this test the term acceptable maximum effort (AME) will refer to the level of exertion the individual is willing to maintain in a series of isometric muscle contractions, for a given set of conditions, over an extended period of time. As such AME is a psychophysical approach to the determination of acceptable static lifting capacity of the individuals.

Recent studies by Karwowski and others have shown that fuzzy modeling is a useful tool in assessing the acceptability of stresses involved in manual lifting (ref.10-ref.16). The present study makes use of fuzzy modeling techniques in testing the hypothesis that AME is a better predictor of lifting capability of workers than MVC.

EXPERIMENTAL METHODS AND PROCEDURES

10 male subjects volunteered to participate in the present study. The physical characteristics of the subjects were as follows: (1) age (mean = 22.5 years, s.d. = 2.6); (2) height (mean = 172.7 cm, s.d.= 5.1); and (3) weight (mean = 69.4 kg, s.d. =13).

A randomized complete block design with three factors experiment was used in this study. Each subject was considered as a block. A total of 18 treatment combinations were performed within each block in a random fashion. Only one treatment was performed by each subject on a given day.

The three factors that were studied were the type of task, frequency of lift and technique of lift. Static and dynamic lifting were the two types of task employed in this study. The frequency levels were set to 2, 4, and 6 lifts/min. Asfour et al. (ref.17) and Karwowski and Yates (ref.18) indicated that the psychophysical methodology can be reliably used to establish safe lifting limits for frequencies below 6 lifts/min. The lifting techniques for the dynamic part of this experiment were back lift (lifting from the floor to 76 cm above the floor with bent back - straight knees), Leg lift (lifting from the floor to 76 cm above the floor with straight back-bent knees), and arm lift (lifting from 76 cm to 127 cm above the floor). The dependent variables were maximum acceptable dynamic effort (MADE) and acceptable maximum effort (AME). The MADE was determined for each subject by adjusting the amount of load lifted according to the psychophysical approach used by Snook (ref.1), and Ayoub et al. (ref.2). The box dimensions used to hold the weights were 38 x 38 x 25 cm and had cut-out handles. All subjects were instructed to perform the muscular effort to the level that they perceived to be acceptable.

In the static part of the experiment, the acceptable maximum effort (AME) was obtained for each subject by pulling against a metal bar provided with two handles 38 cm apart. The bar was connected to a load cell for measurement of the force exerted. The height of the bar was adjusted to correspond to the box handle height at the origin of the lift. The subjects were asked to exert themselves to the level they are willing to maintain in a series of isometric muscle contractions. In both dynamic and static lifting, 60 lifts per treatment were performed. The MADE was considered the final weight of the box, whereas the AME was the average of the last 10 readings of the force transducer. The MVC was defined, for each type of lift, as the maximum force exerted isometrically in a single voluntary contraction. The dynamic equivalent for the MVC is the maximum dynamic effort (MDE). This is the maximum amount of load contained in a box that a person could lift in a single trial for each type of lift.

FUZZY SET MODEL

General model

Karwowski and Ayoub (ref.12) and Karwowski et al. (ref.13) introduced the fuzzy modeling approach in manual materials handling research. In the fuzzy set model, a universal set of X = {x} denotes a collection of all possible levels of the maximum acceptable dynamic effort (MADE) and acceptable maximum effort (AME) recorded. The acceptability measures of "x" are defined in terms of the membership function f(x) of a particular fuzzy subset in X. A fuzzy subset A in X is a set of ordered pairs:

$$A = \{ (x, f_A(x)) \mid x \in X \} \qquad (1)$$

where membership function $f_A(x)$ represents the grade of membership of X in A. It is defined, in this study, for the interval [0,1] as follows:

$$f_A(x) = \begin{cases} x/ME & \text{for } x < ME \\ 1 & \text{for } x = ME \end{cases} \qquad (2)$$

ME denotes a reference point which is the maximum effort for either the static or the dynamic values recorded in the experiment through a single trial.

<u>Dynamic lifting model</u>

A fuzzy subset D in X, named the maximum acceptable dynamic effort (MADE), also known as the lifting capacity, from the dynamic lifting point of view is defined as follows:

$$D = \{ (x, f_D(x)) \mid x \in X \} \qquad (3)$$

with membership function:

$$f_D(x) = \begin{cases} x/MDE & \text{for } x < MDE \\ 1 & \text{for } x = MDE \end{cases} \qquad (4)$$

Where MDE refers to the maximum dynamic effort and is determined by the maximum amount of load contained in a box that an individual can lift in a single trial for each type of lift.

<u>Static lifting model</u>

A fuzzy subset S in X, called the AME from the static lifting point of view, is defined as follows:

$$S = \{ (x, f_S(x)) \mid x \in X \} \qquad (5)$$

with membership function

$$f_S(x) = \begin{cases} x/MVC & \text{for } x < MVC \\ 1 & \text{for } x = MVC \end{cases} \qquad (6)$$

Where MVC is the maximum force exerted isometrically for each type of lift in a single voluntary contraction.

<u>Fuzzy distance and similarity measure for D and S</u>

In order to evaluate the acceptability measure of dynamic lifting with respect to static lifting, the Hamming distance "HD" (ref.19) between D and S was calculated as follows:

$$HD = \frac{1}{n} \sum_{i=1}^{n} |f_D(x_i) - f_S(x_i)| \qquad (7)$$

Also, the similarity index "SI" (ref.10) between the acceptabilities of dynamic and static lifting was computed according to the formula:

$$SI = 1 - \frac{\sum_{i=1}^{n} |f_D(x_i) - f_S(x_i)|}{n + \sum_{i=1}^{n} [\min\{ f_D(x_i), f_S(x_i)\}]} \qquad (8)$$

A similarity index value of 1.0 indicates a complete identity between the fuzzy subsets D and S, while 0 shows lack of any similarity in the respective values of the membership for these types of task.

RESULTS AND DISCUSSION

Table 1 shows the acceptable maximum effort (AME) and acceptability measure for the fuzzy model of static lifting. It is apparent that, on the average, the acceptability measure of AME ranged between 54% and 79% of the MVC values recorded. Table 2 presents the maximum acceptable dynamic effort (MADE) and acceptability measure for the fuzzy model of dynamic lifting. The highest acceptability measure obtained was on the average 46% of the maximum dynamic effort measured.

A close look at tables 1 and 2 shows that, for all treatment combinations, the acceptability measure of static lifting was always higher than that of dynamic lifting (significant at the 0.05 level). This can be explained as follows: When there is no motion involved (static lifting) human beings have a tendency to overestimate their ability to lift an object. When movement is introduced (dynamic lifting), they revise their estimate downward significantly. This indicates that an estimate of acceptable weight to be lifted at one point in the trajectory of motion could be different from that at another in the trajectory of motion. It is not until the entire motion is performed that the human body can make its adjustment to the acceptable level over the entire range.

TABLE 1

ACCEPTABLE MAXIMUM EFFORT AND ACCEPTABILITY MEASURE FOR THE FUZZY MODEL OF STATIC LIFT

Frequency of Lift (lifts/min)	Type of Lift	AME (Newton)		Acceptability Measure	
		$\overline{X}$	S.D.	$\overline{X}$	S.D.
2	Back Lift	503.70	102.99	0.64	0.12
2	Leg Lift	466.40	139.90	0.55	0.17
2	Arm Lift	436.14	87.13	0.79	0.10
4	Back Lift	464.54	85.47	0.61	0.12
4	Leg Lift	455.72	103.38	0.54	0.12
4	Arm Lift	376.03	77.73	0.71	0.17
6	Back Lift	470.80	87.33	0.62	0.17
6	Leg Lift	424.98	131.68	0.51	0.16
6	Arm Lift	339.52	60.21	0.66	0.19

WHERE : X is the mean value
S.D. is the standard deviation

TABLE 2

MAXIMUM ACCEPTABLE DYNAMIC EFFORF AND ACCEPTABILITY MEASURE FOR THE FUZZY MODEL OF DYNAMIC LIFTING

Frequency of Lift (lifts/min)	Type of Lift	MADE (Newton)		Acceptability Measure	
		$\overline{X}$	S.D.	$\overline{X}$	S.D.
2	Back Lift	177.59	46.01	0.35	0.13
2	Leg Lift	176.71	50.91	0.33	0.07
2	Arm Lift	178.86	47.97	0.46	0.14
4	Back Lift	162.02	60.40	0.31	0.10
4	Leg Lift	162.02	44.74	0.32	0.10
4	Arm Lift	158.40	31.92	0.41	0.09
6	Back Lift	160.41	49.64	0.31	0.11
6	Leg Lift	153.11	21.73	0.30	0.04
6	Arm Lift	152.23	48.85	0.39	0.13

WHERE : X is the mean value
S.D. is the standard deviation

TABLE 3

HAMMING DISTANCE (HD) FOR THE PROPOSED FUZZY MODEL

Frequency of Lift (lifts/min)	Type of Lift	HD between AME and MADE	HD between MVC and MADE
2	Back Lift	0.29	0.65
2	Leg Lift	0.22	0.67
2	Arm Lift	0.33	0.54
4	Back Lift	0.30	0.69
4	Leg Lift	0.22	0.68
4	Arm Lift	0.30	0.59
6	Back Lift	0.31	0.69
6	Leg Lift	0.21	0.70
6	Arm Lift	0.27	0.61

TABLE 4

SIMILARITY INDEX (SI) FOR THE PROPOSED FUZZY MODEL

Frequency of Lift (lifts/min)	Type of Lift	SI between AME and MADE	SI between MVC and MADE
2	Back Lift	0.79	0.52
2	Leg Lift	0.83	0.50
2	Arm Lift	0.77	0.63
4	Back Lift	0.77	0.47
4	Leg Lift	0.83	0.52
4	Arm Lift	0.79	0.58
6	Back Lift	0.76	0.47
6	Leg Lift	0.84	0.46
6	Arm Lift	0.81	0.56

Hamming distance (HD) between AME and lifting capacity was always lower than that between MVC and lifting capacity as shown in Table 3 (significant at the 0.05 level). HD was on the average 0.27 between AME and lifting capacity and 0.65 between MVC and lifting capacity. The similarity index (SI) was higher between AME and lifting capacity,than between MVC and lifting capacity as indicated in Table 4 (significant at 0.05 level). SI was on the average 0.80 between AME and lifting capacity and 0.52 between MVC and lifting capacity. The results of Tables 3 and 4 imply that there is a stronger degree of similarity between AME and lifting capacity than between MVC and lifting capacity. These findings support studies by Garg et al. (ref.4), Khalil et al. (ref.8) and Mital et al. (ref.9). Garg et al. (ref.4) and Mital et al. (ref.9) reported poor correlation between MVC and lifting capacity. Khalil et al. (ref.8) found large differences between MVC and MADE, and a better correlation between AME and MADE.

CONCLUSIONS

Several studies have suggested the use of maximum voluntary contraction (MVC) as a means to predict the lifting capacity of workers. Recent studies found poor correlation between MVC and lifting capacity. In the present investigation a new methodology, based on a series of isometric contractions to a psychophysically acceptable level over an extended period of time, was developed to predict the lifting capacity of workers. Our findings showed that there is a stronger degree of similarity between the newly developed index of acceptable maximum effort (AME) and lifting capacity, than between MVC and lifting capacity. Static lifting tests are considered advantagous over dynamic ones because they are easier to administer, require shorter time, and are considered safer to the operator. Thus, it is suggested that AME not MVC be utilized in job design to assess the lifting capacity of workers. It is felt that the use of AME as a submaximal test will help designers of industrial tasks to have a more realistic index than MVC to assess human lifting capacities without exerting workers at maximal levels.

REFERENCES

1 S.H. Snook, Ergonomics, 21 (1978) 963-985.
2 M.M. Ayoub, N.J. Bethea, S. Deivanayagam, S.S. Asfour, G. Bakken, D. Liles, A. Mital and M. Sherif, Determination and Modeling of Lifting Capacity, HEW (NIOSH), Technical Report, Grant #5R010H00545-02.
3 D.B. Chaffin, G.D. Herrin, W.M. Keyserling and A. Garg, Am. Ind. Hyg. Assoc. J. (1977) 662-675.
4 A. Garg, A. Mital and S.S. Asfour, Ergonomics, 23 (1980) 13-27.
5 D.B. Chaffin, Am. Ind. Hyg. Assoc. J., 36 (1975) 505-511.
6 D.B. Chaffin and G. Anderson, Occupational Biomechanics, John Wiley & Sons, New York, 1984.
7 W.M. Keyserling, Isometric Strength Strength Testing in Selecting Workers for Strenuous Jobs, Unpublished Ph. D. Dissertation, The University of Michigan, Ann Arbor, Michigan, U.S.A.
8 T.M. Khalil, S.S. Asfour, S.M. Waly and A.M. Genaidy, A Comparative Study of Static and Dynamic Lifting Tasks, Proceedings of the Human Factors Society 28th Meeting, San Antonio, Texas, U.S.A., 1984, pp. 595-599.
9 A. Mital, W. Karwowski, A. Mazouz and E. Orsarh, Am. Ind. Hyg. Assoc. J., 47 (1986) 288-292.

10 W. Karwowski, A Fuzzy Sets Based Model of the Interaction Between Stresses Involved in Manual Lifting Activities, Unpubblished Ph. D. Dissertation, Texas Tech University, Lubbock, Texas, U.S.A.
11 W. Karwowski, A Pilot Study on the Interaction Between Physiological, Biomechanical, and the Psychophysical Stresses Involved in Manual Lifting Tasks, Proceedings of the Ergonomics Society's Conference, Taylor and Francis LTD, London, 1983, pp. 641-649.
12 W. Karwowski and M.M. Ayoub, Ergonomics, 27(1984) 641-649.
13 W. Karwowski, M.M. Ayoub, L.R. Alley and J.L. Smith, Fuzzy Sets and Systems: An International Journal, 14 (1984) 65-76.
14 W. Karwowski and A. Mital, Validation of a Fuzzy Model for the Assessment of Human Operator Responses to Manual Lifting Tasks, Proceedings of the Human Factors Society 28th Annual Meeting, San Antonio, Texas, U.S.A., (1984), pp. 408-412.
15 H.A. Hafez, An Investigation of Biomechanical, Physiological, and Environmental Heat Stress Associated with Manual Lifting in Hot Environments, Unpublished Ph. D. Dissertation, Texas Tech University, Lubbock, Texas, U.S.A., 1984.
16 M.M. Ayoub, H.A. Hafez and W. Karwowski, in I.D. Brown, R. Goldsmith, K. Coombes, and M. A. Sinclair (Eds.), Ergonomics International 85, Taylor and Francis, London, 1985, pp. 271-283.
17 S.S. Asfour, A.M. Genaidy, T.M.Khalil and E.C.Greco, in R.E. Eberts and C.G. Eberts (Eds.), Trends in Ergonomics/Human Factors II, Elsevier Science Publishers B.V., Amsterdam, 1984, pp. 617-623.
18 W. Karwowski and J.W. Yates, Ergonomics, 29 (1986) 237-248.
19 D. Dubois and H. Prade, Fuzzy Sets and Systems. Theory and Applications. Academic Press, New York, 1980.

Applications of Fuzzy Set Theory in Human Factors, edited by W. Karwowski and A. Mital
Elsevier Science Publishers B.V., Amsterdam, 1986 — Printed in The Netherlands

FUZZY MODELING OF COMBINED MANUAL MATERIALS-HANDLING CAPACITIES

Bernard C. Jiang[1] and James L. Smith[2]

[1]Industrial Engineering Dept., Auburn University, Auburn, Alabama 36849 (USA)

[2]Industrial Engineering Dept., Texas Tech University, Lubbock, Texas 79409 (USA)

ABSTRACT

Fuzzy set theory was applied to develop capacities models for combined manual materials-handling (MHH) activities. These models can be applied to job design/redesign to reduce the risk involved to the human operator.

Twelve male subjects were employed for this study. The capacities for four individual MMH activities (lifting from floor to knuckle height, lifting from knuckle to shoulder height, carrying for 4.3 m, and lowering from knuckle to floor height) and three combined MMH activities, which were the combination of the above four individual activities, were determined psychophysically under three frequency conditions: one time maximum, 1 handling/min, and 6 handlings/min. In all experimental conditions, the pyschophysically determined capacities were the maximum acceptable work loads for a one-hour work period.

The models were developed to combine two individual MMH capacities into a combined MMH capacity using fuzzy set theory. Membership functions and reference points were determined from the previously published literature. These models were tested from the experimental data in this study. The mean absolute error was 5.2%.

INTRODUCTION

"Automation" and "computerization" are the words for this era. The talents and challenges of human beings have combined to make machines capable of replacing manual work. However, a completely "automated world" (manual-work free) is still far beyond reality at the present time. Either economic considerations (e.g., cost for automation equipment) or practical situations (e.g., space limitation or unexpected conditions) make complete automation difficult. Manual materials handling (MMH) is still unavoidable in many work-places. Overexertion in the manual handling of work has been recognized as a major cause of back injuries (ref.1), and the corresponding expenses and the loss of productivity are overwhelming.

Over the years, a great deal of research has been conducted which has contributed to the reduction of back injuries. However, the problem of numerous back injuries and compensation costs associated with MMH is still severe. In

1979, the National Safety Council estimated that 27 percent of all industrial injuries were associated with MMH. This amounted to 590,000 injuries with attendant costs of approximately 10.4 billion dollars. In 1980, the number of injuries increased to 670,000; this, despite improved medical care, increased automation in industry and more extensive use of preemployment examinations (ref.1).

There are basically four approaches used to determine operator capacity for MMH activities: physiological, biomechanical, psychophysical and epidemiological. The merits and demerits of each approach have been compared (ref.1, ref.2). However, very little research has been reported on the determination of operator capacities for combinations of MMH activities.

The physiological approach has been used for the studies on simultaneous activities, i.e., two or more activities performed simultaneously. Whiley and Lind (ref.3), and Kibom and Brundin (ref.4) examined the respiratory and circulatory responses to simultaneous static (hand-grip) and rhythmic (bicycle ergometer) exercises. Sanchez et al. (ref.5) studied the effects of dynamic work (walking), static work (pushing, pulling or holding) and combined static and dynamic work on heart rate and oxygen consumption. A manual carrying task can be categorized as a combined activity in which walking and holding an object are combined. Gordon et al. (ref.6) studied the effects of back load carrying on heart rate and rating of perceived exertion (RPE). Morrissey and Liou (ref.7) examined the metabolic cost of load carrying using three container sizes and five treadmill speeds.

The physiological approach has also been employed in examination of sequential MMH activities, i.e., one activity performed after another. Garg (ref.8) and Garg et al. (ref.9) developed regression models to estimate metabolic energy expenditure rates for MMH tasks composed of various activity elements. These elements were sitting, standing, lifting, lowering, walking, carrying, holding, pushing and pulling. This model was based on the assumption that the net total metabolic cost of a series of activities could be estimated by summing their net steady-state individual metabolic costs as obtained from separate performances. This additivity assumption was not verified experimentally. Asfour (ref.10), based on his study, showed that this assumption was not valid for an energy cost for lifting from floor to 76.2 cm and lifting from 76.2 to 127 cm.

To date, combined MMH activities have not been studied psychophysically. Most of the previous psychophysical studies were centered on individual activities of MMH tasks only, i.e., one of lifting, lowering, carrying, holding, pushing or pulling. The combination of two or more activities has never been examined. These kinds of combined activities are prevalent in industry and in our daily lives, such as packing operations, assembly operations, and loading/unloading activities. Currently, no interaction of task activities is assumed, and the only way to analyze a combined activity is to analyze each component of the combined activity and use the most critical task as the limiting factor for the combined activity.

The objective of this study was to utilize the psychophysical approach to examine the effects of combinations of lifting, carrying and lowering activities in sequence. During the psychophysical testing, physiological response was measured by monitoring the heart rate to ensure that the subject was responding with a "reasonable" level of physical work. The relationship between combined activities and individual activities was determined by using fuzzy set theory.

METHODS AND RESULTS

Subjects

Twelve male students were selected as subjects for this study. The subjects were selected according to the height and weight criteria which constituted a representative sample of the U.S. male population. All subjects were free from back pain and no musculoskeletal abnormalities were noted. The subjects' mean height was 175.7 cm (SD=4.91 cm) and mean weight was 72.5 kg (SD=8.44 kg). They were familiarized with the experimental procedures before the experimental data were collected.

Experimental procedures and results

After a familiarization with the experimental procedure, the data base for each subject's physical condition was collected by conducting strength tests, anthropometric measurements and a physical work capacity (PWC) test. The subject's capacities for simulated MMH activities were tested psychophysically.

Anthropometric measurements and physical work capacity. Height, weight and 24 other anthropometric measurements were taken for each subject. The protocol for anthropometric measurements was adopted from Ayoub (ref.11). PWC in terms of maximum oxygen consumption ($\dot{V}O_2max$) was determined on a bicycle ergometer device using a submaximal technique. The procedure required the subjects to work at three different submaximal loads (e.g., 400, 600 and 900 kpm/min) for 4 minutes each, or until they reached steady state. Heart rate and oxygen consumption were recorded for the last minute of each load. The relationship between heart rate and oxygen consumption for each subject was then computed using the method of least squares regression. Then the PWC in L/min was estimated via extrapolation by using the predicted maximum heart rate (220 - age of the subject), and projecting for $\dot{V}O_2$ max in L/min as described by Kamon and Ayoub (ref.12).

Strength/stamina tests. A set of isoinertial strength tests was selected for this study. Isoinertial strength is a dynamic strength measurement using known weights, in which exertions are performed incrementally until the maximum strength is reached (ref.13,14). Isoinertial strength has been used as an effective predictor variable for individual MMH activities, i.e., lifting, lowering, carrying and holding (ref.15). A comparison of different types of strength has also shown that isoinertial strength is superior to isometric or isokinetic strength in terms of predictability, simplicity and task similarity (ref.16). The isoinertial strength tests conducted in this study were: lift from floor to 1.83 m height (T1), lift from floor to elbow height (T2), lift from floor to knuckle height (T3), and lift from knuckle to shoulder height (T4). An incremental weight lifting machine used by Ayoub et al. (ref.15) was modified for use in this study. The machine can measure a range of weight from 18.2 to 93.2 kg, and it can measure as precisely as 1.1 kg. It consists essentially of a carriage that a person lifts up and down within vertical guardrails. The subjects' strength data are shown in Table 1.

TABLE 1

Mean and standard deviation for strength data*

	T_1 (1.83 m max lift)	T_2 (elbow ht lift)	T_3 (Knuckle ht lift)	T_4 (K-S lift)
mean	60.05	80.14	88.82	50.00
standard deviation	10.64	11.36	7.27	9.68

*All strength units in kg.

Simulated MMH activities. Four individual MMH activities--lifting from floor to knuckle height (LFK), lifting from knuckle to shoulder height (LKS), lowering from knuckle to floor height (LOW) and carrying for 3.4 m (C), and three combined MMH activities--LFK+C (LC), LFK+C+LKS (LCL), LFK+C+LOW (LCLO). Based on a survey conducted by Drury et al. (ref.17), and further discussed by Mital and Ilango (ref.18), the most common three MMH activities in industry are lifting, lowering and carrying, in that order. These three activities were examined in this study. Combined activities were chosen for two reasons: first, they simulate industrial situations, and second, they allow for the examination of the relationship between individual and combined activities. Each activity was conducted under three frequency conditions (one time maximum, 1/min, 6/min[1]) and two replications. One time maximum capacity represented a subject's maximum voluntary capacity. One and six handlings per minute

[1]LFKM represents LFK at frequency of one time maximum, LFK1 represents LFK at frequency of 1/min, and LFK6 represents LFK at frequency of 6/min. Similar notations were used for other activities.

represented an infrequent and a frequent handling in MMH activities. For the experimental MMH activities, a box without handles was constructed using 1.3-cm plywood. The size of the box was 44.5 cm (frontal) x 29.2 cm (sagittal) x 24.8 cm (vertical).

The work duration utilized in this study consisted of a one-hour work bout rather than an eight-hour workday. The one-hour period was selected because it more closely approximates an intermittent work schedule that would be encountered when heavy MMH activities occur only occasionally, such as unloading a truck of boxed material or participating in emergency services (ref.19,20). The psychophysical test protocol was adopted from Ayoub et al. (ref.21). The subjects were not restricted to a particular lifting technique but rather utilized what is called "free-style" lifting (ref.22,23). Previous research in the field of MMH has also shown "free-style" lifting to be more appropriate than a structured lifting style (ref.15,24). The only instructions given to the subjects pertained to a specified grip position to be used on certain simulated activities, otherwise subjects were allowed to lift the load in a manner that was most comfortable to them. For the carrying activity, the subjects carried the load for a distance of 3.4 m. The walking speed was not controlled. The subjects were instructed to select their normal paces. Each subject was given approximately 20 minutes to determine the maximum load he was willing to handle, and then he handled that load for 10 minutes or until the heart rate reached steady state. Heart rate was recorded for at least 4 minutes after the steady state was reached. Heart rate was monitored by using a small heart-rate monitoring device called an "Exersentry." This device is described in detail by Smith and Wilson (ref.25). The subjects' MMH capacities and corresponding heart rate data are shown in Table 2.

MODEL DEVELOPMENT

Since body weight affects the relationship between the subject's MMH capacity and his muscle strength, the body weight was added to the MMH capacity as a dependent variable in this study. There were two major reasons for doing this. First, the body weight plays an important role in MMH activity. For example, to conduct a carrying activity, the operator must move the object as well as his own body weight in order to accomplish the task. For the other activities, such as lifting and lowering, the body weight alone generates a significant component of the MMH activity. Second, the correlations between some of the strength variables and MMH capacities were higher when the body weight was added to the MMH capacities.

The applications of fuzzy set theory in the field of MMH study have been introduced by Karwowski (ref.26,27), Karwowski and Ayoub (ref.28), Karwowski et al. (ref.29), Karwowski and Mital (ref.30), and Hafez (ref.31). The use of fuzzy set theory in determining the relationship between individual activities and combined activity is appropriate because the effect of individual activities on combined activity is imprecise (fuzzy) in nature and therefore, can be associated with fuzzy set.

TABLE 2

Mean capacity and heart rate-data

	Capacity*	Heart Rate***
LFKM	73.1(18.9)**	
LKSM	46.1(7.0)	
LOWM	80.8(21.1)	
CM	79.1(19.6)	
LCM	67.1(17.8)	
LCLOM	64.4(20.4)	
LCLM	46.0(6.8)	
LFK1	40.3(10.2)	98.3(12.4)
LKS1	30.6(4.5)	90.5(11.5)
LOW1	51.7(11.2)	90.1(8.5)
C1	43.6(7.3)	93.5(13.3)
LC1	35.7(7.5)	100.2(10.4)
LCLO1	32.5(6.1)	101.7(11.1)
LCL1	28.6(5.1)	101.0(11.2)
LFK6	26.2(5.7)	126.3(11.9)
LKS6	23.2(4.7)	108.7(11.3)
LOW6	38.8(7.6)	115.7(11.6)
C6	28.7(6.0)	115.7(14.3)
LC6	21.7(4.7)	133.9(14.2)
LCLO6	19.4(3.4)	133.7(13.5)
LCL6	19.1(3.5)	139.4(15.3)

*All capacities units in kg.
**The values in parentheses are standard deviations.
***Heart-rate data were not obtained for "one-time maximum" activities.

Modeling combined MMH capacities from individual MMH capacities using fuzzy set theory which combined two individual MMH capacities into a combined MMH capacity synergistically, rather than statistically. The procedure of applying fuzzy set theory for the modeling of combined MMH capacities from individual MMH capacities is as follows:

<u>Step 1</u> Select a membership function associated with each individual MMH capacity and each combined MMH capacity.

The membership function is a quantitative form of a subjective degree of belief. Based on the model proposed by Karwowski (ref.26), three candidate functions were considered as membership functions for psychophysical lifting capacity: a hyperbolic function, an exponential function and a power function. Through an interactive procedure, Karwowski (ref.26) reported that the hyperbolic function was the best membership function for psychophysical lifting tasks. Hafez (ref.31) adopted the same membership function for psychophysical stress when she combined physiological, biomechanical and environmental heat stresses into psychophysical stress. Therefore, the hyperbolic function is assumed to be an appropriate membership function for psychophysical stress for a lifting activity. It is further assumed that the membership function for psychophysical capacity for a carrying activity and psychophysical capacity for combined activities are also hyperbolic functions.

Let x represent a subject's MMH capacity plus his body weight. Then the membership function f(x) is represented by:

$$f(x) = 1 \quad \text{for } x \leq R, \text{ and} \qquad (1)$$
$$= R/x \text{ for } x > R$$

where R is the reference value which is described in the next step.

Step 2 Defining a reference point

To define a fuzzy set A, a membership value (defined as "acceptability") between 0 and 1 must be assigned to each object x, which represents its grade for membership in A. A reference value should be defined in order to determine that value below which an acceptability of 1 can be assigned to an individual. Therefore, the reference value indicates the weight that a person most likely can handle safely.

Four sources were used to determine a reference value in relation to recommended MMH capacity. Ayoub et al. (ref.2) provided a guideline for manual lifting tasks. This guideline, in turn, had determined a lifting capacity norm from Ayoub et al. (ref.21) and Snook (ref.32). The second source utilized was the NIOSH lifting guide (ref.1). From Snook's data (ref.32) the box size of 36.1 cm in sagittal plane was obtained, and also those capacities for 90% of the male population were adopted as reference values in this study. Finally as indicated by Ciriello and Snook (ref.33), in order to adjust for 90%-tile male population, a normal distribution was assumed based on mean and standard deviation of a certain capacity. Thus,

The capacity for 90%-tile male population
= mean capacity - 1.2816 * standard deviation. (2)

The summary of the previously mentioned four data sets which relate to the present study and to the selected reference values for this study are presented in Table 3.

TABLE 3

Summary of previously recommended MMH capacities and the selected reference points for the study

	(1) Ayoub (1983)*	(2) NIOSH (1981)	(3) Snook (1978)	(4) Ciriello (1983)	Selected Reference Point
LFKM	29.6**	22.9	29.0	40.1	Body Wt. + 27.3
CM			26.0	46.5	Body Wt. + 27.3
LFK1	20.7	21.4	18.0	24.1	Body Wt. + 18.2
C1			19.0	28.5	Body Wt. + 18.2
LFK6	17.9	13.7			Body Wt. + 13.6
C6			14.0	14.5	Body Wt. + 13.6

(1) Box size = 30.5 cm in sagittal plane; for 90% male and 8-hr duration.
(2) Action limit; for 99% male and 1-hr duration.
(3) Box size = 36.1 cm in sagittal plane; for 90% male.
(4) For 90% male; assumed normally distributed (mean cap.-1.2816*std. dev.).
* Based on frequency of 0.1 lift/min.
** All units in kg.

Step 3 Combining individual activitiy capacities into combined activity capacities was investigated.

The conjunction of two fuzzy sets A and B into a fuzzy set C was accomplished through the "connective D" (ref.32) and was expressed as:

$$D(f_A, f_B) = \frac{f_A * f_B}{r + (1-r)(f_A + f_B - f_A*f_B)} \qquad (3)$$

where f_A and f_B are the membership values that correspond to membership functions A and B, respectively, and r is an arbitrary parameter, $r \geq 0$.

The value of connective D is then the membership value for fuzzy set C. The capacity for the combined lifting-and-carrying activity can be predicted through the membership function and the predicted membership value of connective D. In the present study, an r value of 0 was found to produce the best results in terms of minimizing the Hamming distance and minimizing the dissimilarity index (DI). Because of the lack of developed methodology and the lack of previous data, the combining of three capacities, such as combined lifting, carrying and lowering, was not considered in this study.

Step 4 Testing of the model comprises the next procedure.

The closeness, or similarity, between two fuzzy sets can be measured by the dissimilarity index (DI), based on the distance between two sets, and it can be expressed as:

$$DI = \frac{\sum_{j=i}^{n} |D(x_j) - ULC(x_j)|}{n + \sum_{j=i}^{n} \min \{D(x_j), ULC(x_j)\}} \tag{4}$$

where: $D(x_j)$ = connective D; $ULC(x_j)$ = membership value of combined membership function for lifting and carrying; j = the number of element x in set X, j=1,2,...,n; n = number of elements in set X.

The two fuzzy sets are identical if the DI value equals zero. The determination of a value of DI that satisfies the similarity between two fuzzy sets should be based on the required or expected precision.

The fuzzy set-based models were also tested for the mean absolute percent error of predicted capacity. Mean absolute percent error was computed as the difference between the capacity predicted from the fuzzy sets model and the observed capacity from this experiment. The summary of model validation in terms of Hamming distance, DI and mean absolute percent error is presented in Table 4.

TABLE 4
Hamming distance, dissimilarity index (DI) and mean absolute percentage error for fuzzy set models

	Hamming Distance	DI	Mean Absolute % Error
LFKM + CM --> LCM	0.4026	0.0223	6.345
LFK1 + C1 --> LC1	0.3238	0.0163	4.208
LFK6 + C6 --> LC6	0.4442	0.0246	5.204

DISCUSSION

Fuzzy set-based models are appealing because they are data-set independent. In other words, these models were not developed from the experimental sample in the study. These models were validated from the present sample as acceptable models from the predictability point of view. Therefore, the application of these models was not subject to restrictions of the range of predictions or the small sample size as in a regression model. These models can be applied to job design/redesign. Given individual lifting and individual carrying capacities under certain frequencies (one time max, 1/min, or 6/min), the capacity for a combined MMH activity can be determined through fuzzy set models.

Fuzzy set models have two major disadvantages. First, the determination of the membership function for psychophysical capacity as a hyperbolic function was based on previous studies (ref.26,27,31). Although it is a reasonable approach, the foundation to determine the membership function can be improved if there is a methodology available. Second, the reference values for the membership functions were determined according to methodologies established in the previous studies. Because of the discrepancies in the previously recommended data, relatively conservative values have been selected as reference values. More reliable reference values can be obtained when more psychophysical capacity data are generated in the future.

Several carrying activities were timed with a stopwatch. The walking speed had a range of 3.44 to 4.26 km/hr with an average speed of 3.73 km/hr. Most of the past studies on MMH of carrying activities had the subject walking on a treadmill (ref.7,36,37) at a controlled speed. Mital and Ilango (ref.18) conducted a psychophysical study on load carriage. They reported that the subjects were trained to walk at a speed of 4.8 km/hr. However, the actual walking speed was not reported.

For a certain handling frequency, the difference in the length of the resting periods for the individual and the combined MMH activities was obvious. This difference did not affect the capacity for the frequency of one time maximum; however, it became very significant for the frequency of 6/min. It was apparent that, for the frequency of 6/min, a combined activity had very little resting time. Consequently, there was very little time for physiological recovery of muscle fatigue. The subjects also made similar comments on this continuous type of work and mentioned the lack of sufficient recovery time. They felt that it was difficult to adjust the weight in the box so that they could keep a balanced physiological condition. They commented also that it was more stressful doing a combined activity than doing an individual activity at a frequency of 6/min. This situation was reflected in heart-rate readings in which the average heart rate for combined activities was higher than those for individual activities at the same frequency. Physiological limitation plays a more important role in conducting combined activities at a faster pace. It is apparent that strength and physiological capability have a mixed influence on MMH capacities. For low frequencies of handling, such as one time maximum or 0.1 handling/min., the subject's strength is the dominant factor because of the complete physiological recovery obtained from the long resting period. As the frequency becomes faster, the physiological capability is more important. Fortunately, the psychophysical approach used in this study assumed that the subject integrated all the stresses and responded with the weight he was willing to handle. The effects of strength and physiological capabilities on MMH capacities under different frequencies were not examined in this study; however, the effects merit future study.

REFERENCES

1 NIOSH, U.S. Department of Health and Human Services, Center for Disease Control, Work Practices Guide for Manual Lifting, Publication No. 81-122, 1981.

2 M.M. Ayoub, C.F. Gidcumb, H. Hafez, K. Intaranont, B.C. Jiang and J.L. Selan, A Design Guide for Manual Lifting. A Draft Report Submitted to OSHA, 1983.

3 R.W. Whiley and A.R. Lind, Respiratory Responses to Simultaneous Static and Rhythmic Exercises in Humans. Clinical Science and Muscular Medicine, 49 (1975) 427-432.

4 A. Kibom and T. Brundin, Circulatory Effects of Isometric Muscle Contractions Performed Separately and in Combination with Dynamic Exercise, European Journal of Applied Physiology, 36 (1976) 7-17.

5 J. Sanchez, H. Monod and F. Chabaud, Effects of Dynamic, Static and Combined Work on Heart Rate and Oxygen Consumption, Ergonomics, 22 (1979) 935-943.

6 M.J. Gordon, B.R. Goslin, T. Graham and J. Hoare, Comparison Between Load Carriage and Grade Walking on a Treadmill, Ergonomics, 26 (1983) 289-298.

7 S.J. Morrissey and Y.H. Liou, Metabolic Costs of Load Carriage with Different Container Sizes, Ergonomics, 27 (1984) 847-853.

8 A. Garg, A Metabolic Prediction Model for Manual Materials Handling Jobs, Ph.D. Dissertation, University of Michigan, 1976.

9 A. Garg, D.B. Chaffin and G.D. Herrin, Prediction of Metabolic Rates for Manual Materials Handling Jobs, American Industrial Hygiene Association Journal, 1978, pp. 661-674.

10 S.S. Asfour, Energy Cost Prediction Models for Lifting and Lowering Tasks, Ph.D. Dissertation, Texas Tech University, 1980.

11 M.M. Ayoub, Reference Manual for an Anthropometric Survey. Anthropology Research Project, Inc., 503 Xenia Avenue, Yellow Springs, Ohio 45387, 1980.

12 E. Kamon and M.M. Ayoub, Ergonomics Guides to Assessment of Physical Work Capacity, Published by American Industrial Hygiene Association, 1976.

13 K.H.E. Kroemer, An Isoinertial Technique to Assess Individual Lifting Capability, Human Factors, 25 (1983) 493-506.

14 M.M. Ayoub, J.L. Selan and D.H. Liles, An Ergonomics Approach for the Design of Manual Materials-Handling Tasks, Human Factors, 25 (1983) 507-515.

15 M.M. Ayoub, J.D. Denardo, J.L. Smith, N.J. Bethea, B.K. Lambert, L.R. Alley, and B.S. Duran, Establishing Physical Criteria for Assigning Personnel to Air Force Jobs, Final Report, Air Force Office of Scientific Research, Contract No. F49620-79-C-006, 1982.

16 B.C. Jiang, Psychophysical Capacity Modeling of Individual and Combined Manual Materials Handling Activities, Ph.D. Dissertation, Texas Tech University, 1984.

17 C.G. Drury, C.H. Law and C.S. Pawenski, A Survey of Industrial Box Handling, Human Factors, 24 (1982) 553-565.

18 A. Mital and M. Ilango, Load Characteristics and Manual Carrying Capacities, In Proceedings of the Human Factors Society Conference 27th, Norfork, VA, 1983, pp. 274-278.

19 J.L. Smith and B.C. Jiang, A Manual Materials Handling Study of Bag Lifting, American Industrial Hygiene Association Journal, 45 (1984) 505-508.

20 S.J. Legg and C.M. Pateman, A Pyhsiological Study of the Repetitive Lifting Capabilities of Healthy Young Males, Ergonomics, 27 (1984) 259-272.
21 M.M. Ayoub, N.J. Bethea, S. Deivanayagam, S.S. Asfour, G.M. Bakken, D. Liles, A. Mital and M. Sherif, Determination and Modeling of Lifting Capacity. Final Report, HEW (NIOSH) Grant No. 5R01OH-00545-02, 1978.
22 D.F. Jones, Back Injury Research: A Common Thread, American Industrial Hygiene Association Journal, 1972, pp.596-602
23 J.R. Brown, Lifting as an Industrial Hazard, American Industrial Hygiene Association Journal, 34 (1973) 292-297.
24 G.M. Bakken, Lifting Capacity Determination as a Function of Task Variables, Ph.D. Dissertation, Texas Tech University, 1983.
25 L.A. Smith and D. Wilson, An Inexpensive System for Remote Monitoring of Work Heart Rate, Ergonomics, 26 (1983) 719-722.
26 W. Karwowski, A Fuzzy Set Based Model on the Interaction Between Stresses Involved in Manual Lifting Tasks, Ph.D. Dissertation, Texas Tech University, 1982.
27 W. Karwowski, A Pilot Study of the Interaction Between Physiological, Biomechanical, and the Psychophysical Stresses Involved in Manual Lifting Task, Proceedings of the Ergonomics Society's Annual Conference, Taylor and Frances London, 1983, pp. 95-100.
28 W. Karwowski and M.M. Ayoub, Fuzzy Modelling of Stresses in Manual Lifting Tasks, Ergonomics, 27 (1984) 641-649.
29 W. Karwowski, M.M. Ayoub, L.R. Alley and J.L. Smith, Fuzzy Approach in Psychophysical Modeling of Human Operator - Manual Lifting Tasks, Fuzzy Sets and Systems, 14 (1984) 65-76.
30 W. Karwowski and A. Mital, Validation of Fuzzy Model for the Assessment of Human Operator Reponses to Manual Lifting Tasks, Proceedings of the 28th Annual Conference of the Human Factors Society, San Antonio, 1984, pp. 408-412.
31 H.A. Hafez, An Investigation of Biomechanical, Physiological, and Environmental Heat Stresses Associated with Manual Lifting in Hot Environments, Ph.D. Dissertation, Texas Tech University, 1984.
32 S.H. Snook, The Design of Manual Handling Tasks, Ergonomics, 21 (1978) 963-985.
33 V.M. Ciriello and S.H. Snook, A Study of Size, Distance, Height, and Frequency Effects on Manual Handling Tasks, Human Factors, 25 (1983) 473-483
34 H.J. Zimmermann, Fuzzy Programming and Linear Programming with Several Objective Functions, Fuzzy Sets and Systems, 1 (1978) 45-55.
35 V. Dimitrov and O. Cuntchew, Efficient Governing Humanistic Systems by Fuzzy Instructions, In Modern Trends in Cybernetics and Systems (edited by J. Rose and C. Bilciu), Berlin, Heidelberg, New York: Springer-Verlag, 1977, pp. 125-130.
36 B. Givoni and R.F. Goldman, Predicting Metabolic Energy Cost, Journal of Applied Physiology, 30 (1971) 429-433.
37 K.B. Pandolf, B. Givoni and R.F. Goldman, Predicting Energy Expenditure with Loads While Standing or Walking Very Slowly, Journal of Applied Physiology, 43 (1977) 577-581.

Acknowledgment: The authors of this paper wish to thank Angeline Honnell and Barabara Wilson for their efforts in the preparation of this paper.

Applications of Fuzzy Set Theory in Human Factors, edited by W. Karwowski and A. Mital
Elsevier Science Publishers B.V., Amsterdam, 1986 — Printed in The Netherlands

COMBINING STRESS FOR MANUAL MATERIALS HANDLING

M.M. AYOUB[1] and HALA HAFEZ[2]

[1]Industrial Engr. Dept., P.O. Box 4130, Texas Tech University, Lubbock, TX

[2]Engineering Production Dept., Ain Shams University, Cairo, Egypt

ABSTRACT

A wide variety of manual materials handling is performed in hot environments, however, little data exist on the effects of hot environments on lifting capacity. This study was conducted to assess lifting capacity in hot environments, investigate the stresses imposed on the worker under these conditions, a fuzzy model was developed which combined the biomechanical, physiological and thermal stresses. The model was used to test the hypothesis that the combination of the acceptability of these stresses leads to an overall measure of the lifting task acceptability. The results showed that head stress significantly affects the capacity to lift. The fuzzy set model showed that the thermal stress may not be adequately expressed independently when combining these three stresses.

INTRODUCTION

The National Institute for Occupational Safety and Health (ref.1) reports that overexertion in the workplace accounts for a large number, and in some industries the majority, of disabling injuries. Most of these injuries involve the act of manually handling materials. The relative severity of materials handling operations can be evaluated only when all elements of a lifting task are considered together as an integral set.

Most of the previous studies have investigated man's capacity to perform manual handling tasks in a moderate environment. The effect of the physical environment on such tasks has not been sufficiently investigated. Numerous attempts have been made to evaluate the physical stresses of manual lifting in nuetral thermal environments. Recommendations of the maximum loads that individuals should be permitted to lift have been based on four approaches: (1) the epidemiological, (2) the biomechanical, (3) the physiological, and (4) the psychophysical.

The contribution of environmental thermal stresses to the stresses associated with manual lifting activities has seldom been considered. These stresses are very common in many industries and may have a very significant effect on the difficulty of lifting tasks and must be considered as an integral part of the lifting process. Therefore a study was conducted with the following objectives.

1. The assessment of the lifting capabilities of individuals in hot environments, using the psychophysical approach, 2. The determination of the effect of heat load on the compensation made by individuals in their metabolic heat production by reducing the work load, 3. The investigation of the physiological responses of oxygen consumption, heart rate, and body core temperature to lifting, while using the psychophysical approach, 4. The use of the fuzzy sets theory to predict the lifting task acceptability from the combination of the acceptability of the biomechanical, physiological, and thermal stresses, and 5. The development of lifting capacity norms based on the acceptability measure of the combined stress.

METHODS

Six male college students ranging in age from 18 to 27 years participated in a laboratory study. After the subjects underwent a physical examination, they were acclimatized for ten consecutive days using a modified version of the training program reported by Asfour and Ayoub (ref.2). The independent variables used for this study were: (1) the environmental temperature; and (2) frequency of lift. The temperature levels were 22, 27, and 32°C WBGT. Three frequencies of lift were used: .1, 3, and 6 lifts per minute. The dependent variables were: (1) the maximum acceptable weight of lift; and (2) several physiological responses, mainly oxygen consumption, heart rate and rectal temperature. The task consisted of lifting a square tote box with handles 38 cm in length and 25 cm in height from floor to knuckle height using a natural free style lifting technique. The maximum acceptable weight of lift was determined psychophysically.

Three fixed temperature levels and three fixed frequencies were used in this study giving nine basic treatment combinations. Six subjects participated and the experiment was repeated twice (9 x 6 x 2 = 108 runs). Due to the nature of this experiment, it was impractical as well as time consuming to completely randomize the order of experimentation for each subject. Once the environmental chamber was heated to a certain temperature, it required a relatively long time to change to another temperature level. Also, since running each of the psychophysical experiments required a long time (30 minute rest period prior to lifting, 47 minutes or 2 hours 45 minutes of actual lifting depending on the frequency, and an additional 3 minute recovery period in the case of lifting for 47 minutes), it would not have been possible to run the experiment for every subject every day if the additional time for heating the chamber to different temperature levels was included.

The inability to test each subject every day would have presented a problem since the subjects were heat acclimatized prior to the lifting experiment, as discussed earlier, and it was preferrable to expose the subjects to heat every day lest their acclimatization be affected. Furthermore, it was essential to test each subject at basically the same time every day in order to reduce the variations in his physiological functions, mainly body temperature, as a result of time of day.

Therefore, once a temperature was set, all six subjects were exposed to that temperature in a specified order. For each subject, a random order of frequencies was chosen. Another temperature level was then randomly selected and the response again determined from each of the six subjects at all levels of randomly selected frequencies. There were, therefore, two restrictions on randomization since the temperature was chosen, then the subject and then the frequencies were randomized at that particular temperature and subject. This design was then a split split plot design (ref.3, ref.4).

RESULTS

The anova procedure of the Statistical Analysis System Software package (SAS82.3) was used to perform the analyses of variance. Separate analysis were performed for weight selected, O_2 consumption, working body temperature, working heart rate, cardiac cost, sweat rate. Table 1 shows the results of the anova tests. Table 2 shows the overall responses of the experimental lifting tasks. Figures 1, 2, and 3 show the effects of temperature, frequency of lift, temperature x frequency interaction, on the weight selected and oxygen consumption.

Figure 1 shows the effects of temperature and frequency by temperature interaction. the temperature of 22° and 27°C did not significantly affect the psychophysically selected weight of lift (max acceptable weight of lift). However the weight of lift at 32°C was significantly different. Figure 2 shows the classical effect of frequency of lift on the weight of lift. The three frequencies of lift, .1, 3, and 6 lifts/min have significantly different effects on the weight lifted. From Figure 3, it can be seen that the 22°C WBGT, and 27°C WBGT temperature did not significantly differ on their effect on VO_2, however, the 32°C WBGT differed significantly. This seems to follow a similar response to the weight lifted. The effect of frequency of lift on VO_2 is shown in Figure 4 and the response as expected is linear and significantly different at all three frequencies. Table 2 shows a brief summary of the experimental results for all dependent variables used in the experiment.

FUZZY SET MODEL

In addition to investigation of the effects of heat stress on manual lifting, another main aim was the development of a fuzzy sets model to combine the biomechanical, physiological, and thermal stress. The combined stress expressed as a membership function is assumed to be equivalent to the psychophysical membership function (ref.5). In effect, it is hypothesized that under the psychophysical approach the body will combine the stresses imposed,

TABLE 1

Significant main effects and interactions

Dependent Variable	Independent Variable and Interaction
Weight Selected	Temp, SN***, FQ, Temp x freq SN x fQ
O_2 Consumption	Temp**, SN, FQ
Resting Body Temp	Temp, SN
Working Body Temp	Temp, SN*, FQ
Working HR	Temp, SN, FQ
Cardiac Cost	SN, Temp x Req*
Sweat Rate	Temp, FQ

**Significant at the .01 level, otherwise significant at the .05 level

TABLE 2

Average responses of the different lifting tasks

	Temperature in °C WBGT								
	22			27			32		
Freq. (lifts/min)	0.1	3	6	0.1	3	6	0.1	3	6
Weight (lbs)	89.54	56.54	44.92	94.83	53.38	41.46	85.17	46.21	35.42
VO_2 (l/min)	0.455	0.805	1.037	0.431	0.749	1.012	0.465	0.811	1.128
TCRest (°C)	37.13	37.05	37.19	37.19	37.28	37.30	37.25	37.35	37.27
HRRest (bpm)	69	67	70	73	74	75	78	77	78
TCWork (°C)	37.19	37.48	37.60	37.35	37.57	37.65	37.59	37.71	37.79
HRWork (bpm)	79	100	105	86	101	108	89	106	119
CCost (bpm)	10	33	35	13	27	33	11	29	41
SWRate (lb/hr)	0.154	0.458	0.446	0.473	0.740	0.783	0.641	1.083	1.262
Temp. Rating	3.5	3.8	3.9	4.9	5.1	5.6	6.1	6.2	6.7

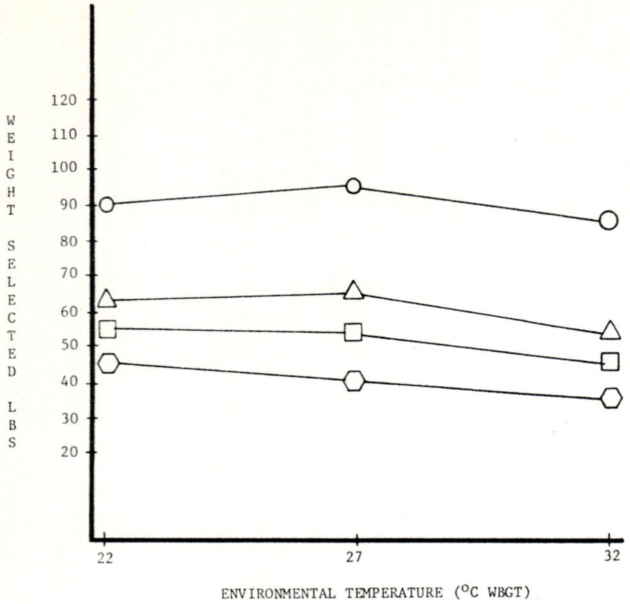

Figure 1. Effect of temperature by frequency interaction on weight selected

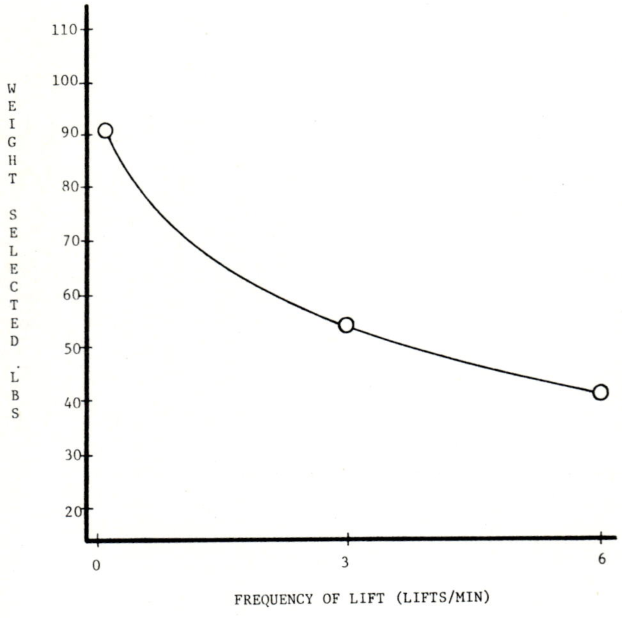

Figure 2. Effect of frequency on weight selected

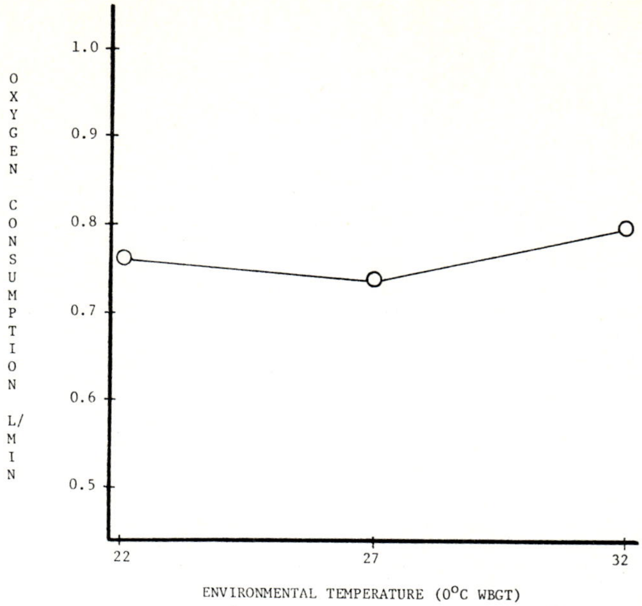

Figure 3. Effect of Temperature on oxygen consumption

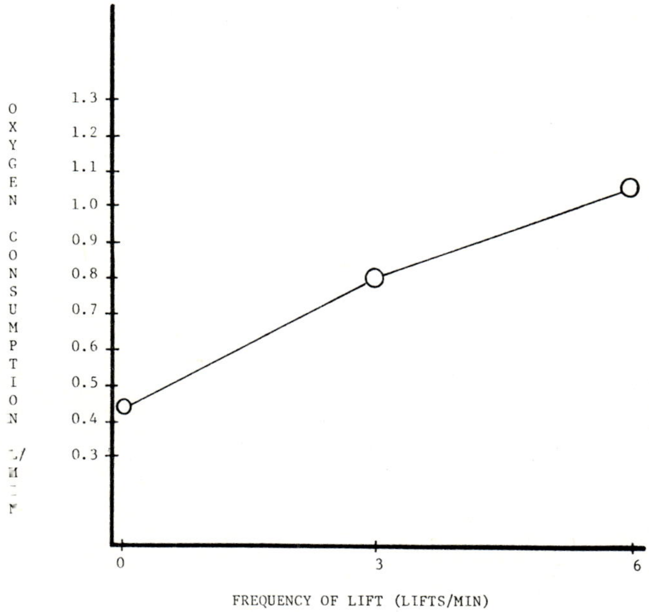

Figure 4. Effect of frequency on oxygen consumption

and based on the perceived exertion, the subject will arrive to the maximum acceptable weight of lift. Therefore, the hypothesis is accepted if the combined membership function is equivalent to the psychophysical membership function (ref.5, ref.6).

In order to test this hypothesis, it was necessary to determine the forms of the membership function and the reference points for each stress component. Based on the results of previous work, the forms of the membership functions used were (ref.6):

Biomechanical:

$f_B(x) = 1$ for $t \leq b$,

and $f_B(x) = b/t$ for $t > b$,

Physiological:

$f_H = 1$ for $v \leq h$,

and $f_H = \exp(k(h-v)/h)$,

$k > 0$ for $v > h$.

Thermal Stress:

$f_T = 1$ for $W \leq C$

$f_T = a - b\, w$ for $W > C$

OR

$f_T = 1$ for $w \leq c$,

and $f_T = c/w$ for $w > c$

Psychophysical:

$f_p(x) = 1$ for $x \leq p$,

and $f_p(x) = p/x$ for $x > p$.

P was based on psychophysical data from Ayoub, 1980.

The model proposed by Karwowski (ref.6) was modified in order to incorporate the third stress considered in this study, viz, the thermal stress. The SUMMARY procedure of the Statistical Analysis System software package (SAS82.3) was the basic procedure used throughout the modification phase of the model.

The first attempt was to use the linear membership function to the thermal stress. According to this function the acceptability of the thermal stress at 22°C WBGT, regardless of the frequency or work load, was equal to 1.0, at 27°C WBGT the acceptability was equal to 0.44. The Hamming distances thus obtained between the grades of acceptability of the combined and psychophysical stresses at the different frequencies and temperatures as a result of using a linear membership function were too large. This indicated that the thermal stress expressed in this way was too severe and resulted in acceptability values of the combined stress that were too low. The fact that the statistical analysis of the results indicated that there were no significant differences between the weight selected at 22 and 27°C WBGT and between oxygen consumption at 22 and 27°C WBGT, and that the values of the response variables at 32°C WBGT were only slightly different than those at 22°C WBGT, suggested that thermal stress may not have been a contributor to the total stress experienced at 27°C WBGT and contributed only mildly at 32°C WBGT.

The linear thermal membership function was thus discarded and the hyperbolic membership function examined. According to this function, the acceptability measures of 22, 27 and 32°C WBGT were 1.0, 0.87 and 0.74, respectively. Despite the fact that the results were considerably improved over the results using a linear membership function, the Hamming distances between the acceptability measures of the combined and psychophysical stresses were still too large.

At this point, two approaches to deal with the problem were developed. One of these approaches was to assume that thermal stress was confounded with the physiological stress and that it could not be assessed independently, which could be justified by the fact that temperature had a significant effect on oxygen consumption. The second approach was to examine other methods for combining the individual stresses. Because of the limited scope of this paper, only the first approach will be presented.

Because of these experimental results, the thermal stress was assumed to be confounded with physiological stress and was, therefore, excluded from the analysis. The original model proposed by Karwowski (ref.6), using the algebraic product as the combining operator between the grades of acceptability of the biomechanical and physiological stresses, was utilized. The application of the model resulted in values of the mean Hamming distances between the acceptability measures of the combined and psychophysical stresses that fell within the range of values reported by Karwowski (ref.6) or were slightly larger (at the 6 lifts/min frequency). This meant that no special modification of the combining process was made, and that the grades of acceptability of the biomechanical and physiological stresses contributed directly to the resulting values of the combined stress.

Table 3 gives a summary of the conditions chosen for the model together with the resulting mean distances between the acceptability measures of the combined and psychophysical stresses at the different frequencies and temperatures and the values of the dissimilarity index (DI).

In this study, it was hypothesized that a combination of the acceptability of the biomechanical, physiological and thermal stresses leads to an overall measure of the lifting task acceptability, expressed by the acceptability of the psychophysical stress. It was demonstrated, however, that thermal stress does not contribute directly to the combined stress, but rather indirectly as a component of the physiological stress.

To measure the similarity between the acceptability of the combined and psychophysical stresses, the dissimilarity index (DI), representing a fuzzy measure of similarity between the two fuzzy sets, was utilized. The resulting values of this index are given in Table 3.

Based on the definition of similarity between two fuzzy sets, it appears that a threshold values equal to 0.07 satisfies the similarity requirements with respect to acceptability of the load lifted. For such a threshold value, it may be said that the combined fuzzy set and the psychophysical fuzzy set are

TABLE 3

Reference points for the model and the resulting similarity measures

Frequency lifts/min	Temperature °C WBGT	Parameters b	h	k	Mean Distance	DI
0.1	22	420	0.5	0.3	0.0464	0.0292
	27	420	0.5	0.3	0.0375	0.0239
	32	420	0.5	0.3	0.0503	0.0317
3	22	400	0.5	0.3	0.0631	0.0406
	27	400	0.5	0.3	0.0550	0.0349
	32	400	0.5	0.3	0.0643	0.0408
6	22	380	1.0	0.5	0.0520	0.0318
	27	380	1.0	0.5	0.0915	0.0560
	32	380	1.0	0.5	0.1110	0.0684

sufficiently close to one another, and since the term sufficiently is fuzzy, it may be said that within the level of precision expected in this type of experiment, these two fuzzy concepts are identical (ref.7).

The model allows the design of lifting tasks based on the acceptability of the combined effect of the biomechanical, physiological and thermal stresses. The accuracy of such procedures with respect to the task and environmental variables used in this study may be assessed through the differences between the reference points for the capacity norms based on the psychophysical methodology and those based on the acceptability of the combined stress. Those differences were expressed in this study as the mean arithmetic differences between the reference points in pounds, and the percentage differences between those points. Table 4 gives the values computed for the differences.

An example of the practical use of such a model would involve assuming a certain weight for the task. For a particular individual, the biomechanical stress could be predicted using the equations developed by El-Bassoussi (ref.8) knowing his characteristic lifting technique (squat or stoop). In a moderate environment, oxygen consumption could either be measured or predicted using equations already developed (ref.9, ref.10). In the hot environments, since no

TABLE 4

Differences between the combined and psychophysical capacity norms

Frequency of Lift (lifts/min)	Temperature Mean (°C WBGT)	Difference (lbs)	Percentage Error
0.1	22	3.346	5.99
	27	1.290	2.31
	32	0.776	1.50
3	22	2.514	7.36
	27	0.699	2.18
	32	0.176	0.64
6	22	-0.813	-2.82
	27	-1.821	-6.87
	32	-1.810	-8.06

models are available for the prediction of oxygen consumption, it would be necessary to measure oxygen consumption. the hypothetical capacity norm, that is the maximum acceptable weight of lift anticipated for the task, can be determined by multiplying the weight initially assumed for the task by the resulting acceptability measures of the combined stress.

Figures 5 and 6 graphically illustrates the results of the model. Figure 5 shows the membership functions and the reference values for both the biomechanical and physiological fuzzy sets. Figure 6 graphically illustrates the membership functions and reference points of the psychophysical fuzzy sets. For example, if a 56 lb (25 Kg) weight is lifted at a frequency of 3 lifts/minute in a 27°C WBGT environment, it will generate a compressive force of approximately 560 Kg (ref.8). Therefore from Figure 5, the acceptability of the biomechanical stress is found to be approximately 0.72. For this load and frequency of lift, the VO_2 is expected to be .81 liters/minute (from the experimental data). From Figure 5, the acceptability of the physiological stress can be found to be .82. Using the fuzzy set model the combined stress is the product of the two acceptability measures. Therefore, the combined acceptability measure is .58 (.72 x .82). Using Figure 6 it can be seen that a 56 lb (25 Kg)

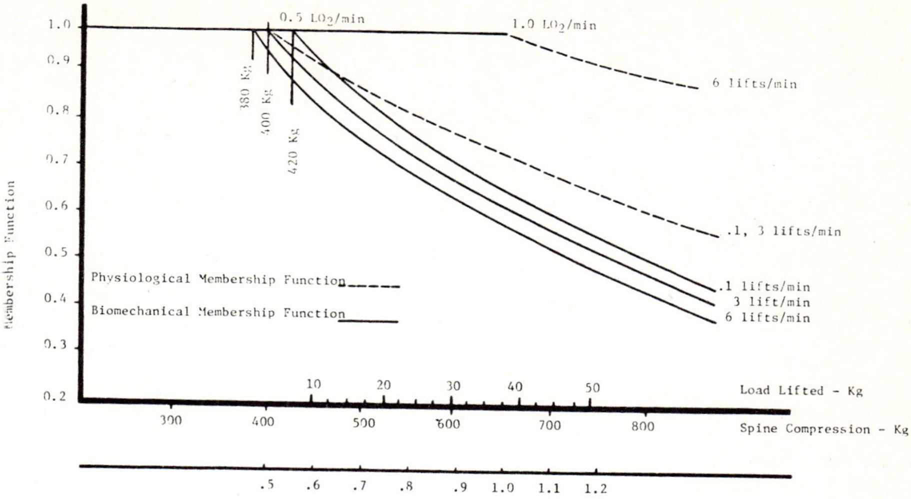

Figure 5. Biomechanical and physiological functions

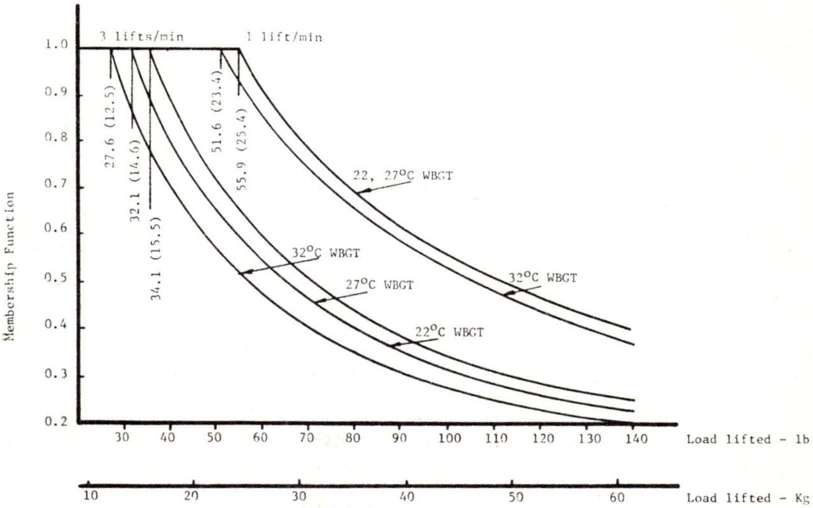

Figure 6. Psychophysical membership function

weight will have an acceptability measure of approximately .58 using a 3 lift/min in a 27°C WBGT environment. To obtain capacity norms using this technique it would be necessary to obtain the product of the used or assigned weight (56 lb or 25 Kg) and the acceptability measure (.58). Therefore the capacity norm for 3 lifts/min for floor to knuckle height in a 27° WBGT environment is 32.5 lb or 14.6 Kg. This is the same as the reference value in Figure 6 for 3 lifts/min and 27°C WBGT environment.

DISCUSSION ON FUZZY SETS MODEL

The results of the study show that the weights selected became lighter, i.e., approached the psychophysical reference weights, the model produced combined stresses which were considerably less acceptable than the psychophysical stresses for all the frequencies and temperatures investigated in this study. This observation was more obvious at the higher frequencies of lift since the actual weights selected tended to approach the psychophysical reference weights and were, therefore, highly acceptable from the psychophysical standpoint.

The relationship between weight selected and the distances between the acceptability measures of the combined and psychophysical stresses reveal that the errors or distances are not randomly distributed. There seems to be a generalized trend of decrease of the distance as the weight selected increased. It appears that when the measures of acceptability of the physiological stress are close to total accpetability, resulting in a flatter relationship, this trend of error distribution becomes more noticeable. Furthermore, by examining the distribution of the acceptability of combined and psychophysical stresses tended to follow the distribution of the measures of acceptability of the biomechanical stress versus weight selected. It therefore apears that, since the biomechanical stress, based on the membership functions used in this study, possesses lower acceptability than the physiological stress, the resulting distances between the acceptability measures of the combined and psychophysical stresses at the different weights selected follow the form of the biomechanical stress membership at the different weights. It may be argued that the biomechanical stress at the higher frequencies should not have such a strong influence on the acceptability of combined or overall stress since the weights lifted at those frequencies are considerably light and, therefore, should possess high acceptability from the biomechanical standpoint. However, it must be remembered that, like physiological stress, a major component of the biomechanical stress is due to the body weight which intuitively is expected to be more severe as the task is repeated more frequently.

The results of this experiment do not suggest that a perfect model has been developed, but that this is another step toward the development of a general model which may accommodate different types of stresses to which a worker is subjected while performing a lifting task in order to predict the overall task acceptability without the need for expensive and time consuming psychophysical experiments.

With the limited information available in this domain, it was possible to demonstrate that the model can predict reasonable values for the overall task acceptability expressed by the acceptability of the psychophysical stress. Nevertheless, several points need to be further investigated such as the reference points for the biomechanical, physiological and psychophysical stresses under heat stress, the membership functions for the different stresses under heat, the possibility of expressing the physiological and thermal stresses as independent attributes, the combining operator for the stresses, and whether it is reasonable to assume that the biomechanical and physiological fuzzy subsets have equally weighted characteristics.

REFERENCES

1 NIOSH. Criteria for a Recommended Standard...Occupational Exposure to Hot Environments. HSM 72-10629, Washington, DC: Government Printing Office, 1972.
2 S.S. Asfour and M.M. Ayoub. Effects of Training on Manual Lifting. Paper presented at the American Industrial Hygiene Association Conference, Houston, Texas, 1980.
3 C.R. Hicks. Fundamental Concepts in the Design of Experiments. Holt, Rinehart and Winston, Inc., New York, 1973.
4 D.C. Montgomery. Design and Analysis of Experiments. John Wiley and Sons, New York, 1976.
5 Hala A. Hafez. An Investigation of Biomechanical, Physiological, and Environmental Heat Stresses Associated with Manual Lifting in Hot Environments. Ph.D. Dissertation (unpublished), Texas Tech University, Lubbock, Texas, 1984.

6 W. Karwowski. A Fuzzy Sets Based Model of the Interaction Between Stresses Involved in Manual Lifting Tasks. Ph.D. Dissertation (unpublished), Texas Tech University, Lubbock, Texas, 1982.
7 M. Nowakowska. Methodological problems of Measurement of Fuzzy Concepts in the Social Sciences. Behavioral Science, 22, 107-115, 1977.
8 M.M. El-Bassoussi. A Biomechanical Model for Lifting in the Sagittal Plane. Ph.D. Dissertation (unpublished), Texas Tech University, Lubbock, Texas, 1974.
9 S.S. Asfour. Energy Cost Prediction Models for Manual Lifting and Lowering Tasks. Ph.D. Dissertation (unpublished), Texas Tech University, Lubbock, Texas, 1980.
10 A. Garg, D.B. Chaffin and G.D. Herrin. Prediction of Metabolic Rates for Manual Materials Handling Jobs. American Industrial Hygiene Association Journal, 39(8), 661-674, 1978.

APPENDIX I.

Applications of Fuzzy Set Theory in Human Factors, edited by W. Karwowski and A. Mital
Elsevier Science Publishers B.V., Amsterdam, 1986 — Printed in The Netherlands

APPLICATION OF FUZZY SET THEORY IN HEALTH CARE SYSTEMS

SU ZAIFU[1], SUN HONGYUAN[2], LI YANJIE[3] and WANG FU[4]

[1]Department of Physics, Tongji Medical University, Wuhan, China
[2]Space Medico-Engineering Institute, Peijing, China
[3]Harbin Teacher's Institute, Harbin, China
[4]Harbin Shipbuilding Engineering Institute, Harbin, China

ABSTRACT

This article presents two different applications of fuzzy set theory in health care systems. The first is an estimation of the grade of pain under acupuncture anesthesia. The second is the evaluation of cardiac function. Within the discussed areas, human subjectivity plays an important role. The applications demonstrate the utility and advantages of fuzzy set theory as a tool in problem solving.

INTRODUCTION

In the People's Republic of China fuzzy set theory has been applied extensively in Chinese traditional medicine, western medicine, acupuncture anesthesia and other health areas (ref.1-6). In some medical colleges, fuzzy set theory is being formally taught to graduate students. One of the reasons for such wide spread application of fuzzy set theory is the realilzation that subjectivity and experience play an important role in medicine. Examples of this include: analysing lab test results; specifying methods of treatment, and even in the diagnosis of disease. A disease such as cardiac coronary or ischemic stroke, seriously threaten the life of the patient. Physicians are puzzled over fore-runners through which some information can be sensed but can't be described precisely, because normal and abnormal lab test results are not sharply defined. Methods used to diagnose such diseases have been proposed by Wang (ref.5) and Le (ref.4).

In Chinese traditional medicine (ref.2,6), information is obtained from a patient by the so-called Four Diagnosis Method (the disease is diagnosed by inspection, ausculation of olfaction, interview, and monitoring the pulse). This information is in the form of verbal descriptors rather than in numbers. Since the verbal descriptors are often vague, fuzzy set theory provides a satisfactory solution. Two examples that demonstrate the application of fuzzy set theory in medical settings are presented.

EXAMPLE 1

Evaluation of anesthetic availability while operating under acupuncture anesthesia

In order to evaluate anaesthestic availability while operating under acupuncture anesthesia, the grade of pain must be estimated first. The present determination availability is based on "Standard of grade of operation under acupuncture anesthesia." According to this standard, the pain in any operation can be classified as: Grade I--no appearance of pain; Grade II--accidental appearance of pain; Grade III--marked pain; Grade IV--distinct appearance of pain and acupuncture anesthesia should be given up. Since this classification of pain is fuzzy, it is possible to use possibility theory which is based on fuzzy set theory to determine the anesthetic availability.

The following step-by-step procedure was applied in order to estimate the anesthetic availability while operating under acupuncture anesthesia.

1. Selection of a numerical attribute which expresses the anesthetic availability. This attribute was defined as the Pain-Index(PI).

2. Classification of related operations under acupuncture anesthesia by dividing them into four graded groupings (Grade-I,...).

3. Drawing up the possibility distribution based on the histogram, of PI.

4. Assignment of qualification which is relevant to the distribution of possibility suggested by (3), and letting the linguistic representation be in agreement with the value of α-possibility. This was used as standard of grading. The grade with the largest possiblity was taken as the grade of estimation.

The operating process, under acupuncture anesthesia, was divided into several intervals with the help of consultation with anethesiologist and the surgeon. The time and the pain response of each interval were carefully recorded. The pain responses were divided into four grades. These grades and a brief description of each follows:

"0" no pain.

"+" slight pain (the appearance of frown, moan, light perspiration, with stable blood pressure (BP) and pulse; the increase in BP and pulse was less than 20 mmHg and 20 beats per minute, respectively).

"++" mild pain (patient holding breath, complaints of pain, making a fist, moving limbs, perspire with fluctuating blood pressure and pulse. The increase in BP was between 20 and 30 mmHg and the increase in pulse between 20 and 30 beats per minute).

"+++" marked pain (patient clenching teeth, loud complaints of pain, heavy perspiration, restlessness and loss of coordination, marked fluctuations in BP and pulse. The respective increases in BP and pulse was more than over 30 mmHg and 30 beats per minute.

Table 1, for example, shows the responses recorded for an appendectomy performed under acupuncture anesthesia.

TABLE 1

Steps of the operating process:	1	2	3	4	5........
The duration of operation	t_1	t_2	t_3	t_4	t_5........
Pain response (PI)r_1	r_1	r_2	r_3	r_4	r_5........
Pain Index (PI)	$10^{r1}t_1$	$10^{r2}t_2$	$10^{r3}t_3$	$10^{r4}t_4$	$10^{r5}t_5$......
Total Pain Index	$P=10^{r1}t_1 + 10^{r2}t_2 + 10^{r3}t_3 + 10^{r4}t_4 + 10^{r5}t_5$				

In each step of the operation, a value of PI is assigned according to the formula $10^{ri}t_i$. Wherein the subscript i represents the operation interval (in Table 1, i=1 to 5), and t and r are the corresponding time interval and pain response. The exponential form of PI in an operation is controlled by the Weber-Fechner's law (ref. 7). The cumulative value of PI in an operation can be written as

$$P = \sum_{i=1}^{5} 10^{r_i} t_i \qquad (1)$$

Several hundred records of appendectomy cases were collected from the district hospital of Hubei and the Teaching Hospital of Wunan Medical College. The list was filled in accordance with the requirements of Table 1. One hundred and four grade II cases and 44 grade III cases were selected to provide the data. These data were used to plot a histogram which showed the distribution of PI with respect to the percentages of cases in each grade (figure 1). The cumulative value of PI also calculated from these data.

As shown in figure 1, the point, log500=2.7, divides the pattern into two categories; the right side indicating grade III, and the left side grade II. Each occupies 80%, and as a boundary, the percentage of case distribution of different sides is higher than the other. This gives a basis for differentiating the grades of availability evaluation.

It is clear that grades I and IV are non-fuzzy. This is simply bacause grade I represents no appearance of pain, and grade IV represents extreme pain. The problem, therefore, is reduced to differentiating between grades II and III.

Here we shall establish the relationship between PI and the possibility attribute.

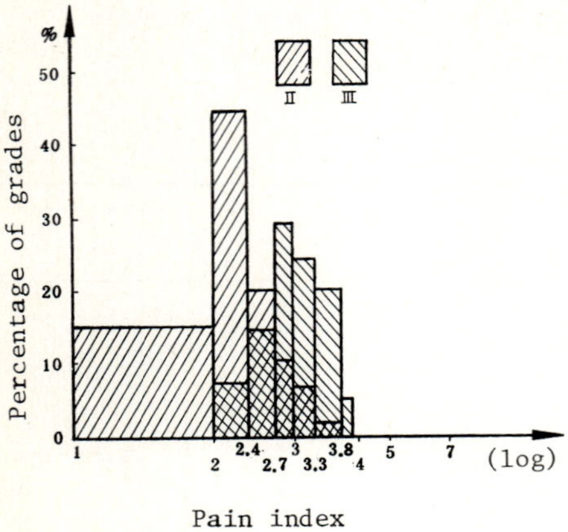

Fig. 1. Histogram showing relation between the logarithm of pain index and the percentages of Grade II and Grade III.

Let PI be represented by a universe of discourse U $[=\{p_1,p_2,\cdots p_n\}]$. Let F be a subset in U with elements of the same grade, such as grade II, grade III, etc. The classification of grade is expressed by an assigned linguistic proposition. Let a proposition represent "The evaluation of availability under acupuncture anesthesia is grade II, i.e., a fuzzy set of grade II," and let its membership function be:

$$f_{II}(P)=1-S(P;a,b,c) \tag{2}$$

where

$S=0$	when $P \leq a$
$S=2(P-a/c-a)^2$	when $a \leq P \leq b$
$S=1-2(P-c/c-a)^2$	when $b \leq P \leq c$
$S=1$	when $P \geq c$

where a, b, and c are parameters defined by the problem, and b (1/2(a+b)) is the cross over point, i.e., $f_{II}(P)=0.5$. Figure 2, a=250, b=500, c=750 show the properties for the possibility distribution pattern of the proposition (the evaluation of availability under acupuncture anesthesia is grade II) are shown. The complement of this proposition is "the evaluation of availability is grade III," so that:

$$f_{III}(P)=1-f_{II}(P)$$

where $f_{III}(P)$ represents the membership function of grade III (i.e., the fuzzy set of grade III). The distribution pattern of $f_{III}(P)$ is shown in Figure 2.

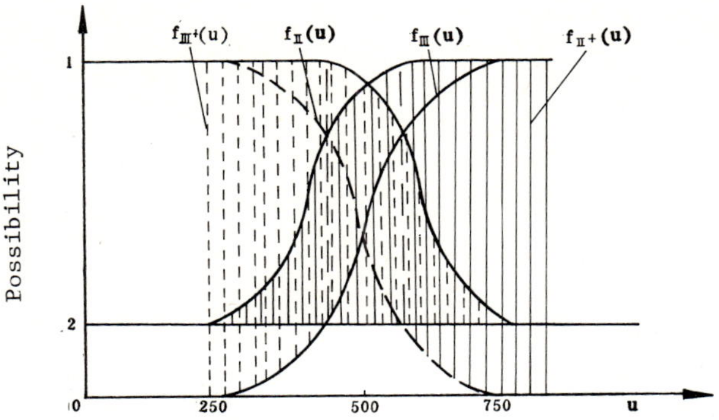

Fig. 2. Possibility qualification distributions based on the information conveyed by Fig. 1, the explanation is found in the text.

If we apply a qualifier to a proposition such as "P is grade II is α-possible," then the possibility distribution of f_{II} (P) is shown as a solid vertical shading in Fig. 2. Similarly we can plot the pattern of the distribution of $f_{III}(P)$.

We further define "1 possible" as "completely possible," "0.9 possible" is "very possible," "0.5 possible" is "possible," "0.1 possible" is "rather possible," etc. We then can take the corresponding PI so that for grade II we get:

TABLE 2

PI	Linguistic
$0 \leq P \leq 250$	completely possible
$250 \leq P \leq 362$	very possible
$362 \leq P \leq 500$	possible
$500 \leq P \leq 638$	rather possible
$638 \leq P \leq 750$	hardly possible
$750 < P$	impossible

then for grade III we have:

PI	Linguistic
$P \geq 750$	completely possible
$750 > P \geq 638$	very possible
$638 > P \geq 500$	possible
$500 > P \geq 362$	rather possible
$362 > P \geq 250$	hardly possible
$250 > P$	impossible

In an appendectomy, if PI=300, we obtain the following two propositions with the possibility qualification:

F^{+} = "the case is grade II" is very possible.

G = "the case is grade III" is hardly possible.

If we have F^{+} G^{+}, then $I(F^{+})$ $I(G^{+})^{8}$, where $II(F^{+})$, $I(G^{+})$ are the magnitudes of information conveyed by the proposition of F^{+} and G^{+}, respectively.

We retain the larger which gives more information as a result of grade. Therefore, PI=300, and the grade of availability of anesthesia in grade II is very possible.

Under linguistic possibility qualification we can evaluate the availability of acupuncture anesthesia. In this case we take a sample of 26 appendectomy cases under acupuncture anesthesia. The one is graded by the method of possibility qualification, the other is by the traditional method. We found that 24 cases correspond with each other, but 2 cases do not correspond. After checking with the original recorder, we found these two cases (no. 1 and no. 26) where two mild pain responses in two intervals of operation for case 1, and in a much longer interval of operation for case 26. It would appear that the traditional method would be the superior in these two cases.

EXAMPLE 2

Evaluation of Human Cardiac Function

With respect to the selection of special task personnel, the measurement of cardiac function is undoubtedly one of the most important physical examination requirements. Since individuals have to perform military or scientific experimental tasks under particular circumstances, it is necessary to evaluate the real-time cardiac function.

Both the prevalence and mortality rates of Computer Aided Design (CAD) are highly correlated with impairment of the human cardiac function. This has attracted attention of medical experts.

So far, there have been some advances in the evaluation of human cardiac function and the diagnosis of CAD by means of a statistical mathematical method in our country and other countries. Fuzzy mathematics has also been applied in this field.

The cardiac function is indicated by multi-parameters. In order to avoid risks noninvasive parameters are usually used. We can take the so-called fine, normal, and weak cardiac functions as a fuzzy subset of the universe of discourse of human cardiac function, which are inturn denoted by A, B and C. To evaluate an individual cardiac function is to determine to what extent it is subordinate to A, B and C. Therefore, the problem of configuring a multivariable membership is involved. There are two methods introduced in the following.

Method 1

Suppose we take a sample from U, where the sample size is n. This sample is composed of professional athletes (n_1), normal subjects (n_2), and patients with CAD (n_3). Therefore, we can say that $n_1 + n_2 + n_3 = n$. The p items of the cardiac parameters of each subject can then be determined. The observed results of the ith subject is denoted by p the dimensional vector:

$$(x_{i1}, x_{i2}, \ldots, x_{in}),\ i=1,2,\ldots,n \tag{3}$$

Mapping the following common set:

A={u|u represents the cardiac function of professional athletes}

B={u|u represents the cardiac function of normal subjects}

C={u|u represents the cardiac function of patients with CAD}

we obtain

$$\begin{aligned} f(A) &\longrightarrow 1 \\ f(B) &\longrightarrow 0 \\ f(C) &\longrightarrow -1 \end{aligned} \tag{4}$$

which is denoted by

$$Y_{n\times 1} \triangleq \begin{bmatrix} 1 \\ \vdots \\ 1 \\ 0 \\ \vdots \\ 0 \\ -1 \\ \vdots \\ -1 \end{bmatrix} \begin{matrix} \left.\vphantom{\begin{matrix}1\\ \vdots\\ 1\end{matrix}}\right\} n_1 \\ \left.\vphantom{\begin{matrix}1\\ \vdots\\ 1\end{matrix}}\right\} n_2 \\ \left.\vphantom{\begin{matrix}1\\ \vdots\\ 1\end{matrix}}\right\} n_3 \end{matrix} \qquad \beta_{(p+1)\times 1} \triangleq \begin{bmatrix} \beta_0 \\ \beta_1 \\ \vdots \\ \beta_p \end{bmatrix} \tag{5}$$

$$X_{n\times(p+1)} \triangleq \begin{bmatrix} 1 & X_{11} & X_{12} & \cdots & X_{1p} \\ 1 & X_{21} & X_{22} & \cdots & X_{2p} \\ \cdots & \cdots & \cdots & \cdots & \cdots \\ 1 & X_{n1} & X_{n2} & \cdots & X_{np} \end{bmatrix}, \qquad \varepsilon_{n\times 1} \triangleq \begin{bmatrix} \varepsilon_1 \\ \varepsilon_2 \\ \vdots \\ \varepsilon_n \end{bmatrix}.$$

To find the expression of f(*), we consider the application of the linear model,

$$y_i = \beta_0 + \beta_1 X_{i1} + \beta_2 X_{i2} + \dots + \beta_1 X_{ip} + \varepsilon_i \qquad (i=1,2,\dots,n) \qquad (6)$$

or write it in its simplified form:

$$Y = X \beta + \varepsilon$$

According to the least square principle, under the condition of rank $(X'X)=p+1$, the least square evaluation of vector β, a parameter to be determined is as follows:

$$\hat{\beta} = (X'X)^{-1} X'Y.$$

When rank $(X'x)<p+1$, we take

$$\hat{\beta}^{+} = (X'X)^{+} X'Y$$

After obtaining $\hat{\beta}$, we can configure the multivariable membership function according to the shape of the Logistic function,

$$f_A(u(x_1,x_2,\dots x_p)) = \left[\frac{1}{1 + \exp[a(\hat{\beta}_0 + \sum_{i=1}^{p} \hat{\beta}_i X_i)]} \right] \qquad (7)$$

where α is a constant, determined according to professional knowledge.

Finally, we use two thresholds ($\lambda_1=0.80$, $\lambda_2=0.60$) to divide $\underset{\sim}{A}$ into three common sets:

$$\begin{aligned} A^* &= \{u \mid f_A(u) \geq 0.80\} \\ B^* &= \{u \mid 0.60 \leq f_A(u) < 0.80\} \qquad (8) \\ C^* &= \{u \mid f_A(u) < 0.60\} \end{aligned}$$

Using this division, the human cardiac function is separated into three levels A*, B*, and C*. Thus, each individual cardiac function can be evaluated on the basis of the extent of membership in $f_{\underset{\sim}{A}}(u)$.

Method 2

Retaining the related assumptions and marks of the first method, we can take the following mapping:

$$\begin{cases} F_1(A) \longmapsto 1 + \frac{-n_1}{n} \\ F_1(B) \longmapsto -\frac{n_2}{n} \\ F_1(C) \longmapsto -\frac{n_3}{n} \end{cases}$$

$$\begin{cases} F_2(A) \longmapsto -\frac{n_1}{n} \\ F_2(B) \longmapsto \frac{1-n_2}{n} \\ F_2(C) \longmapsto -\frac{n_3}{n} \end{cases} \quad (9)$$

$$\begin{cases} F_3(A) \longmapsto -\frac{n_1}{n} \\ F_3(B) \longmapsto -\frac{n_2}{n} \\ F_3(C) \longmapsto 1 + \frac{-n_3}{n} \end{cases}$$

write as follows:

$$\Theta_{(p+1)x3} \triangleq \begin{bmatrix} \theta_{10} & \theta_{20} & \theta_{30} \\ \theta_{11} & \theta_{21} & \theta_{31} \\ \vdots & \vdots & \vdots \\ \theta_{1p} & \theta_{2p} & \theta_{3p} \end{bmatrix} \qquad E_{nx3} \triangleq \begin{bmatrix} e_{11} & e_{12} & e_{13} \\ e_{21} & e_{22} & e_{23} \\ \vdots & \vdots & \vdots \\ e_{n1} & e_{n2} & e_{n3} \end{bmatrix} \quad (10)$$

$$Y_{nx3} \triangleq \begin{bmatrix} 1-\frac{n_1}{n} & -\frac{n_2}{n} & -\frac{n_2}{n} \\ \vdots & \vdots & \vdots \\ 1-\frac{n_1}{n} & -\frac{n_1}{n} & -\frac{n_1}{n} \\ \frac{-n_2}{n} & 1-\frac{n_2}{n} & -\frac{n_2}{n} \\ \vdots & \vdots & \vdots \\ \frac{-n_1}{n} & 1-\frac{n_2}{n} & -\frac{n_2}{n} \\ -\frac{-n_3}{n} & \frac{-n_3}{n} & 1-\frac{-n_3}{n} \\ \vdots & \vdots & \vdots \\ \frac{-n_3}{n} & \frac{-n_3}{n} & 1-\frac{n_3}{n} \end{bmatrix} \begin{matrix} \left.\right\} n_1 \\ \left.\right\} n_2 \\ \left.\right\} n_3 \end{matrix}$$

and consider the following model

$$Y_{nx3} = X_{nx(p+1)} \cdot \Theta_{(p+1)x3} + E_{nx3} \tag{11}$$

In this model when rank $(X'X)=p+1$, the least square estimation of Θ can be obtained,

$$\Theta = (X'X)^{-1}X'Y$$

when rank $(X'X)<p+1$, the estimation of Θ can be obtained

$$\Theta = (X'X)^{+}X'Y$$

From these statements, three membership functions can be derived:

$$\begin{aligned} f_{\underset{\sim}{A}}(u) &\triangleq f_{\underset{\sim}{A}}(u(x_1,x_2,\dots)) = \left[\frac{1}{1+\exp[a_1(\hat{\theta}_{10}+\sum_{i=1}^{p}\hat{\theta}_{1i}X_i)]}\right] \\ f_{\underset{\sim}{B}}(u) &\triangleq f_{\underset{\sim}{B}}(u(x_1,x_2,\dots,)) = \left[\frac{1}{1+\exp[a_2(\hat{\theta}_{20}+\sum_{i=1}^{p}\hat{\theta}_{2i}X_i)]}\right] \\ f_{\underset{\sim}{C}}(u) &\triangleq f_{\underset{\sim}{C}}(u(x_1,x_2,\dots,)) = \left[\frac{1}{1+\exp[a_3(\hat{\theta}_{30}+\sum_{i=1}^{p}\hat{\theta}_{3i}X_i)]}\right] \end{aligned} \tag{12}$$

where α_1, α_2, and α_3 are constants to be determined according to professional knowledge.

Finally, the estimation of the individual cardiac function is taken as the identification of the model. For $u \in U$, if
$f_{\underset{\sim}{B}}(u) = \max f_{\underset{\sim}{A}}(u), f_{\underset{\sim}{B}}(u), f_{\underset{\sim}{C}}(u)$
then $u \in B$, i.e. The cardiac function of u then corresponds to an ordinary normal state, with the rest analogized.

Results

The cardiac function of 200 subjects was studied through application of the two above mentioned methods described earlier. The sample was composed of 70 professional athletes, 70 soldiers and ordinary workers, and 60 patients with CAD. Eight items of noninvasive cardiac function parameters were measured for each individual. These parameters were the ratio of PEP to LVET, stroke volume Index (SVI), isovolumic relaxation time (IRT), ture isovolumic contraction time (TICT), mechanical systol index (CA_2/RR), to ratio of NA_2 to CA_2, total peripheral resistance (TRP), and ejection velocity (10^3 SV/LVET).

With the first method, given $\alpha=-2$, the membership of $\underset{\sim}{A}$ is

$$f_A(u) = \left[\frac{1}{1+\exp[-2(b_0+\sum_{i=1}^{8} b_i X_i)]}\right] \quad (13)$$

where

$b_0 = 2.5$, $b_1 = -2.158$, $b_2 = -0.008$, $b_3 = 0.008$,

$b_4 = -1.087$, $b_5 = -0.0003$, $b_6 = 0.006$, $b_7 = -0.023$,

$b_8 = -3.79$

The extent of each individual membership to $\underset{\sim}{A}$ was calculated according to equation (13), and then differentiated. The results are shown in Table 3, where $\underset{\sim}{A}$, $\underset{\sim}{B}$, $\underset{\sim}{C}$ are the original set and A*, B*, C* are the discriminatory set.

TABLE 3

Results of discrimination

	A	B	C	Total
A*	64	12	0.0	76
B*	6	46	7	59
C*	0	12	53	65
Total	70	70	60	200

As shown in Table 3, the correct rate of differentiation is 81.5% ((64+46+53)/200)=.815 or 81.5%. It was also demonstrated that there is no overlapping and incorrect differentiation between the fine and the weak cardiac function. Perhaps the overlapping and wrong differentiation existing between the fine and the normal groups reflects the actual objectivity of the situation. As a general rule, the cardiac function of a professional athlete is better than that of an ordinary normal subject. The cardiac function of a young soldier is certainly no worse than that of a profesional athlete. Concerning this overlap, there is no objective medical basis for the differentiation. This indicates that by using a multivariable membership function to evaluate the human cardiac function the correct rate is generally not less than 81.5%.

With this second method, we have analyzed the same data and obtained the membership function of the three fuzzy subsets A, B, and C:

$$f_A(u) = \left[\frac{1}{1+\exp[-2(b_{10}+\sum_{i=1}^{8} b_{1i} x_i)]}\right] \quad (14)$$

where

$b_{10} = -0.161,\quad b_{11} = -1.779,\quad b_{12} = 0.004,$
$b_{13} = 0.007,\quad b_{14} = 0.208,\quad b_{15} = 0.016,$
$b_{16} = 0.006,\quad b_{17} = -0.018,\quad b_{18} = -1.05$

$$\underset{\sim}{B}(u) = \left[\frac{1}{1 + \exp[-2(b_{20} + \sum_{i=1}^{8} b_{2i}x_i)]} \right]$$

where

$b_{20} = 2.19,\quad b_{21} = 1.353,\quad b_{22} = 0.002,$
$b_{23} = -0.009,\quad b_{24} = 0.649,\quad b_{25} = -0.056,$
$b_{26} = -0.0005,\quad b_{27} = 0.015,\quad b_{28} = -0.586$

$$f_{\underset{\sim}{C}}(u) = \left[\frac{1}{1 + \exp[-2(b_{30} + \sum_{i=1}^{8} b_{3i}x_i)]} \right]$$

where

$b_{30} = -2.368,\quad b_{31} = 0.497,\quad b_{32} = 0.002,$
$b_{33} = 0.002,\quad b_{34} = 1,\quad b_{35} = 0.046,$
$b_{36} = -0.005,\quad b_{37} = 0.004,\quad b_{38} = 1.908$

The result of the differentiation obtained with the second method is shown in Table 4.

TABLE 4

Result of differentiation

	A	B	C	Total
$\underset{\sim}{A}$	62	4	0	66
$\underset{\sim}{B}$	7	62	6	75
$\underset{\sim}{C}$	1	4	54	59
Total	70	70	60	200

In Table 4, among the 4 subjects who originally belonged to (B) but were identified to be in $\underset{\sim}{C}$, 3 are more than 40 years old. In the group of athletes (A), one subject was identified to be in C by mistake. In the group of patients with CAD, no one was identified to be in $\underset{\sim}{A}$, though 7 subjects were identified to be in $\underset{\sim}{B}$. This case is probably related to the sample we selected. The 60 patients with CAD suffered from neither myocardial infarction nor cardiac failure, although the clinical diagnosis had been made. They were patients with light CAD, who still worked daily.

It can be calculated according to Table 4 that the correct rate of differentiation is 89% ((62+62+54)/200).

DISCUSSION

The first example takes advantage of processing the objects which can only be expressed by verbal descriptions. From the histogram of Fig. 1, it seems that the possibility qualified evaluation possesses a certain stability and gives a more precise classification in clinical practice. This may bring about some convenience in the study of acupuncture anaesthesia. Additionally, with respect to methodology, it is very easy to apply and extend. Although this paper deals with the application to appendectomy, the method may be extended to other operations under acupuncture anesthesia.

The correct rate for the first method was found to be 81.5%, while for the second it was 89%. Although the second method is higher than the first, the computing work load is larger. The correct rate depends not only upon the choice of indices, but it also depends upon the type of the function.

REFERENCES

1 Guo, R., Fuzzy Set Model for Computerized Diagnosis System in Traditional Chinese Medicine, News On 1980 Ann. Rep. Meet. of Beijing Working Group on Fuzzy Sets, Beijing, China.

2 Su Zaifu, The Fuzzy Model in Diagnosis-(Bianzheng) of Chinese Traditional Medicine, J. Huazhong Inst. Tech., Fuzzy Mathematics, No. 2, 1980.

3 Su Zaifu, Application of Fuzzy Relation Equation to Disease-Diagnosis and on Treatment of Chinese Traditional Medicine, Science Exploration, 1, 1981.

4 Li Yanjie et al., An Application of Fuzzy Mathematics in Forecasting Precursor of the Coronary Artery Disease, Fuzzy Mathematics, 3, 1983.

5 Wang Fu, Research on the Fuzzy Mathematical Model of Forecasting Ischemic Stroke (IS), Fuzzy Mathematics, 1, 1983.

6 Su Zaifu et al., Pansystems and First Order Artificial Intelligence Bianzheng System of Chinese Traditional Medicine, Chinese Journal of Biomedical Engineering, Vol. 1, No. 1, 1985.

7 Guyton, C., Physiology of Human Body, Saundrs College Publishing, 1984, pp. 147.

8 Zadeh, L. A., Fuzzy sets a basis for a theory of possiblility, Fuzzy sets and systems, 1, 1978, pp. 3-28.

APPENDIX II.

Applications of Fuzzy Set Theory in Human Factors, edited by W. Karwowski and A. Mital
Elsevier Science Publishers B.V., Amsterdam, 1986 — Printed in The Netherlands

RESEARCH GUIDE TO APPLICATIONS OF FUZZY SET THEORY IN HUMAN FACTORS

JANARDAN KULKARNI[1] and WALDEMAR KARWOWSKI[2]

[1]Laura Kersey Library of Engineering, Physical Sciences and Technology, University of Louisville, Louisville, Kentucky 40292 (U.S.A.)

[2]Department of Industrial Engineering, University of Louisville, Louisville, Kentucky 40292 (U.S.A.)

ABSTRACT

The paper presents the literature related to application of fuzzy set theory to Human Factors and has 784 research papers listed in the bibliography which is given at the end. As in any other collection this literature list has a subjective viewpoint and may be in some sense incomplete. Analysis of these data provides interesting insights regarding the characteristics of articles of the applications of fuzzy set theory, literature trends and the future. This paper consists of two parts. In Part I, the methodology used to gather the information, language, citation, and descriptor analysis, and conclusions are described. Part II is the main bibliography.

INTRODUCTION

Fuzzy Set Theory (FST) is one of the most fascinating areas in the field of mathematics. Notable research contributions to FST and its applications have grown at a fast pace during the last two decades. The literature on FST and its applications is more prolific since the 1970s than before since there has been an immense spurt of research activity. It is scattered over many journals, working papers, conference proceedings, technical reports, theses, dissertations, and monographs. A comprehensive bibliography of all the sources will be a helpful tool even for a researcher or a specialist in this field. The main aim of this paper is to incorporate as many sources as possible, focussing on the application areas of FST, and to consolidate the existing research in just one area, viz. Human Factors. There exist two extensive bibliographies prepared earlier by B. R. Gaines and L. J. Kohut, and by A. Kandel, and R. R. Yager, but they do not explicitly provide an extensive coverage of human factors applications. Our bibliography contains 784 research papers published in 147 journals appearing until 1984 and covering topics given at the end in the subject index.

MODUS OPERANDI (METHODOLOGY)

This section gives a birds-eye view of the resources used in preparing the present paper. The compilation of the bibliography was done by using "DIALOG

Search System," a subsidiary of the Lockheed Corporation. The following table gives the list of the data bases searched and the key words used to search the data bases.

TABLE 1

Name of Data Base	Key Words Used
1. Dissertation Abstracts	Man-Machine Systems, Fuzzy Sets,
2. Conference Paper Index	Human Factors, Uncertainty, Plausibility,
3. Occupational Health and Safety	Behavior Modification, Imprecision,
4. Life Sciences Collection	Vagueness, Information Processing,
5. Social Scisearch	Fuzzy Models, Cognition, Physical Stress,
6. Psycinfo	Problem Solving, Linguistic Approach,
7. Sociological Abstracts	Subjective Judgments, Workload, Decision
8. Compendex	Making, Work Design and Organization,
9. NTIS	Psychophysics, Memory and Learning, Social
	Systems, Psychological Measurements

We have selected only Human Factors areas of applications of FST in this paper for a more in-depth review.

The papers are listed alphabetically by the first author and chronologically within each author's name and include full bibliographic information. The total number of 784 papers in this bibliography represents the work of over 1100 different authors, of which 529 represent single author papers, 188 two authors, 57 three authors, and 16 four or more authors. Figure 1 shows the distribution of the number of authors and the number of titles in this bibliography.

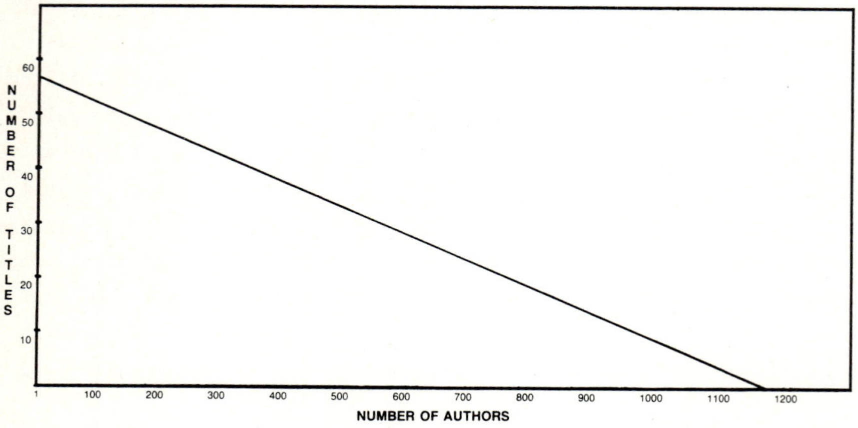

Figure 1. Distribution of the number of papers and their authors.

GROWTH OF THE LITERATURE

Table 2 below reveals some interesting conclusions. Prior to 1970 only a few articles were published. Subsequently interest in FST and its applications grew and the number of publications reached its peak around 1980. The research in FST and its applications in various areas will continue to grow further and therefore there is an apparent need to consolidate the existing research.

TABLE 2

Yearwise growth of research papers on FST and its Applications

Year	All Papers	Year	All Papers
1984	70	1976	65
1983	66	1975	60
1982	68	1974	31
1981	68	1973	28
1980	77	1972	14
1979	50	1971	9
1978	58	1970	10
1977	69		

CITATION ANALYSIS

A list of journals which contain papers on "FST and its applications" is given in Table 3. This is the core list of all the important sources to scan the literature on "FST and its applications" and it consists of 50 journals which were individually searched. The bibliographies given at the end of the papers contained in the revised list were searched to find additional references. This procedure provided a final list of 784 articles, which we hope is as extensive and exhaustive as possible, but an inadvertant exclusion of some references may always be possible.

The journals cited most frequently by the core group are shown in Table 3. They are ranked by frequency, which is the citation rate of the average article when only citations from core journals are considered (ref. 1). The study demonstrated that citation analysis is as effective a tool for defining journal literature of a small field as it is for defining the literature of a large one. Conference literature, theses and dissertations, and technical reports are the next major part of the formal communication system in the FST area. Table 3 shows that the journal most frequently cited by the core group is "Fuzzy Sets and Systems."

The citation data, of course, is an abstraction. Its utility depends upon the types of information that can be derived from it by analysis. The results of citation study led to the formulation of bibliographic laws of concentration (ref. 2) by stating that the tail of the literature of any one discipline (or area of research) consists, in large part, of the cores of the literature of other disciplines.

ABBREVIATIONS OF SOURCE OF PUBLICATION

The abbreviations in the bibliography follow closely the style used by Mathematical Reviews and INSPEC list of journals. Some abbreviations have been expanded to achieve greater clarity. For most sources an indication of the editorship is given in parentheses in the beginning of the abbreviation. No two abbreviations identify the same publication; no two publications, moreover, are given the same abbreviation.

TABLE 3

List of journals cited most frequently in FST and its applications

Title	Number of Papers Published
Fuzzy Sets and Systems	61
International Journal of Man-Machine Studies	44
IEEE Transactions on Systems, Man and Cybernetics	38
Information Sciences	22
Journal of Cybernetics	14
Human Systems Management	10
Journal of Mathematical Analysis and Applications	10
Information and Control	8
International Journal of General Systems	8
Progress in Cybernetics and Systems Research	7
Journal of the American Society for Information Science	7
Cybernetica	6
Ekonomicko-Matematicky Obzor	5
International Journal of Systems Science	5
Pattern Recognition Letters	5
Bulletin of the Psychonomic Society	4
Cybernetics and Systems	4
Engineering Cybernetics	4
Journal of Philosophical Logic	4
Kybernetes	4
Kybernetika	4
Automatica	3
Ergonomics	3
European Journal of Operations Research	3
International Journal of Control	3
Journal of Experimental Psychology - General	3
Information Processing and Management	3
Lecture Notes in Economics and Mathematical Systems	3
Management Science	3
Memory and Cognition	3
Philosophy of Science	3
Proceedings of the American Society for Information Science	3
Proceedings IEEE	3
Synthese	3
Systems and Control (Japan)	3
Transactions of the Institute of Electronic and Communication Engineers, Japan	3

TABLE 3 (continued)

Transactions of the Society of Instrument and Control Engineers, Japan	3
American Philosophical Quarterly	2
Behavior Science	2
IEEE Transactions on Automatic Control	2
IEEE Transactions on Engineering Cybernetics	2
Journal of Experimental Psychology: Human Perception and Performance	2
Journal of the Institution of Electronic and Communication Engineers (Japan)	2
Kybernetika i Diagnostika	2
Mathematical Social Sciences	2
Mathematical Biosciences	2
Philosophical Review	2
Psychological Research	2
Research Reports of the Automatic Control Laboratory, Nagoya University, Japan	2
Theory and Decision	2

LANGUAGE ANALYSIS

All research papers that are not written in English have as the appended word the language in which the paper was written e.g. GERMAN. The percentage distribution of papers in various languages is given in the figure below.

LANGUAGE ANALYSIS OF RESEARCH PAPERS

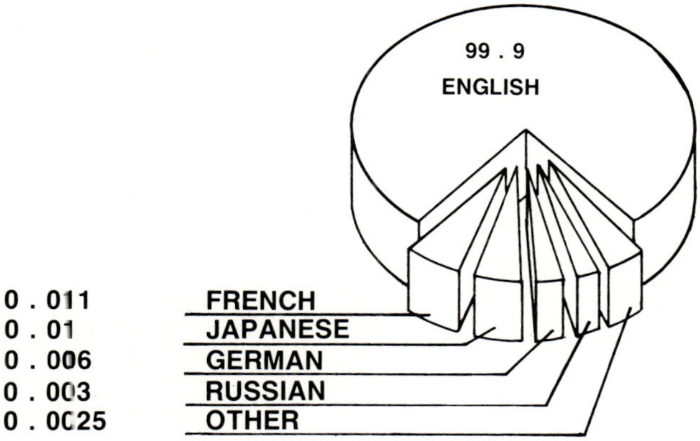

DESCRIPTOR ANALYSIS: CLASSIFICATION APPROACH

In this paper we have attempted to create an information retrieval system for FST and its applications based on indexing by keywords/descriptors/access points/subjects. This approach has a number of benefits, notably--efficiency. The indexing process is guided by the knowledge structure in the field of FST. A "classifier" is any one of a set of characteristics used to substructure or subdivide a given concept in the FST domain. "Classifying/categorizing" refers to a process by which the research papers are grouped to form a subject index of this bibliography. Many classifying/categorizing techniques are available. There are a number of factors that determine the suitability of a classifying/categorizing technique to a given application. The most important of them is, of course, the effectiveness of the classification generated in the context of the application.

CONCLUSION

The 784 articles listed in this bibliography cover a spectrum of areas and fields of applications of FST to Human Factors. Decision Analysis/Making (Fuzzy Decision Making) contains the most articles followed in decreasing order by Fuzzy Sets - Theoretical Applications (Fuzzy Mathematics), Fuzzy Logic (Concepts and Fuzzy Modeling/Measurements). This is certainly an extensive list of application areas in Human Factors.

Errors, particularly omissions, are certain to be prevalent in any bibliography or index of this scope, despite attempts to prevent their occurrence. Readers are invited to comment and to suggest improvements, corrections, or additions for consideration in any updating.

In summary FST and its applications will impact on various fields since it is one of the powerful tools to solve objective problems. Future studies in FST are likely to be devoted to its applications as a new area emerging in "Fuzzy Engineering."

ACKNOWLEDGMENTS

The authors wish to thank Dr. Niraj Kulshreshtha, Geology Department, University of Louisville for inputting the bibliographic data into the computer, editing and making the necessary corrections.

REFERENCES

1 E. Garfield, Citation Indexing, John Wiley, New York, 1979.
2 E. Garfield, The Mystery of the Transposed Journal Lists; wherein Bradford's Law of Scattering is generalized according to Garfield Law of Concentration, Essays of an Information Scientist, vol. 1, p. 222, 1977.

PART II
BIBLIOGRAPHY

1 Aczel, J. and Pfanzgl, J. Remarks on the Measurements of Subjective Probability and Information. Metrika, 5, 91-105, 1966.

2 Adamo, J.M. Fuzzy Decision Trees. Fuzzy Sets and Syst., 4, 207-219, 1980.

3 Adamo, J.M. and Karsky, M. Application of Fuzzy Logic to the Design of a Behavioral Model in an Industrial Environment. Univ. Claude-Bernard Lyon I, 1978.

4 Adams, E.W. Elements of a Theory of Inexact Measurement. Philos. Sci., 32, 205-228, 1965.

5 Adey, W. Ross. Neurophysiological Estimates of Human Performance Capabilities in Aerospace Systems. NTIS Report AD-735 178, 1971 and AFOSR-TR-72-0072 Report 1971, 17 pages.

6 Adlassing, K.P. Survey on Medical Diagnosis and Fuzzy Subsets. In: Gupta, M.M. and Sanchez, E. (eds) - Approximate Reasoning in Decision Analysis. North-Holland Pub., Amsterdam, Netherlands, 203-217, 1982.

7 Adlassing, K.P. A Fuzzy Logic Model of Computer-Assisted Medical Diagnosis. Methods Inf. Med., 19, 141-148, 1980.

8 Alsina, C., Trillas, E., and Valverde, L. On Some Logical Connectives for Fuzzy Sets Theory. J. Math. Ana. and Appl., 93, 15-26, 1983.

9 Amarer, S. On the Goals and Methodologies of Work in Fuzzy Sets Theories. Human Syst. Manage., 4, 309-310, 1984.

10 Amey, L.R. and Bonaert, A.P. The Theory of Fuzzy Subsets - A Review and Application to Planning and Control. Rivista Int. Di Sci. Econom. E. Commerciali, 28, 1-32, 1981.

11 Anderson, D.O. Pruning Fuzziness and Flab from Learning-Disabilities Research. J. Special Edu., 10, 157-161, 1976.

12 Arigoni, A.O. Membership Characteristic Function of Fuzzy Elements: Fundamental Theoretical Basis. Prog. in Cybern. and Syst. Res., 3, 265-269, 1978.

13 Arrow, K.J. Social Choice and Individual Values. New York, 1951.

14 Asai, K. and Tanaka, H. Application of Fuzzy Sets to Decision Making and Control Problems. (Japanese). Syst. and Control (Japan), 19, 235-242, 1975.

15 Asai, K., Tanaka, H. and Okuda, T. Decision Making and its Goal in Fuzzy Environment. In Zadeh, L.A. et al. (eds) - Fuzzy Sets and Their Applications to Cognitive and Decision Processes. Academic Press, New York, USA, 257-277, 1975.

16 Ayyub, Billal M. Structural Safety Analysis of Reinforced Concrete Buildings during Construction (Fuzzy Set). Dissertation, Georgia Inst. of Tech., 367 pages, 1983.

17 Azorin, F. On Fuzzy Sampling from Fuzzy Populations. In: 41st Session of the Int. Statistical Inst., New Delhi, India, Dec. 1977.

18 Baas, S.M. Rating and Ranking of Multiple-Aspect Alternatives Using Fuzzy Sets. In: IFAC Symp. on Large Scale Systems, Theory and Application, Udine, Italy. June 1976. Also in Automatica 13, 47-58, 1977.

19 Baker, W.E. Introduction to Fuzzy Regression Analysis with a Ballastic Application. NTIS Report ADA 146093/0/XAB, 1984. ARBRL - TR - 02575, 1984. AD - F300 476, 1984.

20 Baklien, B. Organizational Climate - A Fuzzy Concept of Organization Psychology. Nordisk Psykologi, 35, 1-13, 1983.

21 Baldwin, J.F. Fuzzy Relational Inference Language for Expert Systems. NTIS Report AD-P002 380/4. Also in Proc. 13th Int. Symp. on Multiple-Valued Logic, Kyoto, Japan, 416-421, 1983.

22 Baldwin, J.F. Fuzzy Logic and Fuzzy Reasoning. In: Mamdani, E.H. and Gaines, B.R. (ed) - Fuzzy Reasoning and Its Applications. Academic Press, New York, USA, 1981, 133-148.

23 Baldwin, J.F. Fuzzy Logic Knowledge Basis and Automated Fuzzy Reasoning. In Lasker, G.E. (ed) - Applied Systems and Cybernetics, 6, 2959-2965, 1981.

24 Baldwin, J.F. and Pilsworth, B.W. Axiomatic Approach to Implication for Approximate Reasoning with Fuzzy Logic. Fuzzy Sets and Syst., 3, 193-219, 1980.

25 Baldwin, J.F. and Guild, N.C.F. Feasible Algorithms for Approximate Reasoning Using Fuzzy Logic. Fuzzy Sets. and Syst., 3, 225-251, 1980.

26 Baldwin, J.F. and Guild, N.C.F. FUZLOG: A Computer Program for Fuzzy Reasoning. In Proc. 9th Int. Symp. Multiple Valued Logic, Bath, England, May 1979, 38-45.

27 Baldwin, J.F. Fuzzy Logic and Approximate Reasoning for Mixed Input Arguments. Int. J. Man-Mach. Stud., 11, 381-396, 1979.

28 Baldwin, J.F. Fuzzy Logic and Fuzzy Reasoning. Int. J. Man-Mach. Stud., 11, 465-480, 1979.

29 Baldwin, J.F. and Pilsworth, B.W. Fuzzy Truth Definition of Possibility Measure for Decision Classification. Int. J. Man-Mach. Stud., 2, 447-463, 1979.

30 Baldwin, J.F. and Pilsworth, B.W. Model of Fuzzy Reasoning Through Multi-Valued Logic and Set-Theory. Int. J. Man-Mach. Stud., 11, 351-380, 1979.

31 Baldwin, J.F. A New Approach to Approximate Reasoning Using Fuzzy Logic. Fuzzy Sets and Syst., 5, 309-325, 1979.

32 Bandler, W. and Kohout, L.J. Four Modes of Interface in Fuzzy Expert Systems. In: Proc. 7th European Meeting on Cybernetics and Systems Research, Vienna, Austria, April 1984.

33 Bandler, W. and Kohout, L. Fuzzy Power Sets and Fuzzy Implication Operators. Fuzzy Sets and Syst., 4, 12-30, 1980.

34 Bandler, W. and Kohout, L.J. Semantics of Implication Operators and Fuzzy Relational Products. Int. J. Man-Mach. Stud., 12, 89-116, 1980.

35 Barnev, P. Fuzzy System Approach to Decision-Making Based on Public Opinion Investigation Through Questionnaires. In: Symp. on Stochastic Control, Budapest, Hungary, Sep. 1974.

36 Barwise, J. and Cooper, R. Generalized Quantifiers and Natural Language. Linguistics and Philosophy, 4, 159-219, 1981.

37 Basu, K. Fuzzy Revealed Preference Theory. J. Econ. Theory, 32, 212-227, 1984.

38 Beachy, K.A. Computer Recognition of Phonemes in the Presence of Cockpit Induced Stress and Noise (Thesis). Air Force Inst. of Tech., Wright Patterson AFB, OH, Report No. AFIT/GE/EE 82D-20, 145 Pages. NTIS Report AD-A135 833/2, 1982.

39 Bedrosian, S.D. Significance of Visual Limitation in Automated Pattern Recognition Application. Pattern Recog. Lett., 2, 395-400, 1984.

40 Bek, R., Pokorny, Z. and Ruzicka, M. Large Variable Systems of Higher Degrees in Fuzification Process - Fuzification of Systems for Technical and Medical Practice II. Kybernetika, 18, 447-454, 1982.

41 Bek, R., Pokorny, Z. and Ruzicka, M. Principle Concepts of Systems Fuzification. Fuzification of Systems for Technical and Medical Practice. Kybernetika, 18, 234-246, 1982.

42 Belke, W., Kappen, B. and Ettrich, H. The Application of Fuzzy Set Theory in Formal Systems of Classification (Russian). Kozel. Mta. Szmita Stech. and Autom. Kut. Intez., 21, 55-71, 1978.

43 Bellman, R.E. Communication, Ambiguity and Understanding. Math. Biosci., 26, 347-356, 1975.

44 Bellman, R.E. Mathematics and the Human Sciences. In: Wilkinson, J., Bellman, R.E. and Garaudy, R. (eds) - The Dynamic Programming of Human Systems. MSS Information Corp., New York, USA, 11-18, 1973.

45 Bellman, R.E. Retrospective Futurology: Some Introspective Comments. In: Wilkinson, J., Bellman, R.E. and Garaudy, R. (eds) - The Dynamic Programming of Human Systems. MSS Information Corp., New York, USA, 35-37, 1973.

46 Bellman, R.E. and Giertz, M. On the Analytic Formalism of the Theory of Fuzzy Sets. Inf. Sci., 5, 149-156, 1973.

47 Bellman, R.E. and Zadeh, L.A. Decision-Making In a Fuzzy Environment. Manage. Sci., 17, 141-164, 1970.

48 Benson, W.H. Application of Fuzzy Set Theory to Data Display. In: Proc. Int. Cong. on Applied Syst. Research and Cybernetics, Acapulco, Mexico, Dec. 1980. NTIS Report LBL-11590. Also in Yager, R.R.--Fuzzy Sets and Possibility Theory. Pergamon Press, New York, USA, 1982.

49 Berndt, R.S. and Caramazza, A. The Development of Vague Modifiers in the Language of Pre-School Children, 1978.

50 Berndt, R.S. and Caramazza, A. The Development of Some Adverbial Modifiers of Dimensional Adjectives. In: Proc. 2nd Ann. Boston Univ. Conf. on Language Development, 1977.

51 Bezdek, J.C. Special Issue on Advances in Fuzzy Information Processing - Editorial. Int. J. Man-Mach. Stud., 19, 1-3, 1983.

52 Bezdek, J.C. Fuzzy Measures of Group Consensus and Uncertainty. In: IEEE Conf. on Decision and Control Including 16th Symp. on Adaptive Processes and a Special Symp. on Fuzzy Set Theory and Applications, New Orleans, LA, USA, Dec. 1977.

53 Bezdek, J.C., Spillman, B. and Spillman, R. Fuzzy Measures of Preference and Consensus in Group Decision-Making. In: Proc. IEEE Conf. on Decision and Control, New Orleans, LA, USA, Dec. 1977, Part II, 1303-1309, 1977. IEEE, New York, USA.

54 Bezdek, J.C. and Castelaz, P.F. Prototype Classification and Feature Selection with Fuzzy Sets. IEEE Trans. Syst. Man and Cybern., SMC - 7, 87-92, 1977.

55 Bezdek, J.C. Physical Interpretation of Fuzzy Isodata. IEEE Trans. Syst. Man and Cybern., SMC - 6, 387-389, 1976.

56 Bezdek, J.C. Cluster Validity with Fuzzy Sets. J. Cybern., 3, 58-73, 1974.

57 Bedzek, J.C. Numerical Taxonomy with Fuzzy Sets. J. Math. Biol., 1, 57-71, 1974.

58 Biswas, P. and Majumdar, A.K. A Supervised Learning Algorithm for Hierarchical - Classification of Fuzzy Patterns. Inf. Sci., 31, 91-106, 1983.

59 Biswas, P. and Majumdar, A.K. A Multistage Fuzzy Classifier for Recognition of Handprinted Characters. IEEE Trans. Syst. Man and Cybern., SMC - 11, 834-838, 1981.

60 Black, M. Margins of Precision. Ithaca, N.Y., Cornell Univ. Press., 1970.

61 Black, M. Reasoning with Loose Concepts. Dialogue, 2, 325-373, 1963.

62 Black, M. Vagueness: An Exercise in Logical Analysis. Philos. Sci., 4, 427-455, 1937.

63 Blackburn, S. Meaning, Reference and Necessity. Cambridge Univ. Press, Cambridge, England, 1975.

64 Blaffert, T. Fuzzy Sets Inverted Searches (German). VDE Facher., 35, 149-154, 1983.

65 Blin, J.M. and Whinston, A.B. Fuzzy Sets and Social Choice. J. Cybern., 3, 28-36, 1975.

66 Blin, J.M. On an Isomorphism Between Aggregation Procedures and Fuzzy Sets. Int. J. Policy Anal. and Inf. Syst., 4, 305-315, 1980.

67 Blin, J.M. et al. Pattern Recognition and Quantitative Political Theory. In: Proc. IEEE Conf. on Decision Control and 11th Symp. on Adaptive Processes, New Orleans, LA, USA, Dec. 1972, 114-117. IEEE, New York, USA.

68 Blishun, A.F. and Shapiro, D.I. Decision Making on the Basis of a Linguistic Representation of Situational Data and Criteria. Eng. Cybern. (USA), 19, 151, 1981.

69 Blockley, D.I. The Nature of Structural Design and Safety. Chichester, England, 1980.

70 Bonissone, P.P. Coping with Uncertainty in Expert Systems: A Comprehensive Study. In: Proc. Amer. Control Conf., San Francisco, CA, USA, June 1983, 3, 1230-1232.

71 Bonissone, P.P. Failure Diagnosis and Decision Making in Industrial Processes: A Fuzzy Set Application. In: Proc. 20th IEEE Conf. on Decision and Control Including the Symp. on Adaptive Processes, San Diego, CA, USA, Dec. 1981, 3, 1087-1093.

72 Bonissone, P.P. A Fuzzy Set Based Linguistic Approach: Theory and Application. In: Proc. Winter Simulation Conf., Orlando, FL, USA, Dec. 1980, A1/99-111. IEEE, New York.

73 Bonissone, P.P. A Pattern Recognition Approach to the Problem of Linguistic Approximation in System Analysis. In: Proc. Int. Conf. on Cybern. and Soc., Denver, CO, USA, Oct. 1979, 793-798. IEEE, New York.

74 Bookstein, A. Relevance of Fuzzy-Set Theory in Information Retrieval. Human Syst. Manage., 4, 323-324, 1984.

75 Bookstein, A. Fuzzy Requests - An Approach to Weighted Boolean Searches. J. Amer. Soc. Inf. Sci., 31, 240-247, 1980.

76 Borisov, A.N. and Osis, J.J. Methods for Experimental Estimation of Membership Functions of Fuzzy Sets. (Russian). In: Kristinkov, D.S., Osis, J.J. and Rastrigin, L.A., (eds) - Kibernetika i Diagnostika, 4, 125-134, 1970, Zinatne, Riga, U.S.S.R.

77 Borisov, A.N., Vulif, G.N. and Osis, J.J. State Prognosis of Complex Systems Using the Theory of Fuzzy Sets (Russian). Kibernetika i Diagnostika, 5, RIGA, 1972.

78 Borkowski, L. (ed) Jan Lukasiewicz: Selected Works. North-Holland Pub., Amsterdam, Netherlands, 1970.

79 Bossel, H., Klaczko, S. and Mueller, N. (eds) - Systems Theory in the Social Sciences. Basel, Stgt., Switzerland, 1976.

80 Bosserman, R.W. and Ragade, R.K. Ecosystem Analysis Using Fuzzy Set Theory. Ecol. Model, 16, 191-208, 1982.

81 Bouchon, B. An Application of the Theory of Questionnaires to Information Theory. In: Topics in Information Theory, Keszthely, Hungary, Aug. 1975, 125-141, North Holland Pub., Amsterdam, Netherlands.

82 Bouchon, B. Fuzzy Questionnaires. Fuzzy Sets and Syst., 6, 1-9, 1981.

83 Bouille, F. Fuzzy Data Processing with the Hypergraph Based Data Structure. In: Int. Conf. on Cybern. and Soc., Tokyo, Japan, Nov. 1978.

84 Bowman, E.H. Consistency and Optimality in Managerial Decision Making. In: Muth, J.F. and Thompson, G.L. (eds) - Industrial Scheduling, Prentice Hall, Englewood Cliffs, New Jersey, USA, 1963.

85 Brainerd, C.J. and Kingma, J. Do children Have to Remember to Reason - A fuzzy - Trace Theory of Transitivity Development. Developmental Review, 4, 311-377, 1984.

86 Brown, C.B. Merging of Fuzzy and Crisp Information. Amer. Soc. Eng. Proc., 106, 123-133, 1980. Discussion 1451-1453, 1980.

87 Brown, C.B. Fuzzy Safety Measure. J. Eng. Mech. Div. - ASCE, 105, 855-872, 1979.

88 Brown, C.B. and Yao, T.P. Fuzzy Sets and Structural Engineering. J. Struct. Eng., 109, 1211-1225, 1983.

89 Brownell, H. and Caramazza, A. Categorizing: A Report. John Hopkins Univ., 1977.

90 Brownell, Hiram H. and Caramazza, Alfonso. Categorizing with Overlapping Categories. Memory and Cognition, 6, 481-490, 1978.

91 Buckley, J.J. Multiple Goal Non Cooperative Conflicts Under Uncertainty: A Fuzzy Set Approach. Fuzzy Sets and Syst., 13, 107-124, 1984.

92 Buckley, J.J. The Multiple Judge, Multiple Criteria Ranking Problem: A Fuzzy Set Approach. Fuzzy Sets and Syst., 13, 25-38, 1984.

93 Buell, D.A. and Kraft, D.H. Evaluation of Fuzzy Retrieval - Systems. Proc. Amer. Soc. Inf. Sci., 18, 298-300, 1981.

94 Buoncristiani, J.F. Probability on Fuzzy - Sets and Epsilon - Fuzzy Sets. J. Math. Anal. and Appl., 96, 24-41, 1983.

95 Buoncristiani, John Faust. Probability on Fuzzy Sets. Dissertation, Boston Univ. 101p, 1980.

96 Capocelli, R.M. and DeLuca, A. Fuzzy Sets and Decision Theory. Inf. and Control, 23, 446-473, 1973.

97 Carlsson, C. An Approach to Handling Fuzzy Problem Structures. Cybern. and Syst., 14, 33-54, 1984.

98 Carlsson, C. Fuzzy Systems: Basis for a Modelling Methodology. In: Proc. 7th European Conf. on Cybernetics and Systems Research II, Vienna, Austria, April 1984.

99 Carlsson, C. Handling of Conflicts in Fuzzy Multiple - Criteria Optimization, Lecture notes in Econ. and Math. Syst., 229, 31-40, 1984.

100 Carlsson, C. Tackling an MCDM - Problem with the Help of Some Results from Fuzzy Set - Theory. European J. Oper. Res., 10, 270-281, 1982.

101 Cavallo, R.E. and Klir, G.J. Reconstruction of Possibilistic Behavior Systems. Fuzzy Sets and Syst., 8, 175-197, 1982.

102 Cerny, M. and Gluckaufova, D. Application of Fuzzy Sets in Multicriterion Decision Problems. Ekon. Mat. Obzor, 14, 1-25, 1978.

103 Chakraborty, M.K. and Das, M. On Fuzzy Equivalence. Fuzzy Sets and Syst., 11, 185-193, 1983.

104 Chanas, S. and Kamburowsk, J. The Use of Fuzzy Variables. Fuzzy Sets and Syst., 5, 11-19, 1981.

105 Chang, C.L. Decision Support in an Imperfect World. In: Proc. of Trends and Appl. Automating Intelligent Behavior Appl. and Frontiers, Gaithersburg, MD, USA, May 1983, 240-247.

106 Chang, C.L. Interpretation and Execution of Fuzzy Programs. In: Zadeh, L. A. et al. (ed) - Fuzzy Sets and Their Application to Cognitive and Decision Processes. Academic Press, New York, USA, 1975.

107 Chang, L.P. and Pavlidis, T. Fuzzy Decision Trees. In: IEEE Int. Conf. on Syst. Man and Cybernetics, Washington, D.C., USA, Nov. 1976, 28-34.

108 Chang, R.L.P. and Pavlidis, T. Applications of Fuzzy Sets in Curve Fitting. Fuzzy Sets and Syst., 2, 67-74, 1979.

109 Chang, Robin Lok Ping. Application of Fuzzy Decision Techniques to Structural Pattern Recognition and Curve Fitting. Dissertation, Princeton Univ., 149 pages, 1976.

110 Chang, S.L. Fuzzy Model and Decision Process. In: Joint National Meeting of The Inst. of Management Science / Oper. Res. Soc. of Amer., New York, USA, May 1978.

111 Chang, S.S.L. On Risk and Decision-Making in a Fuzzy Environment. In: Zadeh, L.A., et al. (ed) - Fuzzy Sets and Their Applications to Cognitive and Decision Processes. Academic Press, New York, USA, 219-226, 1975.

112 Chang, S.S.L. Fuzzy Mathematics, Man and His Environment. IEEE Trans. Syst. Man and Cybern., SMC-2, 92-93, 1972.

113 Chang, S.S.L. and Zadeh, L.A. Fuzzy Mapping and Control. IEEE Trans. Syst. Man and Cybern., SMC-2, 30-34, 1972.

114 Chang, S.Y., Brill, E.D. and Hopkins, L.D. Modeling to Generate Alternatives: A Fuzzy Approach. Fuzzy Sets and Syst., 9, 137-151, 1983.

115 Chatterji, B.N. A Fuzzy Set of Theoretic Approach for Hand Written English Character Recognition. In: Proc. 6th Int. Conf. on Pattern Recog., Munich, Germany, Oct. 1982, 2, 1181. IEEE, New York.

116 Chaudhuri, B.B. and Majumdar, D.D. On Membership Evaluation in Fuzzy Sets. In: Gupta, M.M. and Sanchez, E. - Approximate Reasoning in Decision Analysis, North-Holland Pub., Amsterdam, Netherlands, 3-12, 1982.

117 Chaudhuri, B.B. and Majumdar, D.D. Fuzzy Sets and Possibility Theory in Reliability Studies of Man Machine Systems. In: Proc. Symp. on Policy Analysis and Information Systems, Durham NC, USA, June 1980, 267-273, Plenum Press, New York, USA.

118 Chaudhuri, B.B. and Majumdar D.D. Recognition and Fuzzy Description of Sides and Symmetries of Figures by Computer. Int. J. Syst. Sci., 11, 1435-1445, 1980.

119 Chen, S.C. and Shimura, M. On line recognition of Hand Printed Characters Using Fuzzy Logic. Technical Report on Pattern Recognition and Learning of IECE, PRL74-65, 1974.

120 Chorayan, D.D. Formalization of a Neuronal Ensemble in Terms of the Theory of Fuzzy - Sets. Biofizika, 26, 876-878, 1981.

121 Chrostek, W. Health Hazard Evaluation Determination. Report No. HHE 79-77-605. Towncenter Association Building Rockville, MD. USA. NTIS Report PB 80-147168, 8 p., 1979.

122 Chu, A.T.W., Kalaba, R.E. and Spingarn, K. Comparison of Two Methods for Determining the Weights Belonging to Fuzzy Sets. J. Optimiz. Theory and Appl., 27, 531-538, 1979.

123 Cioffirevilla, C.A. Fuzzy Sets and Models of International - Relations. Amer. J. Political Sci., 25, 129-159, 1981.

124 Clarke, Robert David. Fuzzy Preference Method for Defining Group Preferences. Thesis. Naval Postgrad. School, Monterey, CA, USA, 46 pages, 1980. NTIS Report AD-A084 416/7, 1980.

125 Clements, D. Evaluation of Computer Security Using a Fuzzy Rating Language. ERL Report M 77/41, Univ. of Calif., Berkeley, CA, USA, 1977.

126 Clive, J., Woodbury, M.A. and Siegler, I.C. Fuzzy and Crisp Set - Theoretic Based Classification of Health and Disease; a Qualitative and Quantitative Comparison. In: Proc. 16th Hawaii Int. Conf. on Syst. Sciences, Honolulu, HI, USA, Jan. 1983, 2, 450-462.

127 Cohen, S.A. Fuzziness and Flab - Some Solutions to Research Problems in Learning - Disabilities. J. Special Edu., 10, 129-136, 1976.

128 Cooley, J.W. and Hicks, J.D. A Fuzzy Set Approach to Aggregating Internal Control Judgments. Manage. Sci., 29, 317-334, 1983.

129 Coppo, M. Semantic Support for a Speech Understanding Based on Fuzzy Relations. In: Int. Conf. on Cybern. and Soc., Washington, D.C., USA, May 1976.

130 Cronbach, L.J. Further Evidence on Response Sets and Test Design. Educational Psychol. Measmt., 10, 3-31, 1950.

131 Cumani, A. On a Possibilistic Approach to the Analysis of Fuzzy Feedback - Systems. IEEE Trans. Syst. Man and Cybern., SMC - 12, 417-422, 1982.

132 Czogala, E. and Walichiewicz, L. Synthesis of a Fuzzy and Probabilistic Fuzzy Controllers by Decomposition of Control Rules Derived from Human Operators Actions. In Proc. 1st European Workshop on Real Time Control of Large Scale Systems, Patras, Greece, July 1984.

133 Czogala, E. On Distribution Function Description of Probabilistic Sets and Its Application in Decision Making. Fuzzy Sets and Syst., 10, 21-29, 1983.

134 Czogala, E. and Pedrycz, W. Control Problems in Fuzzy Systems. Fuzzy Sets and Syst., 7, 257-273, 1982.

135 Czogala, E. and Pedrycz, W. Some Problems Concerning the Construction of Algorithms of Decision-Making in Fuzzy Systems. Int. J. Man-Mach. Stud., 15, 201-211, 1981.

136 Dale, A.I. Probability, Vague Statements and Fuzzy-Sets. Philosophy of Science, 47, 38-55, 1980.

137 Damerau, F.J. Fuzzy Adjectives. Linguistics, N196, 57-64, 1977.

138 Dekerf, J. A Bibliography on Fuzzy Sets. J. Comput. and Appl. Maths., 1, 205-212, 1975.

139 Diamond, P. Fuzzy Chaos. Dept. of Math. Univ. of Queensland, Brisbane, Australia, 1975.

140 Dickinson, Dean Berkley. An Expert System Using Fuzzy Set Representations for Rules and Values to Make Management Decisions in a Business Game. Dissertation, Univ. of Arizona, 183 pages, 1984.

141 Dijkman, J.G., Van Haeringen, H. and Delange, S.J. Fuzzy Numbers. J. Math. Anal. and Appl., 92, 301-341, 1983.

142 Dimitrov, V. and Cuntchev, O. Efficient Governing of Humanistic Systems by Fuzzy Instructions. In: Rose, J. and Bilciv, C. (ed) - Modern Trends in Cybernetics and Systems, 2, 125-130, 1977.

143 Dimitrov, V., Welcher, W., and Barnev, P. Optimal Fuzzy Control of Humanistic Systems. In: Proc. Int. Fed. Auto. Control 6th Triennial Congress, Boston and Cambridge, Mass., USA, Aug. 1975.

144 Dimitrov, V.D. Social Choice and Self Organization under Fuzzy Management. Kybernetes, 6, 153-156, 1977.

145 Dimitrov, V.O. Programme System for Social Choice Under Fuzzy Managing. In Proc. IFIP Congress, Toronto, Canada, Aug 1977.

146 Dinnola, A. Sessa, S. and Pedrycz, W. Decomposition Problem of Fuzzy Relations. Int. J. Gen. Syst., 10, 123-133, 1985.

147 Dinola, A. and Ventre, A.G.S. A Relativization of the Concept of Synthesis in Fuzzy Set-Theory. Inf. Sci., 34, 179-186, 1984.

148 Dishkant, H. About Membership Function Estimation. Fuzzy Sets and Syst., 5, 141-147, 1981.

149 Dohnal, M. Fuzzy Simulation of Industrial Problems. Comput. Ind., 4, 347-352, 1984.

150 Dohnal, M. Linguistics and Fuzzy Models. Comput. Ind., 4, 341-345, 1984.

151 Dombi, J. A General Class of Fuzzy Operators, the De Morgan Class of Fuzzy Operators and Fuzziness Measures Induced by Fuzzy Operators. Fuzzy Sets and Syst., 8, 149-163, 1982.

152 Dominowski, R.L. Sources of Fuzziness Regarding Natural Categories. Bull. Psychonomic Soc., 18, 52, 1981.

153 Donnell, M.L. A Theory of Fuzzy Choice. In Joint National Meeting of the Inst. of Manage. Sci. / Oper. Res. Soc. of Amer., New York, USA, May 1978.

154 Dreyfuss, G.R. et al. On the Psycholinguistic Reality of Fuzzy Sets. In Grossman, R.E., San, L.J. and Vance, T.J., (eds) - Functionalism. Univ. of Chicago Press. Chicago, IL, USA, 135-149, 1975.

155 Driankov, D. Dialectics, Fuzzy Sets and Models of Human Actions. In: 5th European Meeting on Cybern. and Syst. Res., Vienna, Austria, April 1980. Also in: Prog. in Cybern. and Syst. Res., 8 and 9, 1980, Hemisphere Pub. Co., Washington, D.C., USA.

156 Dubois, D. and Prade, H. On Distances Between Fuzzy Points and Their Use for Plausible Reasoning. In: Proc. Int. Conf. on Syst. Man and Cybern., Bombay and New Delhi, India, 29th Dec. 1983 - 7th Jan. 1984, 1, 300-303.

157 Dubois, D. and Prade, H. Ranking Fuzzy Numbers in the Setting of Possibility Theory. Inf. Sci., 30, 183-224, 1983.

158 Dubois, D. and Prade, H. A Class of Fuzzy Measures Based on Triangular Norms - A General Framework for the Uncertain - Information. Int. J. Gen. Syst., 8, 43-61, 1982.

159 Dubois, D. Processing Vague Queries in Man Machine Systems: A Fuzzy Set Approach. In: Proc. IFAC Symp. on Theory and Appl. of Digital Control, New Delhi, India, Jan. 1982, 413-416.

160 Dubois, D. and Prade, H. Additions of Interactive Fuzzy Numbers. IEEE Trans. Automatic Control, 26, 926-936, 1981.

161 Dubois, D. and Prade, H. Fuzzy - Sets and Systems - Theory and Applications (Book). Academic Press, New York, USA, 1980.

162 Dubois, D.J. and Prade, H.M. Various Kinds of Interactive Addition to Fuzzy Numbers. Application to Decision Analysis in Presence of Linguistic Probabilities. In: Proc. 18th IEEE Conf. on Decision and Control Including the Symp. on Adaptive Processes, Fort Lauderdale, FL, USA, Dec. 1979, Part II, 783-787. IEEE, New York, USA.

163 Dubois, D. and Prade, H. Fuzzy Data - Processing Shortest Paths Algorithms. (French). Rario - Recherche Operationnelle - Operations Research, 12, 213-227, 1978.

164 Dubois, D. and Prade, H. Operations on Fuzzy Numbers. Int. J. Sys. Sci., 9, 613-626, 1978.

165 Ebanks, B.R. On Measures of Fuzziness and Their Representations. J. Math. Anal. and Appl., 94, 24-37, 1983.

166 Edmonds, E.A. Domains of Interest in Fuzzy - Sets. Int. J. Man-Mach. Stud., 15, 461-468, 1981.

167 Efstathiou, J. and Tong, R.M. Ranking Fuzzy - Sets - A Decision Theoretic Approach. IEEE Tran. Syst. Man and Cybern., SMC - 12, 655-659, 1982.

168 Efstathiou, J. and Rajkovic, V. Multi-Attribute Decision Making a Fuzzy Heuristic Approach. Int. J. Man-Mach. Stud., SMC - 12, 141-156, 1980.

169 Elkins, J. Learning-Disability Research - Structured Reactions to Random Thoughts About a Fuzzy Subject. J. Special Edu., 10, 149-155, 1976.

170 Ernst, C.J. An Approach to Management Expert Systems Using Fuzzy Logic. In: Lasker, G.E. (ed) - Applied Systems and Cybernetics, 6, 2898-2905, 1981.

171 Eschabch, M. and Cunnyngham, J. Logic of Fuzzy Bayesian Inference. In: Proc. 13th Annual Pittsburgh Conf. on Modeling and Simulation, Pittsburgh, PA, USA, April 1982.

172 Eshragh, F. and Mamdani, E.H. A General Approach to Linguistic Approximation. Int. J. Man-Mach. Stud., 11, 501-519, 1979.

173 Esogbue, A.O. Measurement and Valuation of Fuzzy Mathematical Model for Medical Diagnosis. Fuzzy Sets and Syst., 10, 223-242, 1983.

174 Esogbue, A. Fuzzy Sets and the Modeling of Physician Decision Processes. I - The Initial Interview Information Gathering Session. Fuzzy Sets and Syst., 2, 279-291, 1979.

175 Eustace, P. Fuzzy Logic Getting More than Yes - No Answer. Engineer, 255, 44-45, 1982.

176 Ezhkova, I.V. and Pospelov, D.A. Decision - Making on Fuzzy Premises. 2. Deduction Schemes. Eng. Cybern., 16, 1-6, 1978.

177 Felling Out Fuzzy Sets. Electron. Des., 33, 99, May 2, 1985.

178 Fellinger, William Lowell. Specifications for a Fuzzy System Modeling Language. Dissertation, Oregon State Univ., 93 p., 1975.

179 Feron, R. On the Notions of Distance and Derivation in a Fuzzy Structure. In: 5th European Meeting on Cybern. and Syst. Res., Vienna, Austria, April 1980. Also in: Prog. in Cybern. and Syst. Res., 8 and 9, 1980. Hemisphere Pub., Washington, D.C., USA.

180 Fevrier, P. Expressing Measurements by Means of Fuzzy Sets. In: 5th European Meeting on Cybern. and Syst. Res., Vienna, Austria. Also in: Prog. in Cybern. and Syst. Res., 8 and 9, 1980. Hemisphere Pub., Washington, D.C., USA.

181 Fevrier, P. On the Representation of the Results of Measurements by Fuzzy Sets. In Prog. in Cybern. and Syst. Res., 3, 321-324, 1978. John Wiley, New York, USA.

182 Fienberg, S.E. and Gong, G.D. Comment on Graphical Methods for Assessing Logistic Regression Models By Landwehr, Pregibon and Shoemaker. NTIS Report AD-A135684/9, 18 pages, 1983.

183 Fiksel, J. Application of Fuzzy Set and Possibility Theory to Systems Management. In Lasker, G.E. (ed) - Applied Systems and Cybernetics, 6, 2966-2973, 1981. Pergamon Press, New York, USA.

184 Fine, K. Vagueness, Truth and Logic. Synthese, 30, 265-300, 1975.

185 Flachs, G.M. et al. Real Time Video Tracking Concepts. NTIS Report AD-A071 792/6, 45p., 1979.

186 Fox, J. Towards a Reconciliation of Fuzzy Logic and Standard Logic. Int. J. Man-Mach. Stud., 15, 213-220, 1981.

187 Fox, J. Fuzzy Logic and Natural Inference. In: 5th European Meeting on Cybern. and Syst. Res., Vienna, Austria, April 1980. Also in: Prog. in Cybern. and Syst. Res., 8 and 9, 1980, Hemisphere Pub. Co., Washington D.C., USA.

188 Francioni, J.M., Kadel, A. and Eastman, C.M. Imprecise Decision Tables and Their Optimization. In: Gupta, M.M. and Sanchez, E. (ed) - Approximate Reasoning in Decision Analysis, North-Holland Pub., Amsterdam, Netherlands, 89-96, 1982.

189 Franksen, O.I. Fuzzy - Sets Subjective Measurements and Utility. Int. J. Man-Mach. Stud., 11, 521-545, 1979.

190 Freeling, A.N.S. Fuzzy - Sets and Decision - Analysis. IEEE Trans. Syst. Man and Cybern., SMC-10, 341-354, 1980.

191 Friedrich Freksa, Christian. Linguistic Pattern Characterization and Analysis. Dissertation, Univ. California, Berkeley, 1981.

192 Furukawa, M., Nakamura, K. and Oda, M. Fuzzy Varient Process of Memories. Annual Conf. Records of Society of Instrumentation and Control Engineers, Japan, 1973.

193 Furukawa, M., Nakamura, K. and Oda, M. Fuzzy Models of Human Decision - Making Process. Annual Conf. Records of Japanese Automation and Automatic Control Engineers, 1972.

194 Furuta, H. and Shiraishi, N. Fuzzy Importance in Fault Tree Analysis. Fuzzy Sets and Syst., 12, 205-213, 1984.

195 Gaines, B.R. Precise Past -- Fuzzy Future. Int. J. Man-Mach. Stud., 19, 117-134, 1983.

196 Gaines, B.R. Fuzzy and Probability Uncertainty Logics. Inf. and Control, 38, 154-169, 1978.

197 Gaines, B.R. Foundations of Fuzzy Reasoning. In M.M. Gupta, G.N. Saridis and B.R. Gaines (ed) - Fuzzy Automata Decision Processes. North-Holland Pub., Amsterdam, Netherlands, 19-76, 1977.

198 Gaines, B.R. and Kohout, L.J. Fuzzy Decade -- Bibliograhy of Fuzzy Systems and Closely Related Topics. Int. J. Man-Machine Stud., 9, 1-68, 1977.

199 Gaines, B.R. Behavior - Structure Transformation Under Uncertainty. Int. J. Man-Mach. Stud., 8, 337-365, 1976.

200 Gaines, B.R. Foundations of Fuzzy Reasoning. Int. J. Man-Mach. Stud., 8, 623-668, 1976.

201 Gaines, B.R. Understanding Uncertainty. In Proc. Workshop on Discrete Systems and Fuzzy Reasoning, Queens Mary College, London, UK, 1976.

202 Gaines, B.R. Why Fuzzy Reasoning? In: Mamdani, E.H. and Gaines, B.R. (ed) - Discrete Systems and Fuzzy Reasoning, EES-MMS-DSFR-76, Queen Mary College, University of London, (Workshop Proceedings), 1976.

203 Gale, S. Inexactness, Fuzzy Sets and Foundations of Behavioral Geography. Geographical Analysis, 4, 337-349, 1972.

204 Gath, I., Lehmann, D. and Bar-On, E. Fuzzy Clustering of EEG Signals and Vigilance Performance. Int. J. Neurosci., 20, 303-312, 1983.

205 Gerla, G. Sharpness Relation and Decidable Fuzzy Sets. IEEE Trans. Autom. Control, AC-27, 1113, 1982.

206 Gero, J.S. and Volfneuk, M. Building Fuzzy CAD Systems. In: Proc. 4th Int. Conf. and Exhibition on Computers in Design Engineering (CAD 80), Brighton, England, March 1980, 74-79. IPC Sci. and Tech. Press, Guildford, England.

207 Giles, R. The Practical Interpretation of Fuzzy Concepts. Human Syst. Manage., 3, 263-264, 1982.

208 Giles, R. Semantics for Fuzzy-Reasoning. Int. J. Man-Mach. Stud., 17, 401-415, 1982.

209 Giles, R. A Formal System for Fuzzy Reasoning. Fuzzy Sets and Syst., 2, 233-257, 1979.

210 Giles, R. A Logic for Subjective Belief. In: Harper, W. and Hooker, C.A., (ed) - Foundations of Probability Theory, Statistical Inference and Statistical Theories of Science, D. Reidel, Dordrecht, Holland, 41-72, 1976.

211 Giles, R. Lukasiewicz Logic and Fuzzy Set-Theory. Int. J. Man-Mach. Stud., 8, 313-327, 1976.

212 Gitman, Israel. Organization of Data: A Model and Computational Algorithms That Use the Notion of Fuzzy Sets. Dissertation, McGill Univ., Canada, 1970.

213 Goguen, J.A. Objects. Int. J. Gen. Syst., 1, 237-243, 1975.

214 Goguen, J.A. Concept Representation in Natural and Artificial Languages -- Axioms, Extensions and Applications for Fuzzy Sets. Int. J. Man-Mach. Stud., 6, 513-561, 1974.

215 Goguen, J.A. The Logic of Inexact Concepts. Synthese, 19, 325-373, 1969.

216 Gold, D. Social-Research and Social-Problems - Toward a Structural Explanation of a Fuzzy Association. Pacific Sociological Rev., 22, 275-283, 1979.

217 Good, I.J. Subjective Probability as the Measure of a Non-Measurable Set. In: Nagel, E., Suppes, P. and Tarski, A. (eds) - Logic, Methodology and Philosophy of Science, Stanford University Press, Stanford, CA, USA, 319-329, 1962.

218 Gorzakzany, M.B., Kiszka, J.B. and Stachowicz, M.S. Some Problems of Studying Adequacy of Fuzzy Models. In: Yager, R.R. (ed) - Fuzzy Set and Possibility Theory. Pergamon Press, New York, USA, 1982.

219 Gottinger, H.W. On the (Ir)Relevancy of Fuzzy-Sets Theories. Human Syst. Manage., 4, 314-315, 1984.

220 Gottinger, H.W. A Fuzzy Algorithmic Approach to the Definition of Complex or Imprecise Concepts. In Conf. on Systems Theory, Univ. of Biefeld, April 1975.

221 Gottinger, H.W. Toward a Fuzzy Reasoning in Behavioral Science. Cybernetica, 16, 113-135, 1973.

222 Gottwald, S. Fuzzy Sets and Their Applications. An Overview (German). Elektron. Informationsverab. Kybern., 17, 207-235, 1981.

223 Gottwald, S. A Note on Measures of Fuzziness (German). Elektron. Informationsverab. Kybern., 15, 221-223, 1979.

224 Gottwald, S. Set Theory for Fuzzy Sets of Higher Level. Fuzzy Sets and Syst., 2, 125-151, 1979.

225 Gottwald, S. Universes of Fuzzy Sets Closed Under Fuzification. In: Current Topics in Cybern. and Syst., Amsterdam, Netherlands, Aug. 1978, 365-366. Springer-Verlag, Berlin, Germany.

226 Govindaraj, T. An Approach to Modeling the Human Supervisor in a Time-Constrained Environment Using Artificial Intelligence Methodologies. In: Proc. Int. Conf. on Cybern. and Soc., Seattle, WA, USA, Oct. 1982, 511-514. IEEE, New York, USA.

227 Govindaraj, Thiruvenkatasamy. Modelling Human Attention Allocation Strategies in Situations with Competing Criteria. NTIS AD-A108 509/1, 22 pages, 1981.

228 Govindaraj, T. et al. Human Attention Allocation Strategies in Situation with Competing Criteria Development of an Experimental Paradigm and Outlines for Model. In: Proc. 8th Triennial World Cong. of the IFAC, Kyoto, Japan, 7, 3501-3506, 1981.

229 Grattanguinness, I. Fuzzy Membership Mapped Onto Intervals and Many-Valued Quantities. Zeit, Math. Logik und Grund. Math., 22, 149-160, 1976.

230 Green, N.L. Decision Support with Fuzzy Technology. In: Proc. 16th Ann. Meeting of the Amer. Inst. for Decision Sci., Toronto, Ont., Canada, 682 pages, 1984.

231 Grudzewski, W.M. and Pelech, A. Modelling a Structure of the Project in the Fuzzy Sets Language. In: 5th Int. World Cong. on Project Implementation and Management Bridging the Gap, Birmingham, England, 1976, Part IV, 339-349.

232 Gupta, M.M. Fuzzy Information and Decision Processes. In: Proc. IFAC Symp. on Theory and Appl. of Digital Control, New Delhi, India, Jan. 1982, 409-411. Pergamon Press, Oxford, Eng.

233 Gupta, M.M. Fuzzy Concepts and Its Future. In: Joint National Meeting of The Inst. Manage. Sci./Oper. Res. Soc. Amer. New York, NY, USA, May 1978.

234 Gupta, M.M. Fuzziness and Human Behavior in Decision Making. In: Joint National Meeting of Oper. Res. Soc. Amer. and The Inst. Manage. Sci., Miami, FL, USA, Nov. 1976. Paper No. 10.6.

235 Gupta, M.M. Fuzziness and Human Consciousness in Decision Making. In: Joint National Meeting of The Oper. Res. Soc. Amer./The Inst. Manage. Sci., Miami, FL, USA, Nov. 1976.

236 Gupta, M.M. and Sanchez, E. Approximate Reasoning in Decision Analysis. (Book) North-Holland Pub., Amsterdam, Netherlands, 1982, 455 pages.

237 Gupta, M.M. and Sanchez, E. Fuzzy Information and Decision Processes. (Book) North-Holland Pub., Amsterdam, Netherlands, 1982, 451 pages.

238 Gupta, M.M. Ragade, R.K. and Yager, R.R. Advances in Fuzzy Set-Theory and Applications. (Book) North-Holland Pub., Amsterdam, Netherlands, 1979, 753 pages.

239 Gupta, M.M., Sardis, G.N., and Gaines, B.R. Fuzzy Automata and Decision Processes. (Book) North-Holland Pub., Amsterdam, Netherlands, 1977, 496 pages.

240 Gusev, L.A. and Smirnova, I.M. Fuzzy Set Theory and Applications. (Survey) Autom. and Remote Control, 34, 739-755, 1973.

241 Haack, S. Do We Need Fuzzy Logic. Int. J. Man-Mach. Stud., 11, 437-445, 1979.

242 Hacker, W. and Sucro, S.V. Study of Influence of Class Number and Class Size on Formation of Fuzzy-Sets in Configural Objects. Zeit. fur Psych., 188, 26-42, 1980.

243 Hafez, Hala Adel. An Investigation of Biochemical, Physiological and Environmental Heat Stresses Associated with Manual Lifting in Hot Environments (Fuzzy Sets, Acclimatization, Training). Dissertation, Texas Tech. Univ., 327 pages, 1984.

244 Hajnal, M. and Koczy, L.T. A Fuzzy Approach to Texture Analysis. In: 5th European Meeting on Cybernetics and Systems Research, Vienna, Austria, Apr. 1980. Also In: Prog. in Cybern. and Syst. Res., 8 and 9, 1980, Hemisphere Publishing Co., Washington, D.C., USA.

245 Hannan, Edward L. Fuzzy Decision Making with Multiple Objectives and Discrete Membership Functions. Int. J. Man-Mach. Stud., 18, 49-54, 1983.

246 Harris, J.I. Fuzzy Sets: How to be Imprecise Precisely. DOAE Research Working Paper, Ministry of Defense, Byfleet, Surrey, UK, 1974.

247 Hart, W.D. Probability as a Degree of Possibility. Notre Dame J. Formal Logic, 13, 286-288, 1972.

248 Haynor, Anthony Leonard. A Systems Approach to "Personality and Social Structure" With an Application to Selected Writings on Modernity. Dissertation, Rutgers Univ., NJ, 278 pages., 1983.

249 Hendy, R.J., White, E.T. and Hendry, J.E. Supervisory Control Using Heuristic Methods. Chemeca 81. Design for Change. In: Proc. 9th Australasian Chem. Eng. Conf., Christchurch, New Zealand, Sep. 1981, 377-384.

250 Henneman, Richard L. Measures of Human Problem Solving Performance in Fault Diagnosis Tasks. IEEE Trans. on Syst., Man and Cybern., SMC-14, 99-111, 1984.

251 Hersh, H.M. Influence of Population Distributions on Form of Fuzzy Membership Functions. In: Joint National Meeting of The Inst. Manage. Sci./Oper. Res. Soc. Amer., New York, NY, USA, May 1978.

252 Hersh, Harry Morton. Fuzzy Reasoning: The Integration of Vague Information. Dissertation, Johns Hopkins Univ., 105 Pages, 1976.

253 Hersh, H.M. and Caramazza, A. A Fuzzy Set Approach to Modifiers and Vagueness in Natural Language. J. Exp. Psych., 105, 254-276, 1976.

254 Hersh, H.M. and Spiering, J. How Old is Old? In Eastern Psych. Meeting, New York, USA, 1976.

255 Hersh, H. and Caramazza, A. Integrating Verbal Quantitative Information. Bull. Psychonomic Soc., 6, 588-591, 1975.

256 Hersh, H.M. and Caramazza, A. The Quantification of Vague Concepts. In: Psychometric Society Meeting, Iowa City, Iowa, USA, Apr. 1975.

257 Higashi, Masahiko. A systems Modelling Methodology: Probabilistic and Possibilistic Approaches. Dissertation, SUNY at Binghampton, 299 pages, 1984.

258 Higashi, M. and Klir, G.J. On Measures of Fuzziness and Fuzzy Complements. Int. J. Gen. Syst., 8, 169-180, 1982.

259 Hipel, K.W. Fuzzy Set Techniques in Decision Making. Resour. Manage. and Optimiz. (Switzerland), 2, 187-203, 1983.

260 Hirai, H., Asai, K. and Katajima, S. Fuzzy Automation and Its Application to Learning Control Systems. Memo. Fac. of Eng., Osaka City Univ., 10, 67-73, 1968.

261 Hirota, K. On Identification of Fuzzy Systems Under the Existence of Vagueness. In: Proc. IFAC Symp. on Theory and Appl. of Digital Control, New Delhi, India, Jan. 1982, 445-448. Pergamon Press, Oxford, Eng.

262 Hirsch, G. et al. Phonemic Classification Using a Fuzzy Dissimilitude Relation. Fuzzy Sets and Syst., 5, 267-275, 1981.

263 Hirtle, Stephen Craig. Hierarchical Structure in Cognitive Maps. Dissertation, Univ. of Michigan, 119 pages, 1982.

264 Hisdal, E. The If Then Else Statement and Interval-Valued Fuzzy-Sets of Higher Type. Int. J. Man-Mach. Stud., 15, 385-455, 1981.

265 Hogarth, R.M. Cognitive Processes and the Assessment of Subjective Probability Distribution. J. Amer. Stat. Assn., 70, 271-294, 1975.

266 Hohle, U. Measures on Fuzzy Sets. (German). Zeit. fur Wahrscheinl. und ver. Geb., 36, 179-188, 1976.

267 Holstun, J. and Tinker, J. Fuzzy Realpolitik. Nation, 240, 482, 1985.

268 Honda, N. Fuzzy Sets. J. Inst. Electron. Comm. Eng., 54, 1359-1363, 1971.

269 Hormann, A.M. Machine-Aided Value Judgments Using Fuzzy Set Techniques, SP-3590. System Development Corporation, Santa Monica, CA, USA, 1971.

270 Horvath, M.J., Kass, C.E. and Ferrell, W.R. An Example of the Use of Fuzzy Set Concepts in Modeling Learning Disability. Amer. Educat. Res. J., 17, 309-324, 1980.

271 Horvath, Michael John. An Identification Procedure in Learning Disability Through Fuzzy Set Modeling of a Verbal Theory. Dissertation, Univ. of Arizona, 228 pages, 1978.

272 Hunt, R.M. and Rouse, W.B. A Fuzzy Rule-Based Model of Human Problem-Solving. IEEE Trans. on Syst., Man and Cybern., SMC-14, 112-120, 1984.

273 Hunt, Ruston McClellan. Human Pattern Recognition and Information Seeking in Simulated Fault Diagnosis Tasks. Dissertation, Univ. of Illinois, Urbana-Champaign, 101 pages, 1981.

274 Irtem, A. Fuzziness and Consciousness or Goals of Creation. Cybernetica, 20, 253-263, 1977.

275 Ishikawa, A. and Mïeno, H. Design of a Video Information-System Upon Fuzzy Information-Theory. Oper. Res., 23, 52PB, 375, 1975.

276 Ishizuka, M. Inference Methods Based on Extended Dempster and Shafer's Theory for Problems with Uncertainty/Fuzziness. New Gener. Comput., 1, 159-168, 1983.

277 Ishizuka, M., Fu, K.S. and Yao, J.T.P. Inference Procedures Under Uncertainty for the Problem Reduction Method. Inf. Sci., 28, 179-206, 1982.

278 Izumi, K., Tanaka, H. and Asai, K. Fuzzy Logical Construction of Linguistic Model Hedges. (Japanese). Tran. Inst. Electron. and Commun. Eng. Japan., 67D, 655-660, 1984.

279 Jain, R. and Haynes, S.S. Imprecision in Computer Vision. Computer, 15, 39-48, 1982.

280 Jain, R., Fuzzism and Real World Problems. In: Proc. Symp. on Policy Analysis and Information Systems, Durham, NC, USA, June 1980, 129-132. Plenum Press, New York, USA.

281 Jain, R. and Stallings, W. Comments on Fuzzy Set Theory Versus Bayesian Statistics. IEEE Trans. on Syst. Man and Cybern., SMC-8, 332-333, 1978.

282 Jain, R. Analyzing a Real World Scene Sequence Using Fuzziness. In: IEEE Conf. on Decision and Control Including 16th Symp. on Adaptive Processes and a Special Symp. on Fuzzy Set Theory and Applications, New Orleans, LA, USA, Dec. 1977, Part II.

283 Jain, R. Decision Making in Presence of Fuzziness and Uncertainty. In: Proc. 1977 IEEE Conf. on Decision and Control Including 16th Symp. on Adaptive Processes and a Special Symp. on Fuzzy Set Theory and Applications, New Orleans, LA, USA, Dec. 1977, Part II, 1318-1323.

284 Jain, R. Procedure for Multiple-Aspect Decision-Making Using Fuzzy Sets. Int. J. Syst. Sci., 8, 1-7, 1977.

285 Jain, R. Decision Making in Presence of Fuzzy Variables. IEEE Trans. on Syst., Man and Cybern., SMC-6, 698-703, 1976.

286 Jain, R. Outline of an Approach for Analysis of Fuzzy Systems. Int. J. Cont., 23, 627-640, 1976.

287 Jain, R. Tolerance Analysis Using Fuzzy Sets. Int. J. Syst. Sci., 7, 1393-1401, 1976.

288 Jajinski, R., Pelech, A. and Wojcicka, M. An Analysis of Weak-Defined Information from Questionnaire-Based Research with the Use of Fuzzy Sets Languages. KOMM No. 219, INST. Org. i Zarzaldzania Politech. Wroclawskieg, Wroclaw, Poland, 1977.

289 James, M. The Computer Brain: Fuzzy Programs. Electron. and Comput. Mon., 3, 58-60, 1983.

290 Jolles, E. Decision-Making in Successive Stages in Fuzzy Environment - Complete Resolution (French). Rev. d'Econ. Politique, 86, 416-436, 1976.

291 Joly, H. et al. Application of Fuzzy Set Theory to the Evaluation of Cardiac Functions. In: MEDINFO 80 Proc. 3rd World Conf. on Medical Information, Tokyo, Japan, 1980, Part I, 91-95. North-Holland Pub., Amsterdam, Netherlands.

292 Jones, W.T. Fuzzy Set Characterization of Interaction in Scientific-Research. J. Amer. Soc. Inf. Sci., 27, 307-310, 1976.

293 Jumarie, G. New Approach to the Transformation of Random Experiments Involving Fuzzy Observations. In: Cybern. and Syst. Res. 7th European Meeting, Vienna, Austria, Apr. 1984, II.

294 Jumarie, G. A Subjectivistic Calculus. Cybernetica, 23, 47-87, 1980.

295 Kacprzyk, J. Softer Optimization and Control-Models via Fuzzy Linguistic Quantifiers. Inf. Sci., 34, 157-178, 1984.

296 Kacprzyk, J. A Generalization of Fuzzy Multistage Decision-Making and Control via Linguistic Quantifiers. Int. J. Control, 38, 1249-1270, 1983.

297 Kacprzyk, J. Multistage Decision-Making under Fuzziness - Theory and Applications. (Book). Verlag TUV Reheinland, Koln, W. Germany, 1983, 128 pages.

298 Kacprzyk, J. and Staniewski, P. Long-Term Inventory Policy-Making Through Fuzzy Decision-Making Models. Fuzzy Sets and Syst., 8, 117-132, 1982.

299 Kacprzyk, J. Fuzzy Termination Time in Decision-Making in a Fuzzy Environment. In: Current Topics in Cybernetics and Systems, Amsterdam, Netherlands, Aug. 1978, 368-369. Springer-Verlag, Berlin, Germany.

300 Kacprzyk, J. Fuzzy Integral as Model of Operator in Complex Automation Systems. Proc. VII Nat. Conf. Autom. Contr., 1977.

301 Kacprzyk, J. Fuzzy Set Theoretic Approach to Optimal Assignment of Work Places. In: Proc. IFAC Symp. on Large Scale Systems Theory and Applications. Udine, Italy, June 1976, 123-132. Ed. by G. Guardabassi and A. Locatelli, Instrument Soc. of America, Publication.

302 Kaenel, P.A. Von. Fuzzy Codes and Distance Properties. Fuzzy Sets and Syst., 8, 199-204, 1982.

303 Kakoff, G. Hedges: A Study in Meaning Criteria and the Logic of Fuzzy Concepts. J. Philos. Logic., 2, 458-508, 1973.

304 Kalamanson, D. and Stegall, F. Cardiovascular Investigations and Fuzzy Sets Theory. Amer. J. Cardio., 35, 30-34, 1975.

305 Kandel, A. Fuzzy Techniques in Pattern-Recognition. (Book) John Wiley, New York, USA, 356 pages, 1982.

306 Kandel, A. Observations in a Fuzzy Environment. In Proc. 20th IEEE Conf. on Decision and Control, San Diego, CA, USA, Dec. 1981, 3, 1110-1111, IEEE, New York, USA.

307 Kandel, A. and Byatt, W.J. Fuzzy Sets, Fuzzy Algebra, and Fuzzy Statistics. Proc. IEEE, 66, 1619-1639, 1978.

308 Kandel, A. and Davis, H.A. The First Fuzzy Decade: Bibliography on Fuzzy Sets and Their Applications. CSR140 Computer Science Dept., New Mexico Institute of Mining and Technology, Socorro, NM, USA, Apr. 1976.

309 Kandal, A. On Analysis of Fuzzy Logic. In: Proc. 6th Hawaii Int. Conf. on System Sci. with Subconference on Regional and Urban Systems, Modeling Analysis and Decision Making, Honolulu, HI, Jan. 1973. 2 Vols. IEEE, New York, USA.

310 Karwowski, W. et al. Fuzzy Approach in Psychophysical Modeling of Human Operator-Manual Lifting System. Fuzzy Sets and Syst., 14, 65-76, 1984.

311 Karwowski, W. and Ayub, M.M. Fuzzy Modeling of Stresses in Manual Lifting Tasks. Ergonomics, 27, 641-649, 1984.

312 Karwowski, W. A Fuzzy Set Based Model of the Interaction Between Stresses Involved in Manual Lifting Tasks. Dissertation, Texas Tech. Univ., 231 pages, 1983.

313 Kaufmann, A. and Gupta, M.M. Introduction to Fuzzy Arithmatic. (Book). Van Nostrand, New York, USA, 1985, 351 Pages.

314 Kaufmann, A. Progress in Modeling of Human Reasoning of Fuzzy Logic. In: Gupta, M.M. Saridis, G.N. and Gains, B.R. (eds) - Fuzzy Automata and Decision Processes, North-Holland Pub., Amsterdam, Netherlands, 1977.

315 Kaufmann, A. Fuzzy Sets Theory vs. Probability Theory? No! Fuzzy Sets Theory and Probability Theory. In: Joint National Meeting of The Oper. Res. Soc. Amer./The Inst. Manage. Sci., Miami, FL, USA, Nov. 1976.

316 Kaufmann, A. Theory of Fuzzy Sub-Sets and Its Application in Social-Sciences. (French). Economie Appliquee, 29, 469-478, 1976.

317 Kaufmann, A. Theory of Fuzzy Subsets. (Book). Academic Press, New York, USA, 1975.

318 Kaufmann, A. Introduction a la Theorie des Sous-Ensembles Flous, 3: Applications a la Classification et la Reconnaisance des Formes, aux Automes et aux Systems, aus Choix des Critares. (Book). Masson et Cie, Paris, France, 1975.

319 Kaufmann, A. Introduction a la Theorie des Sous-Ensembles Flou, 2: Applications a la Linguistique et a la Semantique. (Book) Masson et Cie, Paris, France, 1975.

320 Kaufmann, A. Introduction to the Theory of Fuzzy Subsets. Vol 1. (Book) Academic Press, New York, USA, 1975.

321 Kaufmann, A. Introduction to a Fuzzy Theory of the Human Operator. Special Interest Discussion Session on Fuzzy Automata and Decision Processes. In 6th IFAC World Congress, Boston, Mass., USA, Aug. 1975.

322 Kaufmann, A. Introduction a la Theorie des Sous-Ensembles Flous, 1: Elements Theoretiques de Base. (Book). Masson et Cie, Paris, France, 1973.

323 Ke, Jyh-Sheng. Database Modeling Techniques with Application to Fuzzy Query Translation. Dissertation, Univ. of Illinois at Chicago Circle, 238 pages, 1978.

324 Kempton, Willett Main. Grading of Category Membership in the Folk Classification of Ceramics: Socially Correlated Cognitive Variation and a Test of the Theory of Fuzzy Subsets. Dissertation, Univ. of Texas at Austin, 398 pages, 1977.

325 Kerridge, D.F. Inaccuracy and Inference. J. Roy. Stat. Soc., (Ser B), 23, 184-194, 1961.

326 Khatchadourian, H. Vagueness, Meaning and Absurdity. Amer. Philos. Quart., 2, 119-129, 1965.

327 Kickert, W.J.M. and Mamdani, E.H. Analysis of a Fuzzy Logic Controller. Fuzzy Sets and Syst., 1, 29-44, 1978.

328 Kickert, W.J. Fuzzy Theories on Decision-Making. (Book). Leiden/Boston, Martinus Nijhoos, Belgium/USA, 1978.

329 Kickert, W.J.M. and Koppelaar, H. Application of Fuzzy Set-Theory to Syntactic Pattern-Recognition of Handwritten Capitals. IEEE Trans. on Syst., Man and Cybern., SMC-6, 148-151, 1976.

330 Kim, Sung. A Fuzzy Set Model of Consumer Behavior. Dissertation, Washington State Univ., 87 pages, 1981.

331 Kimsawaide, Yos. Statistics of Fuzzy Numbers. Masters Thesis, Univ. of Louisville, 89 pages, 1980.

332 King, P.J. and Mamdani, E.H. The Application of Fuzzy Control Systems to Industrial Processes. In: Gupta, M.M. Sardis, G.N. and Gains, B.R. (eds) - Fuzzy Automata and Decision Processes, North-Holland, Amsterdam, Netherlands, 1977, 321-330.

333 Kling, R. Fuzzy Planner: Reasoning with Inexact Concepts in a Procedural Problem Solving Language. J. Cybern., 3, 1-16, 1973.

334 Knopfmacher, J. On Measures of Fuzziness. J. Math. Anal. and Appl., 49, 529-534, 1975.

335 Kobayashi, A. et al. Accelerated Diversity in Measurement (Fuzziness). (Japanese). J. Soc. Instrum. and Control Eng. (Japan), 22, 17-46, 1983.

336 Kochen, M. Is Fuzzy Set-Theory Appropriate for Models of Human-Behavior. Human Syst. Manage., 3, 268-269, 1982.

337 Kochen, M. Enhancement of Coping Through Blurring. Fuzzy Sets and Syst., 2, 37-52, 1979.

338 Kochen, M. Imprecision in Coping and Attending Processes. In: Proc. IEEE Conf. on Decision and Control, New Orleans, LA, USA, Dec. 1977, Part II, 1324-1329.

339 Kochen, M. On Fundamentals, Theory and Application of Fuzzy Sets. In: IEEE Conf. on Decision and Control Including 16th Symp. on Adaptive Processes and a Special Symp. on Fuzzy Set Theory and Applications, New Orleans, LA, USA, Dec. 1977.

340 Kochen, M. Applications of Fuzzy Sets in Psychology. In: Zadeh, L.A. et al. (eds) - Fuzzy Sets and Their Applications to Cognitive and Decision Processes, Academic Press, New York, USA, 1975, 395-408.

341 Kochen, M. and Badre, A.N. On the Precision of Adjectives Which Denote Fuzzy Sets. J. Cybern., 4, 49-59, 1974.

342 Kochen, M. and Badre, A.N. Question-Asking and Shifts of Representation in Problem-Solving. Amer. J. Psych., 87, 369-383, 1974.

343 Koczy, L.T. and Hajnal, M. A New Fuzzy Calculus and Its Application as a Pattern Recognition Techniques. In: Proc. 3rd Int. Cong. of Cybern. and Syst., Bucharest, Rumania, Aug. 1975.

344 Kohout, L.J. Application of Multi-Valued Logic to the Study of Human Movement Control and of Movement Disorders. In Proc. 6th IEEE Int. Symp. Multiple-Valued Logic, Logan, Utah, USA, May 1976, 224-232.

345 Kokawa, M., Nakamura, K. and Oda, M. Fuzzy Theoretic and Concept Formational Approaches to Hint Effect Experiments in Human Decision Processes. Fuzzy Sets and Syst., 2, 25-36, 1979.

346 Kokawa, M., Oda, M. and Nakamura, K. A Study of Handring Method of Fuzzy Data. Trans. Inst. Electron. and Commun. Eng. Japan, 62E, 33-34, 1979.

347 Kokawa, M. Fuzzy Theoretic and Concept-Formational Approaches to Inference and Information Experiments in Human Decision Processes. In: Proc. IEEE Conf. on Decision and Control Including 16th Symp. on Adaptive Processes and a Special Symp. on Fuzzy Set Theory and Applications, New Orleans, LA, USA, Dec. 1977, Part II, 1330-1337.

348 Kokawa, M., Nakamura, K., and Oda, M. Hint Effect and a Jump of Logic in Decision Process. Trans. Inst. Electron. Comm. Eng. (Japan), 580, 256-263, 1975.

349 Kokawa, M., Nakamura, K., and Oda, M. Experimental Approach to Fuzzy Simulation of Memorizing, Forgetting and Inference Process. In Zadeh, L.A. (ed) - Fuzzy Sets and Their Applications to Cognitive and Decision Processes. Academic Press, New York, USA, 409-428, 1975.

350 Kokawa, M., Nakamura, K., and Oda, M. Fuzzy Process of Memory Behavior. Trans. Soc. Instrum. Control Engrs., 10, 385-386, 1974.

351 Kokawa, M., Nakamura, K., and Oda, M. Fuzzy Theoretical and Concept Formational Approaches to Memory and Inference Experiments. Trans. Inst. Electron Comm. Eng. (Japan), 570, 487-493, 1974.

352 Kokawa, M., Nakamura, K., and Oda, M. Fuzzy-Theoretical Approaches to Forgetting Processes and Inference. In: Research Report of Automatic Control Lab., Nagoya Univ., Japan, 21, 1-10, 1974.

353 Kokawa, M., Nakamura, K., and Oda, M. Fuzzy Expression of Human Experience-to-Memory Process. Research Reports of Automatic Control Lab., Nagoya Univ., Japan, 20, 27-33, 1973.

354 Kopstein, Felix F., Kingsley, Edward H. and Siebold, Guy L. Quasi-Algorithm Methods and Techniques for Specifying Objective Job/Task Performance Requirements. NTIS Report AD-A100 435/7, 125 pages, July 1978.

355 Korner, S. Reference, Vagueness and Necessity. Philos. Rev., 66, July 1957.

356 Kramer, U. On the Application of Fuzzy-Sets to the Analysis of the System Driver Vehicle Environment. Automatica, 21, 101-107, 1985.

357 Kramer, Ulrich. Uncertainty and Similarity: Cognitive Aspects of Man-Machine Interaction. In: Proc. of the 3rd European Annual Conf. on Human Decision Making and Manual Controls, Roskide, Denmark, 1983, 313-323.

358 Kramer, U. and Rohr, G. A Model of Driver Behavior. Ergonomics, 25, 891, 1982.

359 Kruse, R. Statistical Estimation with Linguistic Data. Inf. Sci., 33, 197-207, 1984.

360 Kruse, R. On the Entopy of Fuzzy Events. Kybernetes, 12, 53-57, 1983.

361 Kubinski, T. Nazwy Nieostre (Vague Terms). Studia Logica, 7, 115-179, 1958.

362 Kulka, J. and Novak, V. Have Fuzzy Operations A Psychological Correspondence. Studia Psychologica, 26, 131-141, 1984.

363 Kuz'min, V.B. A Parametric Approach to Description of Linguistic Values of Variables and Hedges. Fuzzy Sets and Syst., 6, 27-41, 1981.

364 Kwakernaak, H. Fuzzy Random Variables. 1. Definitions and Theorams. Inf. Sci., 15, 1-29, 1978.

365 Lake, J. Fuzzy Sets and Bald men. Dept. of Mathematics, Polytechnic of the South Bank, Bourough Road, London, UK, 1974.

366 Lakeoff, G. Hedges: A Study in Meaning Criteria and the Logic of Fuzzy Concepts. J. Philos. Logic, 2, 458-508, 1973.

367 Lakeoff, G. Hedges. A Study in Meaning Criteria and the Logic of Fuzzy Concepts, J. Phil. Logic, 2, 458-508, 1973.

368 LeFaivre, Rick. FUZZY: A Programming Language for Fuzzy Problem-Solving. Univ. of Wisconsin, Dept. of Computer Science, Report No. L WIS-CS-74-202, NTIS Report PB-231 813/7, 31 pages, Jan. 1974.

369 Lefaivre, R.A. The Representation of Fuzzy Knowledge. J. Cybern., 4, 57-66, 1974.

370 Leigh, R. and Wetton, M. Thinking Clearly with Fuzzy Logic (Turn Experience of a Human Operator Into a Process Control Machine). Process Eng., 64, 36-37, 1983.

371 Lenker, S.S. The Method for Assigning Membership Grades for Fuzzy Decision Making. In: Joint National Meeting of The Inst. Manage. Sci./Oper. Res. Soc. Amer., New Orleans, LA, USA, 1979.

372 Lerman, I.C. Cross Analysis of Fuzzy Classifications. Biometrics, 36, 369, 1980.

373 Lerner, A.W. and Wanat, J. Fuzziness and Bureaucracy. Public Admin. Rev., 43, 500-509, 1983.

374 Leung, Y. A Fuzzy Set Analysis of Sociometric Structure. J. Math. Sociology, 7, 159-180, 1980.

375 Lewis, Larry Wayne. A Statistical Approach to Fuzzy Data Analysis. Dissertation, Univ. of Louisville, 169 pages, 1981.

376 Lindsay, P.H. and Norman, D.A. Human Information Processing. (Book) Academic Press, New York, USA, 1977, 737 pages.

377 Lockett, G. Computing Aid to a Housing Mix Problem - Planning in a Fuzzy Environment. J. Manage. Stud., 10, 275-281, 1973.

378 Loo, S.G. Measures of Fuzziness. Cybernetica, 20, 210, 1977.

379 Lowen, R. Convex Fuzzy Sets. Fuzzy Sets and Syst., 3, 291-310, 1980.

380 Lusk, E.J. Evaluating Performance Statistics Used to Monitor Performance: A Fuzzy Approach. Fuzzy Sets and Syst., 5, 149-157, 1981.

381 Ma, Mou-Chao and Cao, Zhi-Giang. A Fuzzy Set Model for Category Judgment and Multistage Evaluation Method. Acta. Psychol. Sinica, 15, 198-204, 1983.

382 Machina, K.F. Truth Belief and Vagueness. J. Philos. Logic, 5, 47-77, 1976.

383 Machina, K.F. Vague Predicates. Amer. Philos. Quart., 9, 225-233, 1972.

384 MacVicar-Whelan, P.J. Fuzzy Classification of Human Height. In: Int. Symp. Multiple-Valued Logic, 1980.

385 MacVicar-Whelan, P.J. Fuzzy Sets, Concepts of Height, and Hedge Very. IEEE Trans. on Syst., Man and Cybern., SMC-8, 507-511, 1978.

386 MacVicar-Whelan, P.J. Fuzzy Sets for Man-Machine Interaction. Int. J. Man-Mach. Stud., 8, 687-697, 1976.

387 Maiers, J. and Sherif, Y.S. Applications of Fuzzy Set Theory. IEEE Trans. on Syst. Man and Cybern., SMC-15, 175-189, 1985.

388 Majumder, D.D. On the Relevancy of Fuzzy Mathematical Systems. Hum. Syt, Manage., 4, 315-323, 1984.

389 Majumder, D.D. A Fuzzy Set Theoretic Approach for Recognition of Patterns Generated from Bio-Social Systems. In: 5th European Meeting on Cybern. and Syst. Res., Vienna, Austria, April 1980. Also in Prog. in Cybern. and Syst. Res., 8-9, 1980. Hemisphere Pub., Washington, D.C., USA.

390 Majumder, D.D. and Pal, S.K. On Some Applications for Fuzzy Algorithms in Man-Machine Communication Research. J. Inst. Telecom. Electron. Eng., 23, 117-120, 1977.

391 Malasky, S. W. An Improvement in Cut and Path Set Determination: Safety and Reliability Prediction. In Proc. Ann. Reliability and Maintainability Symp., San Francisco, CA, USA, Jan. 1980, 367-373. IEEE, New York.

392 Malvache, N. and Willayes, D. Contribution of the Fuzzy Sets Theory to Man-Machine Systems. In Gupta, M.M. (ed) - Advances in Fuzzy Sets Theory and Applications, North-Holland, Amsterdam, Netherlands, 1979, 481-489.

393 Malvache, N. Utilization of Fuzzy Sets for System Modeling and Control. In: IEEE Conf. on Decision and Control Including 16th Symp. on Adaptive Processes and a Special Symp. on Fuzzy Set Theory and Applications, New Orleans, LA, USA, Dec. 1977.

394 Malvache, N., Milbred, G. and Vidal, P. Perception Visuelle: Champ de Vision Laterale, Modele de la Fonction du Regard. Rapport de Synthese. Contract DRME No. 71-251, Paris, France, 1973.

395 Mamdani, E.H. and Gaines, B.R. Fuzzy-Reasoning and Its Applications. (Book). Academic Press, New York, USA, 1981, 381 pages.

396 Mamdani, E.H. Special Issue on Fuzzy Reasoning. (Editorial). Int. J. Man-Mach. Stud., 12, 1, 1980.

397 Mamdani, E.H. Application of Fuzzy Logic to Approximate Reasoning Using Linguistic-Synthesis. IEEE Trans. on computers, 26, 1182-1191, 1977.

398 Mamdani, E.H. Advances in Linguistic-Synthesis of Fuzzy Controllers. Int. J. Man-Mach. Stud., 8, 669-678, 1976.

399 Mamdani, E.H. and Assilian, S. Experiment in Linguistic Synthesis with a Fuzzy Logic Controller. Int. J. Man-Machine Stud., 7, 1-13, 1975.

400 Mantaras, R. Lopez De, Salgado, J.G. and Martin, J.A. Classification and Linguistic Characterization of Non-Deterministic Data. Pattern Recog. Lett., 2, 33-41, 1983.

401 Mantas, J. and Heaton, A.G. Handwritten Character Recognition by Parallel Labelling and Shape Analysis. Pattern Recog. Lett., 1, 465-468, 1983.

402 Mantas, J. and Daskalakis, C. Parallel Labelling and Structural Analysis in Handwritten Character Recognition. In: Proc. MELECON '83' Mediterranean Electrochemical Conf., Athens, Greece, May 1983. 2 Vols., IEEE, New York, USA.

403 Mares, M. How to Handle Fuzzy Quantities? Kybernetika, 13, 23-40, 1977.

404 Mares, M. On Fuzzy Quantities with Real and Integer Values. Kybernetika, 13, 41-56, 1977.

405 Markowitz, Judith Ann. A Look at Fuzzy Categories. Dissertation, Northwestern Univ., 313 pages, 1977.

406 Martin, J.K. and Turksen, I.B. A Fuzzy Simulus Response Theory for Information System Design. Working Paper NO. 78-004, Univ. of Toronto, Dept. of Indust. Eng., 1978.

407 Martin, J.K. and Turksen, I.B. Formative Evaluation of Information Need Analysis. Dept. Indust. Eng., Univ. of Toronto, Canada, 1975.

408 Massing, M. Fuzzy Realpolitik - Reply. Nation, 240, 482, 1985.

409 Materna, P. Intensional Semantics of Vague Constants. An Application of Tichy's Concept of Semantics. Theory and Decision, 2, 267-273, 1972.

410 McCall, S.S. Polish Logic 1920-1939. (Book) Clarendon Press, Oxford, England, 1967.

411 McCloskey, M.E. and Glucksberg, S. Natural Categories - Well Defined or Fuzzy Sets. Memory and Cognition, 6, 462-472, 1978.

412 Menges, G. Information, Inference and Decision. D. Reidel, Dordrecht, Holland, 1974.

413 Menges, G. and Skala, H.J. On the Problem of Vagueness in the Social Sciences. In: Menges, G., (ed) - Information, Inference and Decision. D. Reidel, Dordrecht, Holland, 51-61, 1974.

414 Michalski, R.S. and Chilausky, R.L. Knowledge Acquisition by Encoding Expert Rules Versus Computer Induction from Examples: A Case Study Involving Soybean Pathology. Int. J. Man-Mach. Stud., 12, 63-87, 1980.

415 Misra, K.B. and Sharma, A. Performance Index to Quantify Reliability Using Fuzzy Subset Theory. Microelectron. and Reliab, 12, 543-549, 1981.

416 Mizumoto, M. and Zimmerman, H.J. Comparison of Fuzzy Reasoning Methods. Fuzzy Sets and Syst., 253-283, 1982.

417 Mizumoto, M. Fuzzy Reasoning with Various Fuzzy Inputs. Trans. Inst. Electron. and Commun. Eng. Japan, 65E, 562, 1982.

418 Mizumoto, M. and Tanaka, K. Fuzzy-Sets and Their Operations. Inf. and Control, 48, 30-48, 1981.

419 Mizumoto, M. Fuzzy-Sets and Their Operations. 2. Inf. and Control, 50, 160-174, 1981.

420 Mizumoto, M., Toyoda, J. and Tanaka, K. Fuzzy Languages. Systems, Computers, Controls, 1, 36, 1970.

421 Montazemi, Ali Reza. A Framework for Information Requirement Analysis and Decision of Exception Reporting Systems. Dissertation, Univ. of Waterloo (Canada), 1984.

422 Moray, Neville and Waterson, K. Fuzzy Model of Rather Heavy Workload. In: Proc. 18th Annual Conf. on Manual Control, Dayton, OH, June 1982. NTIS Report ADA-131256, 120-126, 1983.

423 Morita, Y. and Oka, Y. On a Loop in Fuzzy Evaluation and Measurement. In: Summary of Papers on General Fuzzy Problems, The Working Group on Fuzzy Systems, Tokyo, Japan, 1976, 87-100.

424 Morita, Y. and Iida, H. Measurement, Information and Human Subjectivity Described by an Order Rèlationship. In: Summary of Papers on General Fuzzy Problems, The Working Group on Fuzzy Systems, Tokyo, Japan, Nov. 1975, 34-39.

425 Mourant, R.R. and Rockwell, T.H. Mapping Eye Movement Patterns to the Visual Scene in Driving: An Exploratory Study. Human Factors, 12, 81, 1970.

426 Murray, T.J., Pipino, L.L. and Van-Gigch, J.P. A Pilot-Study of Fuzzy Set Modification of Delphi. Human Syst. Manage., 5, 76-80, 1985.

427 Murray, T.J. and Van-Gigch, J.P. A Pilot Study of Fuzzy Set Modification of Delphi. In: Proc. 15th Hawaii Int. Conf. on Syst. Sci., Honolulu, HI, USA, Vol. I, 897-905, 1982.

428 Murray, T.J., Pipino, L.L. and Van Gigch, J.P. Modification of the Delphi Technique by Fuzzy Set Theory. In Joint Meeting of The Inst. Manage. Sci./Oper. Res. Soc. Amer. Washington, D.C., USA, May 1980.

429 Nahmias, S. Fuzzy Variables. Fuzzy Sets and Syst., 1, 97-110, 1977.

430 Nakamura, K. and Iwai, S. Introduction of Topological Structure into Relations Between Elements of Fuzzy Sets and Modeling of Human Learning Behavior Through Inference by Analogy (Japanese). Syst. and Control, 24, 278-286, 1980.

431 Nakamura, K. and Yoshioka, M. Some Macroscopic Models of Collective Human Behavior. Summary of Papers on General Fuzzy Problems, The Working Group on Fuzzy Systems, Tokyo, Japan, 79-86, 1976.

432 Negoita, C.V. Fuzzy Systems. (Book). ABACUS Press, London, England, 1981.

433 Negoita, C.V. Review of Fuzzy Sets and Their Applications to Cognition and Decision Processes. IEEE Trans. Syst. Man and Cybern., SMC-7, 122-123, 1977.

434 Negoita, C.V. Fuzzy Models for Social Processes. In: Bossel, H., Klaczko, S., Muller, N. (eds) - Systems Theory in the Social Sciences, Birkhauser Verlag, Basel, Switzerland, 1976, 283-291.

435 Negoita, C.V. Introduction to Fuzzy Set Theory for Systems Analysis. John Wiley, New York, USA, 1976.

436 Negoita, C.V. and Ralescu, D.A. Applications of Fuzzy Sets to Systems Analysis. (Book). John Wiley, New York, USA, 1975.

437 Negoita, C.V. and Ralescu, D.A. Fuzzy Systems and Artificial Intelligence. Kybernetes, 3, 173-178, 1974.

438 Neustupny, J.V. On the Analysis of Linguistic Vagueness. Travaux Linguistiques de Prague, 2, 39-51, 1966.

439 Newman, R.A. Test of a Fuzzy Set Based Decision Model as an Aid in Solving Human Factors Design Problems. In: Proc. Human Factors Soc. 25th Annual Meeting, Rochester, NY, USA, 306-310, Oct. 1981. Human Factors Soc., Santa Monica, CA, 1981.

440 Newstead, S.E. and Griggs, R.A. Fuzzy Quantifiers as an Explanation of Set Inclusion Performance. Psych. Res., 46, 377-388, 1984.

441 Nguyen, H.T. Note on Extension Principle for Fuzzy Sets. J. Math. Anal. and Appl., 64, 369-380, 1978.

442 Nguyen, H.T. Fuzziness and Linguistic Probabilities. J. Math. Anal. and Appl., 61, 658-671, 1977.

443 Nguyan, H.T. Some Mathematical Tools for Linguistic Probabilities. In Proc. IEEE Conf. on Decision and Control, New Orleans, LA, USA, Dec. 1977. Part II, 1345-1350. IEEE, New York, USA.

444 Nias, J. Less Fuzziness on Openness. Elementary School J., 75, 79-86, 1974.

445 Nijkamp, P. Conflict Patterns and Compromise Solutions in Fuzzy Choice Theory. J. Peace Science, 4, 67-90, 1979.

446 Nishiwaki, V. et al. Possible Application of Fuzzy Set Concept in Risk Assessment, Subjective Perception and Public Attitude Study on Nuclear Energy. In: 6th Int. Radition Protection Asso. Cong., Berlin, W. Germany, May 1984.

447 Nojiri, H. Fuzzy Team Decision-Problems in a Sales Organization. European J. Oper. Res., 8, 256-263, 1981.

448 Nojiri, H. Ultra Fuzzy Team Decision in a Dynamic Environment. Inf. Sci., 23, 105-122, 1981.

449 Nojiri, H. On the Fuzzy Team Decision in a Changing Environment. Fuzzy Sets and Syst., 3, 137-150, 1980.

450 Noma, K., Tanaka, H. and Asai, K. Fuzzy Fault Tree Analysis for Accident Caused by Human Errors. Ergonomics, 25, 543, 1982.

451 Novak, V. Assessing of Weights of Parameters Using Calculus of Fuzzy-Sets. Ekon. Mat. Obzor, 16, 1-12, 1980.

452 Novak, V. Fuzzy Weights. (Czech) Ekon. Mat. Obzor, 14, 377-388, 1978.

453 Nowakowska, M. Methodological Problems of Measurement of Fuzzy Concepts in Social-Sciences. Behavioral Sci., 22, 107-115, 1977.

454 Nurmi, H. Modelling Impreciseness in Human Systems. In Prog. in Cybern. and Syst. Res., 7, 225-230, 1980, Hemisphere Pub. Washington, D.C., USA.

455 Oda, M. Structure Analysis and Applications of Fuzzy and Uncertain Reasoning Methods. In: IEEE Conf. on Decision and Control Including 16th Symp. on Adaptive Processes and a Special Symp. on Fuzzy Set Theory and Applications, New Orleans, LA, USA, Dec. 1977.

456 Oden, G.C. Integration of Fuzzy Linguistic Information in Language Comprehension. Fuzzy Sets and Syst., 14, 29-41, 1984.

457 Oden, G.C. Cups and Bowls Revisited - Categorization of Exemplars of Fuzzy Concepts. Bull. Psychonomic Soc., 14, 244, 1979.

458 Oden, G.C. Fuzzy Logical Model of Letter Identification. J. Exp. Psych. Human Perception and Performance, 5, 336-352, 1979.

459 Oden, G.C. Fuzzy Propositional Approach to Psycholinguistic Problems and Application of Fuzzy Set Theory in Cognitive Science. In: Gupta, M.M. (ed) - Advances in Fuzzy Set Theory and Applications. North-Holland Pub., Amsterdam, Netherlands, 1979, 409-420.

460 Oden, G.C. Fuzziness in Semantic Memory - Choosing Exemplars of Subjective Categories. Memory and Cognition, 5, 198-204, 1977.

461 Oden, G.C. Integration of Fuzzy Logical Information. J. Exp. Psych. - Human Perception and Performance, 3, 565-575, 1977.

462 Oden, G.C. and Anderson, N.H. Integration of Semantic Constraints. J. Verbal Learning and Behavior, 13, 138-148, 1974.

463 Oguntade, O.O. and Beaumont, P.E. Opthalmological Prognosis Via Fuzzy Subsets. Fuzzy Sets and Syst., 7, 123-128, 1982.

464 Oguntade, O.O. Pragmatic Aspect of Fuzzy-Programming. Int. J. Syst. Sci., 12, 1335-1349, 1981.

465 Oguntade, O.O. Psychophysics of Humanistic Systems. Cybernetica, 24, 157-168, 1981.

466 Oguntade, O.O. Towards the Psychophysics of Humanistic Systems. Cybernetica, 23, 299-323, 1980.

467 Ohsuga, Setsuo. Knowledge Based Man-Machine Systems. Automatica, 19, 685-691, 1983.

468 Okuda, T. Tanaka, H. and Asai, K. Formulation of Fuzzy Decision Problems with Fuzzy Information Using Probability Measures of Fuzzy Events. Inf. and Control, 38, 135-147, 1978.

469 Okuda, T., Tanaka, H. and Asai, K. Decision-Making and Information in Fuzzy Events. Bull. Univ. Osaka Prefecture, 23A, 1974.

470 Okuma, A. Software and Fuzzy Logic: Let any Good Programmer Design an Expert System (Language Reveal). Electron. Des., 33, 173-178, 1985.

471 Ollero, A. and Freire, E. The Structure of Relations in Personnel Management. Fuzzy Sets and Syst., 5, 115-125, 1981.

472 Otsuki, S. A Model for Learning and Recognizing Machine. Inf. Processing, 11, 664-671, 1970.

473 Pal, S.K., Majumder, D.D. and Chaudhury, B.B. Fuzzy Sets in Handwritten Character Recognition. In: Proc. All India Interdisciplinary Symp. on Digital Tech. and Pattern Recog., ISI, Calcutta, India, 1980.

474 Pal, S.K., Datta, A.K. and Duttamajumder, D. Adaptive Learning Algorithm in Classification of Fuzzy Patterns - Application to Vowels in CNC Context. Int. J. Syst. Sci., 9, 887-897, 1978.

475 Pal, S.K. and Duttamajumader, D. Effects of Fuzzification on Plosive Cognition System. Int. J. Syst. Sci., 9, 873-886, 1978.

476 Parui, S.K. and Majumder, D.D. Estimation of Fuzzy Membership Functions. In 6th Conf. on Pattern Recognition, Munich, Germany, Oct. 1982.

477 Pask, G. Conversation, Cognition, and Learning. (Book). Elsevier, Amsterdam, Netherlands, 1975.

478 Pask, G. The Cybernetics of Human Learning and Performance. (Book). Hutchison, London, England, 1975.

479 Pavlidis, T. Applications of Fuzzy Sets in Curve Fitting. In: Proc. IEEE Conf. on Decision and Control Including 16th Symp. on Adaptive Processes and a Special Symp. on Fuzzy Set Theory and Applications, New Orleans, LA, USA, Dec. 1977, Part II, 1396-1400.

480 Pedrycz, W. Construction of Fuzzy Relational Models. In: Proc. 7th European Meeting on Cybernetics and Systems Research, Vienna, Austria, Apr. 1984, 545-549. North-Holland Pub., Amsterdam, Netherlands, 1984.

481 Pedrycz, W. An Approach to the Analysis of Fuzzy Systems. Int. J. Control, 34, 403-421, 1981.

482 Phelphs, R.H. and Shanteau, J. Livestock Judges: How Much Information Can an Expert Use? Organizational Behavior and Human Performance, 21, 204-219, 1978.

483 Pierce, J.E. Culture - Collection of Fuzzy Sets. Human Organization, 36, 197-200, 1977.

484 Pinkava, V. Fuzzification of Binary and Finite Multivalued Logical Calculi. Int. J. Man-Mach. Stud., 8, 717-730, 1976.

485 Pipkin, J.S. Fuzzy-Sets and Spatial Choice. Annals Assoc. Amer. Geographers, 68, 196-204, 1978.

486 Pipino, L.L., Van-Gigch, J.P. and Tom, G. Experiments in the Representation and Manipulation of Labels of Fuzzy-Sets. Behavioral Sci., 26, 216-228, 1981.

487 Pipino, L.L. and Van-Gigch, J.P. Potential Impact of Fuzzy Sets on the Social Sciences. Cybern. and Syst., 12, 21-35, 1981.

488 Pipino, L.L. Epistemology of Inexact Sciences Revisited in the Light of Fuzzy Sets. In: Joint Meeting of The Inst. Manage. Sci./Oper. Res. Soc. Amer., San Francisco, CA, USA, May 1977.

489 Ponsard, C. An Application of Fuzzy Subsets Theory to the Analysis of the Consumer's Spatial Preferences. Fuzzy Sets and Syst., 5, 235-244, 1981.

490 Progress in Cybernetics and Systems Research. Vol. 8: General System Methodology, Mathematical Systems Theory and Fuzzy Sets. Hemisphere Pub. Co., Washington, D.C., USA, 1980.

491 Pun, L. Experience in the Use of Fuzzy Formalism in Problems with Various Degrees of Subjectivity. In: Special Interest Discussion Session on Fuzzy Automata and Decision Processes, 6th IFAC World Congress, Boston, MA, USA, Aug. 1975.

492 Radecki, T. Outline of a Fuzzy Logic Approach to Information-Retrieval. Int. J. Man-Mach. Stud., 14, 169-178, 1981.

493 Radecki, T. An Evaluation of the Fuzzy Set Theory Approach to Information Retrieval. In: 5th European Meeting on Cybernetics and Systems Research, Vienna, Austria, Apr. 1980. Also in Progress in Cybern. and Syst. Res., 8 and 9, 1980. Hemisphere Pub. Co. Washington, D.C., USA.

494 Radecki, T. Fuzzy Set Theoretical Approach to Document-Retrieval. Inf. Processing and Manage., 15, 247-259, 1979.

495 Radecki, T. Level Fuzzy Sets. J. Cybern., 7, 189-198, 1977.

496 Radecki, T. Mathematical-Model of Time-Effective Information-Retrieval System Based on Theory of Fuzzy Sets. Inf. Processing and Manage., 13, 109-116, 1977.

497 Ragade, R.K. Fuzziness in Multidimensional Scaling. In: Joint National Meeting of The Inst. Manage. Sci./Oper. Res. Soc. Amer., New York, N.Y., USA, May 1978.

498 Ragade, R.K. Fuzzy Graphs in Societic Modeling. In IEEE Conf. on Decision and Control, New Orleans, LA, USA, Dec. 1977.

499 Ragade, R.K. and Gupta, M.M. Fuzzy Set Theory: Introduction. In: Gupta, M.M., Sardis, G.N., Gains, B.R. (eds) - Fuzzy Automata and Decision Processes. North-Holland Pub., Amsterdam, Netherlands, 1977, 105-132.

500 Ragade, R.K. Fuzzy Interpretive Structural Modeling. J. Cybern., 6, 189-211, 1976.

501 Ragade, R.K. Fuzzy Models in Multi-Objective Conflict Analysis. In: Joint National Meeting of The Inst. Manage. Sci./Oper. Res. Soc. Amer., Miami, FL, USA, Nov. 1976.

502 Ragade, R.K. Fuzzy Sets in Communications Systems and Consensus Formation Systems. J. Cybern., 6, 21-38, 1976.

503 Ragade, R.K. Incertitude Characterization of the Retriever-System Communication Process. In Proc. 37th Annual Meeting Amer. Soc. Inf. Sci., Atlanta, GA, USA, Oct. 1974.

504 Ragade, R.K. A Multiattribute Perception and Classification of Visual Similarities. Systems Research and Planning, Bell-Northern Research, Ottawa, Canada, Paper S-001-73, 1973.

505 Ragade, R.K. On Some Aspects of Fuzziness in Communication: I. Fuzzy Entropies. Systems Research and Planning, Bell-Northern Research, Ottawa, Canada, Paper W-002-73, 1973.

506 Ragade, R.K. On Some Aspects of Fuzziness in Communication: II. A Note on Fuzzy Entropies Associated with a Fuzzy Channel. Systems Research and Planning, Bell-Northern Research, Ottawa, Canada, Paper W-006-73, Nov. 1973.

507 Ragade, R.K. On Some Aspects of Fuzziness in Communication: III. Fuzzy Concept Communication. Systems Research and Planning, Bell-Northern Research, Ottawa, Canada, Paper W-005-73, Dec. 1973.

508 Ralescu, A.L. and Ralescu, D.A. Probability and Fuzziness. Inf. Sci., 34, 85-92, 1984.

509 Ralescu, D. Fuzzy Subobjects in a Category and The Theory of B-Sets. Fuzzy Sets and Syst., 1, 193-202, 1978.

510 Rao, M. Bhaskara and Rashed, A. Some Comments on Fuzzy Variables. Fuzzy Sets and Syst., 6, 285-292, 1981.

511 Reball, S. Evaluation of Information Collections by Theory of Fuzzy Sets. Int. Classification, 5, 152-155, 1978.

512 Richardson, Gregory. Information, Fuzzy Choice and The Manipulation of Social Decisions. Dissertation, Cornell Univ., 68 pages, 1975.

513 Rieger, B.B. Fuzzy Representation Systems in Linguistic Semantics. In: 5th European Meeting on Cybernetics and Systems Research, Vienna, Austria, Apr. 1980. Also in: Prog. in Cybern. and Syst. Res., 8 and 9, 1980. Hemisphere Pub., Washington, D.C., USA.

514 Robertson, S.E. Fuzzy-Sets-Reply. J. Amer. Soc. Inf. Sci., 30, 357-358, 1979.

515 Robertson, S.E. Nature of Fuzz: A Diatribe. J. Amer. Soc. Inf. Sci., 29, 304-307, 1978.

516 Robinson, E.J. and Turner, S.J. Improving Library Effectiveness - A Proposal for Applying Fuzzy Set Concepts in the Management of Large Collections. J. Amer. Soc. Inf. Sci., 32, 458-462, 1981.

517 Rose, A. Triumphs of Fuzzy Fogtop. (Book). Dial Press, New York, USA, 1979, 32 pages.

518 Rosen, R. Planning Management, Policies and Strategies - Four Fuzzy Concepts. Int. J. Gen. Syst., 1, 245-252, 1974.

519 Rouse, W.B. and Hunt, R.M. A Fuzzy Rule-Based Model of Human Problem Solving in Fault Diagnosis Tasks. In: Proc. 8th Triennial World Cong. of the IFAC on Control Sci. and Tech. for the Prog. of Soc., Kyoto, Japan, Aug. 1981, 7, 3507-3513.

520 Rouse, W.B. Pilot Interaction with Automated Airborne Decision Making Systems. NASA-CR-164729, Aug. 1981, 28 pages.

521 Rouse, W.B., Rouse, H. and Pellegrino, S.J. Rule-Based Model of Human Problem Solving Performance in Fault Diagnosis Tasks. IEEE Trans. Syst. Man Cybern., SMC-10, 366-376, 1980.

522 Rouse, W.B. Systems Engineering Models of Human-Machine Interaction. North-Holland, Amsterdam, Netherlands, 1980.

523 Rouse, W.B. A Model of Human Decision Making in Fault Diagnosis Tasks That Include Feedback and Redundancy. IEEE Trans. Syst. Man Cybern., SMC-9, 237-241, 1979.

524 Rouse, W.B. A Model of Human Decision Making in Fault Diagnosis Task. IEEE Trans. Syst. Man Cybern., SMC-8, 357-361, 1978.

525 Rubin, D.C. Measuring Fuzziness - Comment on a Fuzzy Set Approach to Modifiers and Vagueness in Natural - Language. J. Exp. Psych. - General, 108, 486-489, 1979.

526 Saaty, T.L. Measuring the Fuzziness of Sets. J. Cybern., 4, 53-61, 1974.

527 Sachs, W.M. Approach to Associative Retrieval Through Theory of Fuzzy Sets. J. Amer. Soc. Inf. Sci., 27, 85-87, 1976.

528 Sambuc, R. et al. Fuzzy Functions Application to Medical Diagnosis. In: MEDINFO 80, Proc. 3rd World Conf. on Medical Informatics, Tokyo, Japan, Oct. 1980. Part 2, 784-788. North-Holland Pub., Amsterdam, Netherlands.

529 Sanatani, S. Market Penetration of New Products in Segmented Populations a System Dynamics Simulation with Fuzzy-Sets. Tech. Forecasting and Social Change, 19, 313-329, 1981.

530 Sanchez, E. Combination of Fuzzy Quantifiers in Natural Languages. In: American Control Conf., San Diego, CA, USA, June 1984.

531 Sanchez, E. Fuzzy Information, Knowledge Representation and Decision Analysis. In: Proc. IFAC Symp. Marseille, France, July 1983. Pergamon Press, Oxford, England.

532 Sanchez, E., et al. Fuzzy Sets and Knowledge Representation. In: Proc. 8th Int. Conf. on Artificial Intelligence. Karlsruhe, FRG, Aug. 1983. William Kaufmann, Los Altos, CA, USA.

533 Sanchez, E. Resolution of Composite Fuzzy Relation Equations. Inf. and Control, 30, 38-47, 1976.

534 Sandell, Henrick Stein Hugo. A Knowledge Engineering Tool for Creating Frame and Rule Based Expert Systems. Dissertation, Vanderbilt Univ., 133 pages, 1984.

535 Sanford, D.H. Notes on Logics of Vagueness and Some Applications. In: Proc. Symp. on Policy Analysis and Information Systems, Durham, NC, USA, June 1980. 123-126. Plenum Press, New York, USA.

536 Sanford, D.H. Infinity and Vagueness. Philos. Rev., 84, 520-535, 1975.

537 Saridis, G.N. Fuzzy Decision-Making in Prosthetic Devices and Other Applications. In: Special Interest Discussion Session on Fuzzy Automata and Decision Processes, 6th IFAC Congress, Boston, MA, USA, Aug. 1975.

538 Schefe, Peter. On Foundations of Reasoning with Uncertain Facts and Vague Concepts. Int. J. Man-Mach. Stud., 12, 35-62, 1980.

539 Schmucker, K.J. Fuzzy Sets, Natural Language Computation and Risk Analysis. (Book). Computer Sci. Press, Rockville, MD, USA, 192 pages, 1984.

540 Schrader, Alvin Marvin. Toward a Theory of Library and Information Science. Dissertation, Indiana Univ., 1016 pages, 1983.

541 Schwartz, Daniel Guy. Appropriate Reasoning, Logics for Self Reference, and the Use of Non-Classical Logics in Systems Modeling. Dissertation, Portland State Univ., 289 pages, 1981.

542 Seif, A. and Aguilar-Martin, J. Multigroup Classification Using Fuzzy Correlation. Fuzzy Sets and Syst., 3, 109-122, 1980.

543 Sekita, Y. and Tabata, Y. Health Status Index Model Using a Fuzzy Approach. European J. Oper. Res., 3, 40-49, 1979.

544 Sembi, B.S., Baker, N.J.C. and Efstathiou, J. Report on the Panel Session at the Workshop on Fuzzy Reasoning. Int. J. Man-Mach. Stud., 11, 407-409, 1979.

545 Seo, F. and Sakawa, M. Fuzzy Multiattribute Utility Analysis for Collective Choice. IEEE Trans. on Systems, Man and Cybern., SMC-15, 45-53, 1985.

546 Seo, F. and Sakawa, M. Fuzzy Assessment of Multiattribute Utility - Functions. Lecture Notes in Economics and Mathematical Systems, 229, 97-104, 1984.

547 Seo, F. Fuzzy Extension of Multiattribute Utility Analysis for Collective Choice. In: Cybern. and Syst. Res. Proc. 7th European Meeting, Vienna, Austria, Apr. 1984. 195-200. North-Holland Pub., Amsterdam, Netherlands.

548 Shahinpoor, M. Mathematical Modelling of the Future for Complex Systems. Math. Modelling, 3, 153-160, 1982.

549 Sheng, Richard Li-Cheng. A Linguistic Approach to Temporal Information Analysis. Dissertation, Univ. California, Berkeley, 159 Pages, 1980.

550 Sheridan, T.B. Interaction of Human Cognitive Models and Computer Based Models in Supervisory Control. NTIS Report AD-A142 677/4, 55 Pages, Mar. 1984.

551 Shimura, M. An Approach to Pattern Recognition and Associative Memories Using Fuzzy Logic. In: Zadeh, L.A. et al. (ed) - Fuzzy Sets and Their Applications to Cognitive and Decision Processes, Academic Press, New York, USA, 1975, 449-476.

552 Shimura, M. Fuzzy Sets Concept in Rank - Ordering Objects. J. Math. Anal. and Appl., 43, 717-733, 1973.

553 Shine, J.A. Bayesian, Evidence, Fuzzy; Which Theory Works Best When Reasoning with Uncertain Knowledge? In Amer. Soc. for Photogrammetry and Amer. Cong. on Surveying and Mapping Convention, Washington, D.C., USA, Mar. 1985.

554 Shortliffe, E.H. Computer-Based Medical Consultation: MYCIN (Book). Elsevier, New York, USA, 1976.

555 Silva, O.D.O., Derenna, E. and Souza, C. Induction of Fuzzy Decision Rules. (Portuguese). NTIS Report N85-13525/9/XAB, 7 Pages, Apr. 1983.

556 Silva, O.D.O. and Erthal, G.J. Algorithms for Minimization of Fuzzy Functions and Incompletely Specified Fuzzy Functions. (Portuguese). NTIS Report N81-30885/0, 81 pages, Aug. 1981.

557 Simcox, W. and Kreifeldt, J. Human Modeling of the Operator Process Interface. In: Proc. Tech. Assoc. of Pulp and Paper Industry Eng. Conf., Boston, MA, USA, Sep. 1984, 231-239.

558 Singh, J.P. and Thiel, C.C. Ice Groups Risk Determination Using Fuzzy Possibility. In: Proc. 4th Int. Symp. on Offshore Mechanics and Arctic Eng., Dallas, TX, USA, Feb. 1985.

559 Sirlin, S.W. and Longman, R.W. Fuzzy Concepts of the Degree of Controllability and Degree of Observability. Adv. in Astronaut. Sci., 54, 1983. Also in: Proc. AAS/AIAA Astrodynamics Conf., Lake Placid, New York, USA, Aug. 1983. 2, 777-795.

560 Siy, P. and Chen, C.S. Fuzzy Logic for Handwritten Numerical Character Recognition. IEEE Trans. Syst. Man and Cybern., SMC-4, 570-575, 1974.

561 Siy, Pepe. Fuzzy Logic for Handwritten Character Recognition. Dissertation, Univ. of Akron, 173 pages, 1973.

562 Siy, P. and Chen, C.S. Fuzzy Logic Approach to Handwritten Character Recognition Problem. In: Proc. IEEE Conf. Syst. Man and Cybern., Anaheim, CA, USA; 113-117, 1971.

563 Sjoberg, L. Aided and Unaided Decision Making: Improving Intuitive Judgment. J. Forecasting, 1, 349-363, 1981.

564 Skala, H.J. Opinion Pooling. In: Proc. 14th Int. Symp. on Multiple Valued Logic, Winnipeg, Man., Canada, May 1984, 264-268. IEEE Computer Soc. Press, New York, USA, 1984.

565 Skala, H.J. Fuzzy Concepts: Logic, Motivation, Application. In: Bossel, H., Klaczko, S. and Muller, N. (eds) - Systems Theory In the Social Sciences. Birkhauser Verlag, Basel, Switzerland, 1976, 292-306.

566 Skala, H.J. On the Problems of Imprecision. D. Reidel, Dordrecht, Netherlands, 1974.

567 Slack, J.M.V. A fuzzy Set-Theoretic Approach to Semantic Memory: A Resolution to the Set-Theoretic versus Network Model Controversy. In: Mamdani, E.H. and Gaines, B.R. (ed) - Discrete Systems and Fuzzy Reasoning. EES-MMS-DSFR-76. Queen Mary College, Univ. of London (Workshop Proceedings), 1976.

568 Slack, J.M.V. Possible Applications of the Theory of Fuzzy Sets to the Study of Semantic Memory. In: Mamdani, E.H. and Gaines, B.R. (ed) - Discrete Systems and Fuzzy Reasoning. EES-MMS-DSFR-76. Queen Mary College, Univ. of London, (Workshop Proceedings), 1976.

569 Smeaton, A.F. Relevance Feedback and a Fuzzy Set of Search Terms in an Information-Retrieval System. Inf. Tech. Res. and Dev., 3, 15-23, 1984.

570 Smets, P. Probability of Fuzzy Events and Axiomatic Approach. Fuzzy Sets and Syst., 7, 153-164, 1982.

571 Smets, P. The Degree of Belief in a Fuzzy Event. Inf. Sci., 25, 1-19, 1981.

572 Smith, T.A. Keyword in a Fuzzy Context. Byte, 8, 389-397, 1983.

573 Smithson, M. Fortran-IV Programs for Fuzzy Set Analysis. Behavior Res. Methods Instrum. and Computers, 16, 553-556, 1984.

574 Smithson, M. Multivariate Analysis using 'AND' and 'OR.' Math. Soc. Sci., 7, 231-251, 1984.

575 Smithson, M. Applications of Fuzzy Set Concepts to Behavioral Sciences. Math. Social Sci., 2, 257-274, 1982.

576 Smithson, M. Models for Fuzzy Nominal Data. Theory and Decision, 14, 51-74, 1982.

577 Sobolewski, M. Classification System Semantics in Terms of Fuzzy Sets. 3rd Eur. Meeting on Cybern. Syst. Res., Vienna, Austria, 1976.

578 Sommer, G. Fuzzy Bayes-Decision Making. In: 5th European Meeting on Cybern. and Syst. Res., Vienna, Austria, Apr. 1980. Also in: Prog. in Cybern. and Syst. Res., 8 and 9, 1980. Hemisphere Pub., Washington, D.C., USA.

579 Spal, J. Application of Modified Fuzzy Sets. Probl. Control and Info. Theory (Hungary), 11, 117-126, 1982.

580 Spal, J. Fundamentals of a Mathematical Theory of Fuzzy Sets. Appl. Math., 27, 326-340, 1982.

581 Spal, J. Methods of Incomplete Description of Systems. (Czech). Automatizace, 23, 170-175, 1980.

582 Spragge, G.L. Canadian Planners Goals - Deep Roots and Fuzzy Thinking. Canadian Public Administration, 18, 216-234, 1975.

583 Stallings, W. Fuzzy Set - Theory Versus Bayesian Statistics. IEEE Trans. Syst. Man and Cybern., SMC-7, 216-219, 1977.

584 Stefik, M. et al. The Organization of Expert Systems: A Tutorial. Artificial Intelligence, 18, 135-173, 1982.

585 Stein, W.E. and Talati, K. Fuzzy Random Variables. In: Joint National Meeting of The Inst. Manage. Sci./Oper. Res. Soc. Amer. Washington, D.C., May 1980. Paper NO. TD12.5.

586 Steinberg, E. and Rinks, D. Application of the Blin - Whinston Algorithm for Resolving Fuzzy Group Preference. Manage. Sci., 25, 159-170, 1979.

587 Sugeno, M. and Takagi, T. Multidimensional Fuzzy Reasoning. Fuzzy Sets and Syst., 9, 313-326, 1983.

588 Sugeno, M. Application of Fuzzy Set Theory (Japanese). J. Inst. Electron. and Commun. Eng. Japan, 65, 1264-1266, 1982.

589 Sugeno, M. and Terano, T. A Model of Learning Based on Fuzzy Information. Kybernetes, 6, 157-166, 1977. Also in: Prog. in Cybern. and Syst. Res., 3, 325-336, 1978. John Wiley, New York, USA.

590 Sugeno, M. Development of Fuzzy Sets Theory (Japanese). Syst. and Control (Japan), 19, 229-234, 1975.

591 Sugeno, M. Fuzzy Decision Making Problems. Trans. Soc. Instrum. and Control Eng., Japan, 11, 709-714, 1975.

592 Sugeno, M., Tsukmoto, Y. and Terano, T. Subjective Evaluation of Fuzzy Objects. In IFAC Symp. on Stochastic Control, 1974.

593 Sugeno, M. and Terano, T. An Approach to the Identification of Human Characteristics by Applying Fuzzy Integrals. In: Proc. of 3rd IFAC Symp. on Identification and System Parameter Estimation, Hague, Netherlands, 1973.

594 Sunderesan, N. and Chatterji, B.N. Application of Fuzzy Sets for Recognition of Handwritten English Characters. In: Proc. IFAC Symp. on Theory and Appl. of Digital Control, New Delhi, India, Jan. 1982. 439-443. Pergamon Press, Oxford, England.

595 Svensson, O. On use of Fuzzy Concepts in Description. In: Proc. 2nd Int. Joint Conf. on Pattern Recognition, Lyngby, Denmark, Aug. 1974. IEEE, New York, USA.

596 Swierniak, A. A Unified Approach to Controllers Design for Uncertain Systems. NTIS Report N82 - 32005/2, 12 pages, 1982.

597 Szulc, T. Fuzzy Frameworks. New Republic, 179, 12, 1978.

598 Tahani, V. Conceptual - Framework for Fuzzy Query Processing - Step Toward Very Intelligent Database Systems. Inf. Processing and Manage., 13, 289-303, 1977.

599 Tamburrini, G. and Termini, S. Some Foundational Problems in the Formalization of Vagueness. In: Gupta, M.M. and Sanchez, E. (eds) - Fuzzy Information and Decision Processes. North-Holland Pub., Amsterdam, Netherlands, 1982.

600 Tanaka, H., Asai, K. and Uejima, S. A Linear Regression Model with Fuzzy Functions. (Japanese). J. Oper. Res. Soc. Japan, 25, 162-174, 1982.

601 Tanaka, H., Uejima, S. and Asai, K. Linear Regression Analysis with Fuzzy Model. IEEE Trans. Syst. Man and Cybern., SMC-12, 903-907, 1982.

602 Tanaka, H., Okuda, T. and Asai, K. Decision Making and Its Goal in a Fuzzy Environment. In: U.S. - Japan Seminar on Fuzzy Sets and Their Applications. Berkeley, CA, USA, July 1974.

603 Tanaka, K. and Mizumato, M. Fuzzy Programs and Their Execution. In: Zadeh, L.A. et al. (eds) - Fuzzy Sets and Their Applications to Cognitive and Decision Processes. Academic Press, New York, USA, 41-76, 1975.

604 Tasnadi, A. Fuzzy Systems and Learning Models. In: Proc. 4th Int. Conf. on Current Topics in Cybern. and Systems, Amsterdam, Netherlands, Aug. 1978, 382-383. Springer Verlag, Berlin, Germany.

605 Tatzagi, T. and Sugeno, M. Derivation of Fuzzy Control Rules from Human Operator's Control Actions. In: Proc. IFAC Symp. on Fuzzy Information, Knowledge Representation and Decision Analysis, Marseille, France, 1983.

606 Tazaki, E.O. Structural Modelling in a Class of Systems by Fuzzy Sets Theory. Fuzzy Sets and Syst., 2, 87-103, 1979.

607 Tazaki, E. Fuzzy Entropy and Its Application to Pattern Recognition. In: 2nd Formator Symp. on Mathematical Methods for Analysis of Large Systems, Prague, Czechoslovakia, June 1974.

608 Terano, T., Murayama, Y. and Akiyama, N. Human Reliability and Safety Evaluation of Man-Machine Systems. Automatica, 19, 719-722, 1983.

609 Terano, T. and Sugeno, M. Conditional Fuzzy Measures and Their Applications. In: Zadeh, L.A. et al. (eds) - Fuzzy Sets and Their Applications to Cognitive and Decision Processes. Academic Press, New York, USA, 151-170, 1975.

610 Terano, T. Fuzziness and its Concept. In: Proc. of a Symp. on Fuzziness in Systems and Its Processing, Professional Group of System Engineering of Soc. of Instrumentation and Control Engineers, Japan, 1971.

611 Thole, U., Zimmermann, H.J. and Zysno, P. On the Suitability of Minimum and Product Operators for the Intersection of Fuzzy Sets. Fuzzy Sets and Syst., 2, 167-180, 1979.

612 Thomas, John C. A Design - Interpretation Analysis of Natural English with Applications to Man - Computer Interaction. NTIS Report AD-A056 121/7, 38 pages, June 1977.

613 Thomas, Sidney Fitz-Ralph. A Theory of Semantics and Possible Inference with Application in Decision Analysis. Dissertation, Univ. of Toronto, Canada, 1980.

614 Thomason, M.G. and Marinos, P.N. Fuzzy Logic Relations and Their Utility in Role Theory. In: Proc. IEEE Int. Conf. on Cybern. and Society, 1972, Washington, D.C., USA. IEEE, New York, USA.

615 Thurston, L.L. A Law of Comparative Judgment. Psychol. Rev., 34, 273-286, 1927.

616 Togai, Masaki. Principles and Applications of Fuzzy Inference: A New Approach to Decision - Making Processes in Ill - Defined Systems. Dissertation, Duke Univ., 155 pages, 1982.

617 Tong, R.M. and Bonissone, P.P. Fuzzy Information - Processing. Inf. Sci., 34, 83-84, 1984.

618 Tong, R.M. An Annotated Bibliography of Fuzzy Control. Technical Memo TM6002-1, Advanced Information and Decision Systems, Mountain View, CA, USA, 1983.

619 Tong., R.M. Fuzzy - Sets and Multi - Criteria Decision - Making. Human Syst. Manage., 3, 266-268, 1982.

620 Tong, R.M. The Evaluation of Fuzzy Models Derived from Experimental Data. Fuzzy Sets and Syst., 4, 1-12, 1980.

621 Tong, R.M. and Bonissone, P.P. A Linguistic Approach to Decision Making with Fuzzy - Sets. IEEE Trans. Syst. Man and Cybern., SMC-10, 716-723, 1980.

622 Tong, R.M. The Construction and Evaluation of Fuzzy Models. In: Gupta, M.M. (ed) - Advances in Fuzzy Set Theory and Applications. Plenum Press, New York, USA, 1979.

623 Tong, R.M. and Bonissone, P.P. Linguistic Decision Analysis Using Fuzzy Sets. Memo UCB/ERL M79/72, Berkeley, 1979.

624 Tong, R.M. Analysis of Fuzzy Control Algorithms Using Relation Matrix. Int. J. Man-Mach. Stud., 8, 679-686, 1976.

625 Tress, P.H. New Models of Social Mobility: Fuzzy Sets and Catastrophe Theory. In Proc. 9th Annual Pittsburgh Conf. on Modeling and Simulation, Pittsburgh, PA, USA, Apr. 1978.

626 Tsukamoto, Y. A Reasoning Method Using Fuzzy Information (Japanese). Trans. Soc. Instrum. and Control Eng. (Japan), 17, 479-484, 1981.

627 Turner, S.J. Fuzzy Set - Theory as Applied to Serial Binding and Tattletaping Decisions. Proc. Amer. Soc. for Inf. Sci., 18, 192, 1981.

628 Uhr, L. Toward Integrated Cognitive Systems Which Must Make Fuzzy Decisions About Fuzzy Problems. In: Zadeh, L.A. et al. (ed) - Fuzzy Sets and Their Applications to Cognitive and Decision Processes, Academic Press, New York, USA, 353-393, 1975.

629 Umano, M., Mizumoto, M. and Tanaka, K. FSTDS - System - Fuzzy Set - Manipulation System. Inf. Sci., 14, 115-159, 1978.

630 Unwin, S.D. and Hayne, M.R. Rational Quantitative Safety Goals: A Summary. In: Proc. Int. ANS/ENS Topical Meeting on Probabilistic Safety Methods and Applications, San Francisco, CA, USA, Feb. 1984. 7 pages. NTIS Report DE-84017058/XAB, Aug. 1984.

631 Usechak, David. Application of Fuzzy Theory to Pattern Recognition. Dissertation, New Jersey Inst. of Tech., 188 pages, 1983.

632 Van Gigch, J.P. A Fuzzy Set Theory Interpretation of Delphi. In: Joint National Meeting of The Inst. Manage. Sci./Oper. Res. Soc. Amer., New Orleans, LA, USA, Apr-May 1979.

633 Van Gigch, J.P. Fuzzy Set Theory and Decision Making in Inexact Sciences. In: Proc. Soc. for Gen. Syst. Res. 21st Annual North American Meeting, Denver, CO, USA, Feb. 1977.

634 Van Amc Rogen, J. and Van Nauta-Lemke, R. An Autopilot for Ships Designed with Fuzzy Sets. In Proc. 5th IFAC Conf., The Hague, Netherlands, June 1977. North Holland, Amsterdam, Netherlands.

635 Van Velthoven, G.D. Fuzzy Models in Personnel Management. In: Proc. 3rd Int. Cong. of Cybern. and Syst., Bucharest, Romania, Aug. 1975. 15 pages.

636 Van Velthoven, G.D. Application of Fuzzy Sets Theory to Criminal Investigation. Thesis, Univ. Louvain, Belgium, 1974.

637 Verma, P.R. Vagueness and the Principle of the Excluded Middle. Mind, 79, 66-77, 1970.

638 Vila, M.A. and Delgado, M. On Medical Diagnosis Using Possibility Measures. Fuzzy Sets and Syst., 10, 211-222, 1983.

639 Vrba, J. and Nekola, J. Possibilities of Fuzzy Set Theory Application for Innovation Process Modeling. (Czech). Ekon-Mat. Obz. (Czech), 19, 188-202, 1983.

640 Wagner, W. A Fuzzy Model of Concept Representation in Memory. Fuzzy Sets and Syst., 6, 11-26, 1981.

641 Wang, P.P. Advances in Fuzzy Sets, Possibility Theory and Applications. (Book). Plenum Press, New York, USA, 1983, 421 pages.

642 Wang, P.P. and Chang, S.K. Fuzzy - Sets - Theory and Applications to Policy Analysis and Information - Systems. (Book). Plenum Press, New York, USA, 1980, 413 pages.

643 Ward, L.M. Fuzzy Theory of Psychophysical Judgment. Bull. Psychonomic Soc., 8, 264, 1976.

644 Watanabe, Satosi. Creative Learning and Propensity Automation. IEEE Trans. on Syst., Man and Cybern., SMC-5, 603-609, 1975.

645 Wechler, W. Modeling Humanistic Systems by Fuzzy Logic. In: IEEE Conf. on Decision and Control Including 16th Symp. on Adaptive Processes and a Special Symp. on Fuzzy Set Theory and Applications, New Orleans, LA, USA, Dec. 1977.

646 Wee, W.G. and Fu, K.S. A Formulation of Fuzzy Automata and Its Application as a Model of Learning Systems. IEEE Trans. Syst. Sci. and Cybern., SSC-5, 215-223, 1969.

647 Weiss, J.J., Sticha, P.J. and Donnell, M.L. Preliminary Investigations Into the Psychological Foundations of Fuzzy Reasoning. NTIS Report AD-A071 812/2, June 1979, 47 pages.

648 Wenstop, Fred. Quantitative Analysis with Linguistic Values. Fuzzy Sets and Syst., 4, 99-115, 1980.

649 Wenstop, F. Verbal Formulation of Fuzzy Dynamic Systems. In: Proc. 4th Int. Conf. on Current Topics in Cybern. and Syst., Amsterdam, Netherlands, Aug. 1978, 386-387. Springer Verlag, Berlin, Germany.

650 Wenstop, F. Fuzzy Set Simulation Models in a System Dynamics Perspective. Kybernets, 6, 209-218, 1977.

651 Wenstop, F. Deductive Verbal Models of Organizations. Int. J. Man-Mach. Stud., 8, 293-311, 1976.

652 Wenstop, F. Application of Linguistic Variables in the Analysis of Organizations. Dissertation, Univ. of California, Berkeley, 1975.

653 Wenstop, F. Evaluation of Verbal Organizational Models. NOAK 75, Oslo, Norway.

654 Whalen, T. Fuzzy Knowledge Based Systems in Management. Human Syst. Manage., 4, 263-274, 1984.

655 Whalen, T. and Schott, B. Issues in Fuzzy Production Systems. Int. J. Man-Mach. Stud., 19, 57-71, 1983.

656 Whalen, T. and Schott, B. Fault Diagnosis in a Fuzzy Network. In: Proc. Int. Conf. on Cybern. and Soc., Seattle, WA, USA, Oct. 1982, 35-39. IEEE New York, USA.

657 Wheless, Deborah Anne. A Normative Model of the Audit Judgment Process: A Fuzzy Set Theory Approach. Dissertation, Univ. Georgia, 1981, 166 Pages.

658 Wierzchon, S.T. Applications of Fuzzy Decision Making Theory to Coping with Ill Defined Problems. Fuzzy Sets and Syst., 7, 1-18, 1982.

659 Willaeys, D. and Malvache, N. The Use of Fuzzy Sets for the Treatment of Fuzzy Information by Computer. Fuzzy Sets and Syst., 5, 323-327, 1981.

660 Willaeys, D. and Malvache, N. Contribution of the Fuzzy Sets Theory to Man-Mach. Systems. In: Gupta, M.M. (ed) - Advances in Fuzzy Set Theory and Applications. North-Holland Pub., Amsterdam, Netherlands, 481-499, 1979.

661 Willaeys, D., Malvache, N. and Hammad, P. Utilization of Fuzzy Sets for System Modelling and Control. In: Proc. IEEE Conf. on Decision and Control, New Orleans, LA, USA, Dec. 1977. Part II, 1435-1439. IEEE, New York, USA.

662 Willmott, R. Two Fuzzier Implication Operators in the Theory of Fuzzy Power Sets. Fuzzy Sets and Syst., 4, 31-36, 1980.

663 Wojciechowski, K. Optimal Tracking Problem for a Fuzzy Model. In: Proc. Int. Symp. on Syst. Analysis and Simulation, Berlin, Germany, Sep. 1980, Akademie Verlag, Berlin, Germany, 163.

664 Woodbury, M.A., Clive, J. and Garson, A. Mathematical Typology: A grade of Membership Technique for Obtaining Disease Definition. Computers and Biomed. Res., 11, 277-298, 1978.

665 Yager, R.R. Approximate Reasoning as a Basis for Rule Based Expert Systems. IEEE Trans. Syst. Man and Cybern., SMC-14, 636-643, 1984.

666 Yager, R.R. Fuzzy Subsets with Uncertain Membership Grades. IEEE Trans. Syst. Man and Cybern., SMC-14, 271-275, 1984.

667 Yager, R.R. A Note on Measure of Autonomy in Drenick's Theory of Organizations. IEEE Trans. Syst. Man and Cybern., SMC-14, 313-315, 1984.

668 Yager, R.R. On the Selection of Objects Having Imprecise Qualities. IEEE Trans. Syst., Man and Cybern., SMC-14, 755-761, 1984.

669 Yager, R.R. Probabilities from Fuzzy Observations. Inf. Sci., 32, 1-31, 1984.

670 Yager, R.R. Linguistic Approach to the Construction of Expert Systems. In: 'Management of Risk and Uncertainty in Systems Acquisition: Proc. 1983 Defense Risk and Uncertainty Workshop, Fort Belvoir, VA, USA, July 1983. NTIS Report AD-A136 230, 214-216, 1983.

671 Yager, R.R. Membership in a Compound Fuzzy Subset. Cybern. and Syst., 14, 173-184, 1983.

672 Yager, R.R. On the Implication Operator in Fuzzy - Logic. Inf. Sci., 31, 141-164, 1983.

673 Yager, R.R. Querying Knowledge Base Systems with Linguistic Information Via Knowledge Trees. Int. J. Man-Mach. Stud., 19, 73-95, 1983.

674 Yager, R.R. Some Relationship Between Possibility, Truth and Certainty. Fuzzy Sets and Syst., 11, 151-156, 1983.

675 Yager, R.R. Binary Answers to Imprecise Questions. Kybernetes, 11, 285-295, 1982.

676 Yager, R.R. Fuzzy Prediction Based on Regression - Models. Inf. Sci., 26, 45-63, 1982.

677 Yager, R.R. Fuzzy - Sets and Possibility Theory. Recent Developments. (Book). Pergamon Press, Oxford, England, 1982. 672 pages.

678 Yager, R.R. An Introduction to Applications of Possibility Theory. Hum. Syst. Manage., 3, 246-269, 1982.

679 Yager, R.R. Linguistic Hedges: Their Relation to Context and Their Experimental Realization. Cybern. and Syst., 13, 357-374, 1982.

680 Yager, R.R. Measuring Tranquility and Anxiety in Decision Making: An Application of Fuzzy Sets. Int. J. Gen. Syst., 8, 139-146, 1982.

681 Yager, R.R. and Rubinson, T.C. Linguistic Summaries of Data Bases. In: Proc. 20th IEEE Conf. on Decision and Control Including the Symp. on Adaptive Processes, San Diego, CA, USA, Dec. 1981, 3, 1094-1097. IEEE, New York, USA.

682 Yager, R.R. A Procedure for Ordering Fuzzy Subsets of the Unit Interval. Inf. Sci., 24, 143-161, 1981.

683 Yager, R.R. An Approach to Inference in Approximate Reasoning. Int. J. Man-Mach. Stud., 13, 323-338, 1980.

684 Yager, R.R. Finite Lineraly Ordered Fuzzy - Sets with Applications to Decisions. Int. J. Man-Mach. Stud., 12, 299-322, 1980.

685 Yager, R.R. Fuzzy Sets in Decisions in Complex Systems. In: Joint Automatic Control Conf., San Francisco, CA, USA, Aug. 1980.

686 Yager, R.R. Fuzzy Sets, Probabilities and Decision. J. Cybern., 10, 1-18, 1980.

687 Yager, R.R. Fuzzy Thinking as Quick and Efficient. Cybernetica, 23, 265-298, 1980.

688 Yager, R.R. Logical Online Bibliographic Searcher - An Application of Fuzzy - Sets. IEEE Trans. on Syst., Man and Cybern., SMC-10, 51-53, 1980.

689 Yager, R.R. On Modeling Interpersonal Communications. In: Wang, P.P. and Chang, C.K. (eds) - Fuzzy Set Theory and Applications to Policy Analysis and Information Systems. Plenum Press, New York, USA, 309-320, 1980.

690 Yager, R.R. On the Measure of Fuzziness and Negation. II Lattices. Inf. and Control, 44, 236-260, 1980.

691 Yager, R.R. Some Observations on Probabilistic Qualification in Approximate Reasoning. Inf. Sci., 22, 217-234, 1980.

692 Yager, R.R. Using Levels Sets to Extract Fuzzy Membership Grades. In: 5th European Meeting on Cybern. and Syst. Res., Vienna, Austria, Apr. 1980. Also in: Prog. in Cybern. and Syst. Res. 8 and 9, 1980.

693 Yager, R.R. Measure of Fuzziness and Negation. 1. Membership in the Unit Interval. Int. J. Gen. Syst., 5, 221-229, 1979.

694 Yager, R.R. Measurement - Informational Discussion of Fuzzy Union and Intersection. Int. J. Man-Mach. Stud., 11, 189-200, 1979.

695 Yager, R.R. Note on Fuzziness in a Standard Uncertainty Logic. IEEE Trans. on Syst., Man and Cybern., SMC-9, 387-388, 1979.

696 Yager, R.R. Solving Fuzzy Mathematical Relationships. Inf. and Control, 41, 29-55, 1979.

697 Yager, R.R. Fuzzy Decision Making Including Unequal Objectives. Fuzzy Sets and Syst., 1, 87-95, 1978.

698 Yager, R.R. Linguistic Models and Fuzzy Truths. Int. J. Man-Mach. Stud., 10, 483-494, 1978.

699 Yager, R.R. On the Need for Membership Grades in Fuzzy Sets. In: Proc. 4th Int. Conf. on Current Topics in Cybernetics and Systems, Amsterdam, Netherlands, Aug. 1978. 389-390, Springer - Verlag, Berlin, Germany.

700 Yager, R.R. Validation of Fuzzy - Linguistic Models. J. Cybern., 8, 17-30, 1978.

701 Yager, R.R. Building Fuzzy Systems Models. In: Proc. 1st Conf. on Applied General Systems Res.: Recent Developments and Trends, Binghamton, NY, USA, Aug. 1977, Plenum Press, New York, USA.

702 Yager, R.R. Multiple Objective Decision - Making Using Fuzzy Sets. Int. J. Man-Mach. Stud., 9, 375-382, 1977.

703 Yager, R.R. Fuzzy Decision Making Including Unequal Importance of Goals and Constraints. In: The Joint National Meeting of The Oper. Res. Soc. Amer./The Inst. Manage. Sci., Miami, FL, USA, Nov. 1976.

704 Yager, R.R. and Basson, D. Decision - Making with Fuzzy Sets. Decision Sciences, 6, 590-600, 1975.

705 Yao, J.T.P. et al. NAFIP-1 Panel Discussion on Introduction of Fuzzy Sets to Undergraduate Engineering and Sciences Curricula. Int. J. Man-Mach. Stud., 19, 5-6, 1983.

706 Yeh, Yean-Chuan. Alternative Methods for Fire Risk Assessment and Applications to Nuclear Power Plants. Dissertation, Renselaer Polytech. Inst., 1984.

707 Yezhkova, I.V. and Pospelov, D.A. Decision - Making on Fuzzy Grounds. 1. Universal Scale. Eng. Cybern., 15, 1-7, 1977.

708 Yudin, D.B. Decision - Making in Complex Situations. IEEE Trans. Eng. Cybern., 209-222, 1970.

709 Zadeh, Lofti, A. Syllogistic Reasoning in Fuzzy Logic and Its Application to Usuality and Reasoning with Dispositions. IEEE Trans. on Syst., Man and Cybern., SMC-15, 754-763, 1985.

710 Zadeh, L.A. Fuzzy Probabilities. Inf. Processing and Manage., 20, 363-372, 1984.

711 Zadeh, L.A. Making Computers Think Like People. (Fuzzy Set Theory) IEEE Spectrum, 21, 26-32, 1984.

712 Zadeh, L.A. A Theory of Commonsense Knowledge. In: Skala, H.J. (ed) - Issues of Vagueness. D. Reidel, Dordrecht, Netherlands, 257-296, 1984.

713 Zadeh, Lofti A. Commonsense Knowledge Representation Based on Fuzzy Logic. ERL Memo M83/26, Univ. of California, Berkeley, 1983. Also in: Computers, 16, 61-65, 1983.

714 Zadeh, L.A. A Computational Approach to Fuzzy Quantifiers in Natural Languages. Comput. and Math. with Appl., 9, 149-184, 1983.

715 Zadeh, L.A. Linguistic Variables Approximate Reasoning and Dispositions. Med. Inf., 8, 173-186, 1983.

716 Zadeh, L.A. The Role of Fuzzy Logic in the Management of Uncertainty in Expert Systems. Fuzzy Sets and Syst., 11, 199-227, 1983.

717 Zadeh, L.A. Fuzzy Probabilities and Their Role in Decision Analysis. In: Proc. IFAC Symp. on the Theory and Appl. of Digital Control, New Delhi, India, Jan. 1982, 15-21, Pergamon Press, Oxford, England.

718 Zadeh, L.A. Linguistic Variables and Approximate Reasoning. In: Proc. 6th Ann. Symp. on Computer Applications in Medical Care, Washington, D.C., USA, Oct. 1982, 787-791.

719 Zadeh, L.A. A Note on Prototype Theory and Fuzzy - Sets. Cognition, 12, 291-297, 1982.

720 Zadeh, L.A. Fuzzy Logic and the Theory of Evidence. In: Joint National Meeting of The Inst. Manage. Sci./Oper. Res. Soc. Amer., Washington, D.C., USA, May 1980. Paper No. TD17.1.

721 Zadeh, L.A. Fuzzy Sets Versus Probability. Proc. IEEE, 68, 421, 1980.

722 Zadeh, L.A. Possibility Theory as a Basis for a Mathematical Model of Human Communication. UNESCO, Vienna, Austria, 1980.

723 Zadeh, L.A. Fuzzy Sets and Soft Data. In: Amer. Assoc. for the Adv. of Sci. 145th National Meeting, Houston, TX, USA, Jan. 1979.

724 Zadeh, L.A. Fuzzy Sets as a Basis for Information Analysis. NTIS Report AD-A067 211/3 Mar. 1979. 9 pages.

725 Zadeh, L.A. A Theory Approximate Reasoning. Tech. Memo M77/58, Univ. California, Berkeley, 1977. Also in: Machine Intelligence, 9, 149-194, 1979.

726 Zadeh, L.A. Fuzzy Logic as a Model for Human Reasoning. In: Proc. Int. Conf. on Cybern. and Soc., Tokyo, Japan, Nov. 1978, IEEE, New York, USA.

727 Zadeh, L.A. Fuzzy Sets as a Basis for a Theory of Possibility. Fuzzy Sets and Syst., 1, 3-28, 1978.

728 Zadeh, L.A. Fuzzy System Theory - a Frame Work for the Analysis of Humanistic Systems. In: Proc. IEEE Conf. on Circuit and Syst., 1978.

729 Zadeh, L.A. Fuzzy Logic as a Model for Human Reasoning. In: Proc. Int. Conf. on Appl. Gen. Syst. Res. Recent Developments and Trends, Binghamton, NY, USA, Aug. 1977. Plenum Press, New York, USA.

730 Zadeh, L.A. Fuzzy Sets as a Basis for Approximate Reasoning and Knowledge Representation. In First World Conf. on Math. at the Service of Man, Barcelona, Spain, July 1977.

731 Zadeh, L.A. PRUF and Its Applications to Inference from Fuzzy Propositions. In: Proc. IEEE Conf. on Decision and Control, New Orleans, LA, USA, 1977.

732 Zadeh, L.A. Theory of Fuzzy Reasoning and Probability Theory vs Possibility Theory in Decision Making. In: Proc. IEEE Conf. on Decision and Control Including 16th Symp. on Adaptive Processes and a Special Symp. on Fuzzy Set Theory and Applications, New Orleans, LA, USA, Dec. 1977.

733 Zadeh, L.A. Fuzzy - Algorithmic Approach to Decision Analysis. In: Joint National Meeting of The Oper. Res. Soc. Amer./The Inst. Manage. Sci., Miami, FL, USA, Nov. 1976.

734 Zadeh, L.A. A Fuzzy Algorithmic Approach to the Definition of Complex or Imprecise Concepts. In: Bossel, H., Klaczko, S. and Muller, N. (eds) - Systems Theory In the Social Sciences, Birkhauser Verlag, Basel, Switzerland, 202-282, 1976.

735 Zadeh, L.A. Fuzzy Algorithmic Approach to Definition of Complex or Imprecise Concepts. Int. J. Man-Mach. Stud., 8, 249-291, 1976.

736 Zadeh, L.A. Information and Fuzzy Sets. Proc. Amer. Soc. for Inf. Sci., 13, 83, 1976.

737 Zadeh, L.A. The Linguistic Approach and Its Application to Decision Analysis. In: Ho, Y.C. and Mitter, S.K. (ed) - Directions in Large - Scale Systems, Plenum Press, New York, USA, 1976.

738 Zadeh, L.A. Semantic Inference from Fuzzy Data by Mathematical Programming. In: Proc. IEEE Int. Conf. on Cybern. and Soc., Washington, D.C., USA, Nov. 1976.

739 Zadeh, L.A. Theory of Fuzziness and Its Applications to Information Processing and Decision Making. NTIS Report AD-A064 598/6, Oct. 1976, 5 pages.

740 Zadeh, L.A. Calculus of Fuzzy Restrictions. In: Zadeh, L.A. et al. (eds) - Fuzzy Sets and Their Applications to Cognitive and Decision Processes, Academic Press, New York, USA, 1-39, 1975.

741 Zadeh, L.A. Fuzzy Logic and Approximate Reasoning. Synthese, 30, 407-428, 1975.

742 Zadeh, L.A. et al. Fuzzy Sets and Their Applications to Cognitive And Decision Processes. (Book). Academic Press, New York, USA, 1975.

743 Zadeh, L.A. Inference from Fuzzy Data. In: 9th Annual Asilomar Conf. on Circuits, Systems and Computers, Western Periodicals Co., North Hollywood, CA, USA, 1975.

744 Zadeh, L.A. Linguistic Cybernetics. In: Rose, J. (ed) - Adv. in Cybern. and Syst., 3, 1607-1615, 1975. Gordon and Breach, New York, USA.

745 Zadeh, L.A. Local Logics - Fuzzy Logic and Approximate Reasoning. In: Proc. Int. Symp. on Multiple - Valued Logic, Indiana Univ., Bloomington, IN, USA, 1975.

746 Zadeh, L.A. A Relational Model for Approximate Reasoning. In: Proc. IEEE Int. Conf. on Cybern. and Soc., San Francisco, CA, USA, Sep. 1975.

747 Zadeh, L.A. The Concept of a Linguistic Variables and Its Application to Approximate Reasoning. In: Fu, K.S. and Tu, J.T. (ed) - Learning Systems and Intelligent Robots, Plenum Press, New York, USA, 1-10, 1974.

748 Zadeh, L.A. Fuzzy Logic and Its Application to Approximate Reasoning. In: Inf. Processing: Proc. IFIP Cong., North-Holland, Amsterdam, Netherlands, 591-594, 1974.

749 Zadeh, L.A. A New Approach to System Analysis. In: Marois, M. (ed) - Man and Computer, North-Holland, Amsterdam, Netherlands, 55-94, 1974.

750 Zadeh, L.A. Outline of a New Approach to the Analysis of Complex Systems and Decision Processes. IEEE Trans. Syst. Man and Cybern., SMC-3, 28-44, 1973.

751 Zadeh, L.A. A System - Theoretic View of Behavior Modification. In: Wheeler, H. (ed) - Beyond the Punitive Society, W.H. Freeman, San Francisco, CA, USA, 160-169, 1973.

752 Zadeh, L.A. Fuzzy Languages and Their Relation to Human Intelligence. In: Proc. Int. Conf. on Man and Computer, Bordeaux, France, 1972. S. Karger, Basel, Switzerland, 130-165.

753 Zadeh, L.A. A Fuzzy Set Interpretation of Linguistic Hedges. J. Cybern., 2, 4-34, 1972.

754 Zadeh, L.A. A Fuzzy Set Theoretical Interpretation of Linguistic Hedges. Univ. California, Berkeley, Memo ERL-M335, 1972. J. Cybern. 2, 4-34, 1972.

755 Zadeh, L.A. A Rationale for Fuzzy Control. J. Dynamic Syst., Measurements and Control, G94, 3-4, 1972.

756 Zadeh, L.A. Human Intelligence vs. Machine Intelligence. In: Proc., Int. Conf. on Sci. and Soc., Belgrade, Yugoslavia, 127-133, 1971.

757 Zadeh, L.A. Quantitative Fuzzy Semantics. Inf. Sci., 3, 159-176, 1971.

758 Zadeh, L.A. Similarity Relations and Fuzzy Orderings. Inf. Sci., 3, 177-200, 1971.

759 Zadeh, L.A. Towards a Theory of Fuzzy Systems. In: Kalman, R.E. and DeClaris, R.N. (eds) - Aspects of Network and Systems Theory. Holt, Rienhart and Winston, New York, USA, 1971.

760 Zadeh, L.A. Towards Fuzziness in Computer Systems - Fuzzy Algorithms and Languages. In: Boulaye, G. (ed) - Architecture and Design of Digital Computers. Dunod, Paris, France, 9-18, 1971.

761 Zadeh, L.A. Biological Applications of the Theory of Fuzzy Sets and Systems. In: Proctor, L.D. (ed) - Biocybernetics of the Central Nervous System, Little Brown and Co., Boston, MA, USA, 199-212, 1969.

762 Zadeh, L.A. Fuzzy Algorithms. Inf. and Control, 12, 94-102, 1968.

763 Zadeh, L.A. Probability Measures of Fuzzy Events. J. Math. Anal. and Appl., 23, 421-427, 1968.

764 Zadeh, L.A. Shadows of Fuzzy Sets. (Russian). Problems in Transmission of Information, 2, 37-44, 1966.

765 Zadeh, L.A. Fuzzy Sets and Systems. In: Fox, J. (ed) - System Theory. Microwave Res. Inst. Symp. Series XV, Polytechnic Press, Brooklyn, New York, USA, 29-37, 1965.

766 Zadeh, L.A. Fuzzy Sets. Inform. and Control, 8, 338-353, 1965. NTIS Report AD-608981, 1964.

767 Zeleny, M. On the (IR) Relevancy of Fuzzy - Sets Theories. Human Syst. Manage., 4, 301-306, 1984.

768 Zeleny, M. Qualitative Versus Quantitative Modeling in Decision Making. Hum. Syst. Manage., 4, 39-42, 1983.

769 Zeleny, M. Membership Functions and Their Assessment. In: Proc. 4th Int. Conf. on Current Topics in Cybern. and Syst., Amsterdam, Netherlands, Aug. 1978, 391-392. Springer Verlag, Berlin, Germany.

770 Zeleny, M. The Theory of Displaced Ideal. In: Zeleny, M. (ed) - Multiple Criteria Decision Making, Kyoto, Japan, 1975. Also In: Lecture Notes in Economics and Mathematical Systems, 123, 153-206, 1976, Springer Verlag, Berlin, Germany.

771 Zhurid, B.A. and Silov, V.B. A Method for Constructing Logical - Linguistic Models of Intellectual Robots. Eng. Cybern, (USA), 21, 57-61, 1983.

772 Zimmer, Alf. C. A Model for the Interpretation of Verbal Predictions. Int. J. Man-Mach. Stud., 20, 121-134, 1984.

773 Zimmer, A.C. Verbal vs. Numerical Processing. In: Scholz, R. (ed) - Individual Decision Making Under Uncertainties, North Holland, Amsterdam, Netherlands, 1983.

774 Zimmer, Alf C. What Really is Turquoise? A Note on the Evolution of Color Terms. Psychol. Res., 44, 213-230, 1982.

775 Zimmermann, H.J. Fuzzy Set Theory and Its Applications (Book). Academic Press, New York, USA, 1985.

776 Zimmermann, H.J. and Zysno, P. Latent Connectives in Human Decision Making. Fuzzy Sets and Syst., 4, 37-51, 1980.

777 Zimmermann, H.J. Theory and Applications of Fuzzy Sets. In: Proc. Conference on Oper. Res., Toronto, CA, USA, June 1978, 1017-1033. North Holland, Amsterdam, Netherlands, 1979.

778 Zimmermann, H.J. Results of Empirical Work in Fuzzy Set Theory. In: Proc. Int. Conf. on Applied General Syst. Res., Recent Developments and Trends, Binghamton, NY, USA, Aug. 1977. Plenum Press, New York, USA.

779 Zimmermann, H.J. Description and Optimization of Fuzzy Systems. Int. J. Gen. Syst., 2, 209-215, 1976.

780 Zimmermann, H.J. Bibliography: Theory and Applications of Fuzzy Sets. Lehrstuhl for Unternehmensforschung, RWTH, Aachen, Germany, Oct. 1975.

781 Zimmermann, H.J. The Potential of Fuzzy Decision - Making in the Private and Public Sector. SOAK 75, Lidingo, Sweden, 1975.

782 Zimmermann, H.J. Optimization in Fuzzy Environments. In: Joint National Meeting of the 46th Meeting of Oper. Res. Soc. Amer./Inst. Manage. Sci., San Juan, Puerto Rico, USA, Oct. 1974.

783 Zweifel, P. Fuzzy Measurement of Output - a Problem of Evaluation for Health - Policy. Social Sci. and Med., 16, 1291-1300, 1982.

784 Zwick, M. Fuzziness and Catastrophe. In: Proc. Int. Conf. on Cybern. and Soc., Tokyo, Japan, Nov. 1978. Part II, 1237-1241, IEEE, New York, USA.

SUBJECT INDEX TO BIBLIOGRAPHY

BIBLIOGRAPHICAL NOTES

BIOGRAPHICAL NOTES

ASFOUR, S. S. is an Associate Professor of Industrial Engineering and Neurological Surgery at the University of Miami. He is also the Associate director of the Ergonomic Division, the Comprehensive Pain Center and the Department of Neurological Surgery. Dr. Asfour holds a Ph.D. in Industrial Engineering from Texas Tech University. He is a member of the Human Factors Society, Ergonomics Research Society, Institute of Industrial Engineers, Society of Manufacturing Engineers, Alpha Pi Mu, Tau Beta Pi, Sigma Xi Research Society, Phi Kappa Phi, and Omicron Delta Kappa. His research interests are in work physiology, psychophysics, biomechanics, and rehabilitation engineering.

AYOUB, M. M. PH.D., is a Paul Whitfield Horn Professor of Industrial Engineering and the Director of the Institute for Ergonomics Research at Texas Tech. He is a recognized authority in the area of biomechanics and work physiology an has extensive research experience in occupational biomechanics and work physiology. He is a member of Institute of Industrial Engineers, the Ergonomics Society, American Industrial Hygiene Association, the American Society of Biomechanics, The Human Factors Society, and American Society of Engineering Education. He is a Registered Professional Engineer in Texas.

BOY, Guy A. is currently a Research Engineer at the Centre d'Etudes et de Recherches de Toulouse, France (Research Center of the Office National d'Etudes et Recherches Aerospatiales) where he is developing a group in Cognitive Science. He is also an Assistant Professor of Decision Theory and Artificial Intelligence at the Ecole Nationale Superieure de l'Aeronautique et de l'Espace, Toulouse. His chief research interest is in the application of control and decision theories, and of artificial intelligence principles and techniques, to problems of human-machine interactions. In particular, he is interested in process control and expert systems in aeronautics and space applications. From September 1984 to January 1986, he was a Research Associate in the Aerospace Human Factors Research Division at the NASA-Ames Research Center, California. During this time, he also gave a series of lectures in artificial intelligence at NASA. He holds an engineering degree (equivalent to Master of Science in U.S.) in Physics and Chemistry from the National Polytechnic Institute of Toulouse, France; an equivalent degree in Control Theory and System Analysis from the Ecole Nationale Superieure de l'Aeronautique et de l'Espace, Toulouse, France; a doctoral degree in Control Theory and System Analysis from the Ecole Nationale Superieure de l'Aeronautique et de l'Espace; and a masters degree in Experimental Psychology from the University of Toulouse, France.

CZOGALA, Ernest was born in 1941 and received his Ph.D. and Habilitation degrees from Silesian Technical University (Poland), where he is now an associate professor. He has published more than 100 scientific papers and several books. Most of them deal with the fields of probability theory and fuzzy sets theory, and their applications especially in identification and control of ill-defined processes, decision making methods and some aspects of systems analysis.

DRURY, Colin G. is a professor of Industrial Engineering and Director of the Human Factors Laboratory at the State University of New York at Buffalo. His chief research interests are in applying mathematical models of human performance to the design of industrial jobs. Particular areas of interest are industrial inspection and the control of industrial processes, emergency medical services, and manual materials handling. He was previously manager of ergonomics at Pilkington Brothers, Ltd., Glass manufacturers, England. He is a Fellow of the Human Factors Society and the Ergonomics Society. In 1980 he received the Bartlett Medal of the Ergonomics Society. He holds a bachelors degree in Physics from the University of Sheffield, England, and a Ph.D. in Work Design in Ergonomics from the University of Birmingham, England.

EVANS, Gerald W. is an Assistant Professor in the Industrial Engineering Department at the University of Louisville. He received a B.S. in Mathematics, and an M.S. and Ph.D. in Industrial Engineering, all from Purdue University. Prior to his present position, he was a senior research engineer at General Motors Research Laboratories. His publications include articles that have appeared in IIE Transactions, Management Science, the International Journal of Production Research, and IEEE Transactions on Engineering Management. Dr. Evans' current research interests include the application of mathematical programming, multi-objective decision theory, and fuzzy sets to quality control, layout design, production planning, and the design of expert systems. He is a Senior Member of IIE, and a member of ORSA, TIMS, and DSI.

FARRENY, Henri is a Professor of Computer Science at the Polytechnic National Institute of Toulouse (I.N.P.T., France). He received the Engineer and the Doctor-Engineer degrees from the "Ecole Nationale Superieure d'Electronique, Electrotechnique, Informatique et Hydraulique de Toulouse" respectively in 1969 and 1971. He obtained the "Doctorat d'Etat" degree in 1980 with a thesis entitled "A Problem-Solving System Oriented to the Control of Robots." He taught Artificial Intelligence at several universities for about ten years. He is the author of two books: "Introduction a LISP" (Masson, Paris, 1984) and "Les Systemes-Experts. Principes et Exemples" (CEPADUES, Toulouse, 1985) and over thirty technical papers. His current research interests are in artificial intelligence, expert systems, and natural language communication systems. He is co-leader of the French National Program of Concerted Research in Artificial Intelligence (P.R.C.-I.A.).

FRANUS, Edward is a professor of Psychology at the Jagiellonian University of Krakow, Poland. He is also Director of the Institut of Work Psychology, and in the past was Vice-Chairman of the Psychology Department at the Jagielonian University. He holds M.S. and Ph.D. degrees in psychology (both from Jagielonian University). His main research interests include psychology of work and technology, human factors psychology and ergonomics. In particular, he is interested in methodology of ergonomics research, studies of the scientific and technological thought processes, and psychology of education. He is a member of the Polish Ergonomics Association, Polish Psychological Society, and Ergonomics and Psychological Committees of the Polish Academy of Sciences. He is also a member of the Editorial Board of Ergonomia, the scientific journal published by the Polish Ergonomics Association.

GENAIDY, A. M. is a graduate student working towards his doctorate in human factors at the University of Miami. He holds a B. D. in Biomedical Engineering from Cairo University, and an M.S. in Biomedical Engineering from the University of Miami. Mr. Genaidy is a student member of the Human Factors Society, Institute of Industrial Engineers, and Society of Manufacturing Engineers. He is also a member of Alpha Pi Mu, Tau Beta Pi, and Sigma Xi. His research interests are in work physiology, psychophysics, biomedical instrumentation, and computer applications in industry.

GROBELNY, Jerzy is an Assistant Professor at the Institute of Production Engineering and Management in Technical University of Wroclaw, Poland. His chief research interests are in applying formal methods and model to the work station design process. Particular areas of interest are computer aided design in work space planning, expert systems for man-machine system design and computer simulation models in ergonomics. He holds a MS degree and a Ph.D. in Industrial Engineering from the Technical University of Wroclaw, Poland.

HAFEZ, Hala A. is an Assistant Professor of Design and Production Engineering at Ain Shams University, Cairo, Egypt. Dr. Hafez received her B.S. in 1972 and M.S. in 1973 in Production Engineering from Ain Shams University. In 1983, she received a Ph.D. at Texas Tech University. Her research interest is in Ergonomics. She is a member of Tau Beta Pi and Alpha Pi Mu honor societies. She is a member of Institute of Industrial Engineers, Human Factors Society and AIHA.

HIROTA, Kaoru is an Associate Professor of Hosei University. He was born in 1950 and received the Doctor-Engineer degree from Tokyo Institute of Technology in 1979. He is the author of "Image Pattern Recognition," "Fuzzy Controlled Intelligent Robot," and "Introduction to Software Engineering" (all were written in the Japanese language) and over fifty technical papers. He is a secretary of IFSA, Japan chapter. His current research interests are image pattern recognition, intelligent robots, artificial intelligence, and fuzzy systems.

JIANG, Bernard C. is an Assistant Professor of Industrial Engineering at Auburn University. He holds a B.E. (1976) in Industrial Engineering from Chun Yuan University (Taiwan), and M.S. (1981) and a Ph.D. (1984) in Industrial Engineering from Texas Tech University. His research interests are in applied ergonomics, human/computer/environment interaction, and human factors in manufacturing. He is a senior member of IIE. He is also a member of Human Factors Society, SME, AIHA and the Ergonomics Society.

KACPRZYK, Janusz is a research associate at the Systems Research Institute, Polish Academy of Science, Warsaw, Poland. Currently he is Visiting Professor of Computer Science at the University of Tennessee, Knoxville, Tennessee, USA. His main interests are in knowledge-based decision support systems, particularly issues related to their "human-friendliness", and in the application of fuzzy and possibilistic tools for "softening" decision making and knowledge engineering models. He is the author of 3 books on those subjects, two in Polish and one in English, the author and co-author of more than 50 papers, and the co-editor of one edited volume. He is on the editorial board of the International Journal of Intelligent Systems. He holds a M.S. degree in Computer Science and Automatic Control from the Technical University of Warsaw, Poland, and a Ph.D. in Systems Analysis from the Systems Research Institute, Polish Academy of Sciences, Warsaw, Poland.

KARWOWSKI, Waldemar is an Assistant Professor of Industrial Engineering and Engineering Management at the University of Louisville. He holds a Masters degree in Industrial Management and Information Sciences (1978) from the Technical University of Wroclaw, Poland, and a Ph.D. in Industrial Engineering (1982) from Texas Tech University, Lubbock, Texas. He is a member of the Human Factors Society, the Ergonomics Society, the American Industrial Hygiene Association, the Institute of Industrial Engineers, North American Fuzzy Information Processing Society, and Tau Beta Pi. Professor Karwowski teaches in the area of applied ergonomics, industrial safety engineering, and design of man-machine systems. His main research interests concentrate on workplace design; as well as applications of fuzzy methodologies in human performance measurement and safety engineering; and development of knowledge-based expert systems for ergonomics applications. Dr. Karwowski is an Associate Editor of the International Journal of Industrial Ergonomics.

KHALIL, T. M. is professor and chairman of the Department of Industrial Engineering at the University of Miami. He is also the Director of the Ergonomic Division of the Comprehensive Pain Center and the Department of Neurological Surgery, and is the Director of the Environmental Health and Safety Program of University of Miami. He holds professorships in the Departments of Biomedical Engineering, Neurological Surgery, and Epidemiology and Public Health. Dr. Khalil is a member of the Human Factors Society, IIE, SME, ASSE, and AIHA. He received a fellow award from IIE, 1st Ergonomics Division Award from IIE, and the Human Factors Society's Jack A. Kraft Award.

MA, Mou-chao is an Associate Research Professor at the Institute of Psychology, Academia Sinica, The People's Republic of China. His chief research interests are in applying fuzzy sets theory to the evaluation associated with psychological property. Particular areas of interest are empirical investigation of fuzzy sets theory, methodology of psychophysics and psychological scale, human factors in ergonomics, and social psychological problems. He is a member of the board of directors of the Fuzzy Systems and Fuzzy Mathematics Society, a fellow of the editorial board of the Journal of Fuzzy Mathematics and of the Journal of Psychology communication, and an advisor to the Journal of China Advertising.

MAREK, Tadeusz is an Associate Professor of Psychology and Head of Mathematical Psychology Group the Institute of Psychology at Jagiellonian University at Krakow, Poland. His main research interests are in the mathematical and statistical modeling of human information processes in rested and mentally fatigued men. He is an author of six books and about eighty articles. These topics include theory of mental fatigue and its measurement, man-computer interaction, the temporal dimension of vision, and application of sequential and cluster analysis in behavioral sciences and mathematical modeling of visual perception. He is a fellow of the Polish Ergonomics Society, the Psychological Society, the Secretary of Science for Ergonomics Committee of Polish Academy of Science, and the International Seminar Macinter (Man-Computer Interaction Research). He holds a Masters Degree and a Ph.D. in Psychology from the Jagiellonian University at Krakow, Poland.

MITAL, Anil is an Associate Professor of Industrial Engineering and Director of the Ergonomics Research Laboratory at the University of Cincinnati, Cincinnati, Ohio, U.S.A. He is the Editor-in-Chief of the International Journal of Industrial Ergonomics and also serves as Chairman of the International Foundation for Industrial Ergonomics and Safety Research. He holds a Bachelor of Engineering degree in Mechanical Engineering (1974) from Allahabad University, India (with highest honors); an M.S. in Industrial Engineering from Kansas State University, U.S.A. (1976); and a Ph.D. in Industrial Engineering from Texas Tech University, U.S.A. (1980). His research interests are in Applied Ergonomics and the application of Systems Methodologies to Ergonomics. He is a member of several professional and honor societies including the Institute of Industrial Engineers, Human Factors Society, Sigma Xi, Tau Beta Pi, and Phi Kappa Phi. He has received several awards including the 1984 Sigma Xi Distinguished Research Award and the Society of Automotive Engineers' 1985 Ralph R. Teetor Award for significant contributions to teaching, research, and student development. In 1984, he was recognized as The Young Engineer Of The Year by Engineers and Scientist of Cincinnati.

NOWOROL, Czeslaw is an Assistant Professor of Mathematics, Statistics and Human Factors at the Agricultural University and an Assistant Professor of Statistics and Methodology in the Institute of Psychology (Group of Mathematical Psychology and Statistics) at the Jagiellonian University at, Krakow, Poland. His main research interest are in applying mathematical and statistical models to human factors/ergonomics and behavioral sciences. Particular areas of interest are building of mathematical and statistical models, computer programming and techniques, ergonomical problems of a man-computer system (especially design and testing of new keyboards and computer workstands), measurement of mental load and fatigue, measurement of mental load and fatigue of operators and mathematical psychology. He is a Fellow of the Polish Ergonomics Society and Psychological Society and an International Seminar MACINTER (Man-Computer Interaction Research). He holds a Master of Science degree in Mathematics from the Jagiellonian University and a D.Sc. in Technical Sciences from the Institute of Construction, Techniques and Energy of Agriculture at Warsaw, Poland.

NOWAKOWSKI, Marek is an Associate Professor of Production Engineering and Director of Ergonomics Laboratory at the Technical University of Wroclaw at Wroclaw. His main research areas are modeling of human performance, analysis and designing of human-computer interaction and applications of fuzzy sets and possibility theory in human factors. Particular areas of interest are creating symbiotic human-computer systems, artificial intelligence and expert systems and also technologies in knowledge engineering. He is a member of Polish Ergonomics Society. He holds a M.A. degree in Production Engineering and Computer Sciences and Ph.D. in modeling of human operator and reliability analysis from the Technical University of Wroclaw.

PACHOLSKI, Leszek M. holds a Professors State Nomination in Technical Sciences and works as a Professor of Industrial Engineering and Director of the Institute of Organization and Management at the Technical University of Poznan, Poland. He also manages the Ergonomics Laboratory at the above Institute. The basic subjects of his research interests are ergonomics methodology and engineering of human interactive systems. Particular areas of interest are in applying of mathematical modelling in ergonomic diagnostics and design of man-machine industrial systems. He is also well-known as an author of programs and modern methods for studying ergonomics/human factors in Poland. He holds a M.Sc. degree in Mechanical Engineering, D.Sc. in Industrial Engineering, and Ph.D. in Ergonomics from the Technical University of Poznan. At present, he acts as the Vice-president of the Polish Ergonomics Society, and is a member of the Ergonomics Committee of Polish Academy of Sciences. He is also an expert of the International Organization for Standardization cooperating with the ergonomics subcommittee in Poland.

PEDRYCZ, Witold received his Ph.D. and Habilitation degrees in Computer Engineering from Silesian Technical University (Poland), where he is now an associate professor. Currently, he is also with the Research and Development Center of Electronic Medical Instruments, Zabrze, Poland. His research interests include fuzzy systems, computer science, pattern recognition, and computer control. He is an associate editor of Pattern Recognition Letters and a co--author of an Introduction into Fuzzy Set Theory (in Polish, Gliwice, 1980).

PRADE, Henri is a Charge de Recherche at the National Center for Scientific Research (C.N.R.S.). He received the Engineer and the Doctor-Engineer degree from the Ecole Nationale Superieure de l'Aeronautique et de l'Espace respectively in 1975 and 1977, and the "Doctorat d'Etat" degree from the University of Toulouse in 1982. He is the author (with Dr. D. Dubois) of two books: "Fuzzy Sets and Systems: Theory and Applications" (Academic Press, New York, 1980) and "Theorie des Possibilites" (Masson, Paris, 1985), and over eighty technical papers. He is co-editor of the fuzzy set bulletin, BUSEFAL, which has appeared quarterly since 1979. He is also a member of the editorial organizations of the international journal "Fuzzy Sets & Systems," the Chinese journal "Fuzzy Mathematics" and the "International Journal of Intelligent Systems." His current research interests are in fuzzy set and possibility theory, approximate reasoning and non-classical logics, artificial intelligence and expert systems, natural language communication systems, incomplete and uncertain information data base management systems, and operations research.

RAMSEY, Jerry D. is currently a Professor of Industrial Engineering and Associate Vice President for Academic Affairs at Texas Tech University, Lubbock, Texas. He is a registered Professional Engineer and a Certified Safety Executive. He received his B. S. and M. S. at Texas A&M University and his Ph.D. in 1967 at Texas Tech University. His teaching and Research interests include ergonomics, occupational safety and health, product safety, human factors engineering, and occupational environments especially heat stress. He has been named as a member or chairman of national committees for several professional groups and is a contributing author to six books and over 50 professional publications.

SHARIT, Joseph is an Assistant Professor of Industrial Engineering at the State University of New York at Buffalo. He received his Masters and Ph.D. Degrees in Industrial Engineering from Purdue University. His chief research interests are in the areas of human-computer and human-systems interaction, with emphasis on human decision making, human reliability, and decision-support systems. Other interests include analysis and evaluation of mental workload, especially in computer-based work environments.

SHIMOMURA, Takeshi was born in Tokyo, Japan on November 6, 1943. He received a B.S. degree from Osaka Electro-Communication University in 1966. He joined the Department of Industrial Engineering at Osaska Electro-Communication University, and is now a senior lecturer there. His interests are possibilistic regression analysis and applications. He is especially concerned with analysis of learning phenomenon in the possibilistic approach.

SMITH, James L. is an Associate Professor of Industrial Engineering at Texas Tech University. He received a BSIE from Northwestern University, an MSIE from the University of Texas at Arlington, and a Ph.D. from Auburn University. His primary research interests are in the areas of applied ergonomics, work physiology and biomechanics. He is a member of the Human Factors Society, the Ergonomics Society, the American Society of Biomechanics, the Institute of Industrial Engineers, the American Industrial Hygiene Association, and the Research Society of Sigma Xi.

SMITHSON, Michael J. is a Senior Lecturer in Sociology at James Cook University in Queensland, Australia. His primary research interests include mathematical models and statistical methods for the social sciences, social psychological theories of exchange behavior, and the sociology of knowledge. He has a long--standing involvement in developing methods applying fuzzy set theory to behavioral and social science research problems, and is currently writing a book on the topic. Dr. Smithson holds a B.Sc. in Mathematics from Harvey Mudd College in Claremont, California, and a M.Sc. and Ph.D. in Sociology from the University of Oregon, in Eugene.

TANAKA, Hideo was born in Osaka, Japan on September 28, 1938. He received a B.S. degree from the University of Kobe in 1962 and his M.S. and Ph.D. degrees from the Osaka City University in 1966 and 1969, respectively. After completing his doctoral work, he joined the faculty of the Department of Industrial Engineering, University of Osaka Prefecture. In 1972-1973 he was a visiting research associate at the University of California, Berkeley, and in 1975-1977 he was an Alexander Von Humboldt Foundation Fellow at the University of Aachen in West Germany. In 1981-1982 he was visiting Research Associate at Kansas State University. He is now an Associate Professor in the Department of Industrial Engineering at the University of Osaka Prefecture. His areas of interest are fuzzy systems and its application to Operations Research.

WALY, S. M. is a graduate student working towards his doctorate in the interdepartmental Ph.D. program at the University of Miami. He holds a B.S. in Biomedical Engineering from Cairo University, and an M.S. in Biomedical Engineering from the University of Miami. Mr. Waly is a member of the Egyptian Society of Biomedical Engineering, Human Factors, SME, Tau Beta Pi, and the Editorial Board of the Egyptian Journal of Biomedical Engineering. Waly's areas of research interest are rehabilitation engineering, work physiology, digital signal processing, and computer applications in clinical and industrial settings.

WANG, Mao-Juin James received his Ph.D. at State University of New York, Buffalo. He is an Assistant Professor in the Division of Industrial Engineering, Alfred University. His research interests include industrial inspection, mathematical modelling of human performance, occupational safety, and biomechanics.

WATADA, Junzo is an Associate Professor in the Faculty of Business Administration at Ryukoku University in Osaka, Japan. His research interests are possibilistic models and their applications to industrial engineering. Recently he has been concerned with expert systems and decision support systems. Until 1979 he was with Fujitsu Ltd. In 1983-1984 he was a visiting research associate at the School of Electrical Engineering, Purdue University. He received a BSEE and an MSEE from Osaka City University, and a Ph.D. in Industrial Engineering from the University of Osaka Prefecture.

YAGER, Ronald R. is a professor of Information Systems and Director of the Machine Intelligence Institute at Iona College in New Rochelle, New York. His main research interests are in the management of uncertainty for knowledge based systems and in the representation of knowledge. He is also interested in the foundations of fuzzy set theory. Dr. Yager has previously taught at Pennsylvania State University. He has published over 120 articles and has edited four books. He is editor-in-chief of the International Journal of Intelligent Systems and a member of the editorial board of a number of other journals including Fuzzy Sets and Systems. He holds a bachelors degree in Electrical Engineering from the City College of New York and a Ph.D. in systems science from the Polytechnic Institute of Brooklyn.

ZAIFU, Su is an associate professor of department of physics of Tongji Medical University, Wuhan, China. His chief research interests are in artificial intelligence and medical expert systems. He is a member of the Chinese Medicine Processing Society, the editor of the fuzzy Mathematics and the Science Exploration.

ZIMMER, Alf C. is Professor of Experimental and Applied Psychology at the University of Regensburg in the Fed. Rep. of Germany. His main research interests are in mundane (or everyday) reasoning of lay people and experts and in qualitative judgments of uncertainty about repetetive or singular events and sequences of events (scenarios). One fact of particular interest for him is the design of user friendly interfaces for expert systems based on the above mentioned research. He is a fellow of the Cognitive Science Society and editor of "Gestalt Theory" (an international and interdisciplinary journal). He holds a diploma in Psychology (approx. M.A.), or Dr. phd. (Ph.D.) in Psychology and Statistics and a Dr. phil. habil in Psychology, all from the Westfalische Wilhelms-Universitat Munster, FRG.

AUTHOR INDEX

AUTHOR INDEX